METHODS IN MOLECULAR BIOLOGY

For further volumes:
http://www.springer.com/series/7651

Histone Variants

Methods and Protocols

Edited by

Guillermo A. Orsi

Institut Curie, PSL Research University CNRS, UMR3664, Equipe Labellisée Ligue contre le Cancer, Paris, France; Sorbonne Universités, UPMC Univ Paris 06 CNRS, UMR3664, Paris, France

Geneviève Almouzni

Institut Curie, PSL Research University CNRS, UMR3664, Equipe Labellisée Ligue contre le Cancer, Paris, France; Sorbonne Universités, UPMC Univ Paris 06 CNRS, UMR3664, Paris, France

Humana Press

Editors
Guillermo A. Orsi
Institut Curie
PSL Research University CNRS, UMR3664
Equipe Labellisée Ligue contre le Cancer
Paris, France

Sorbonne Universités
UPMC Univ Paris 06 CNRS, UMR3664
Paris, France

Geneviève Almouzni
Institut Curie
PSL Research University CNRS, UMR3664
Equipe Labellisée Ligue contre le Cancer
Paris, France

Sorbonne Universités
UPMC Univ Paris 06 CNRS, UMR3664
Paris, France

ISSN 1064-3745 ISSN 1940-6029 (electronic)
Methods in Molecular Biology
ISBN 978-1-4939-8662-0 ISBN 978-1-4939-8663-7 (eBook)
https://doi.org/10.1007/978-1-4939-8663-7

Library of Congress Control Number: 2018949156

Cover illustration: Confocal microscopy image of HeLa cell mitotic chromosomes with immunofluorescent staining against phosphorylated serine 10 of histone H3 (H3S10P). Credit: David Sitbon.

Printed on acid-free paper

This Humana Press imprint is published by the registered company Springer Science+Business Media, LLC part of Springer Nature.
The registered company address is: 233 Spring Street, New York, NY 10013, U.S.A.

Preface

Histone variants—non-allelic isoforms of histones—have emerged in the past decades as key players in chromosome organization and function. As histones, they are among the most conserved proteins in eukaryotes. By composing the nucleosome, they provide the first basic level of chromatin organization while ensuring the versatility of this module and its regulatory functions in the epigenomic landscape.

Recently, thanks to the development of a series of new technologies, much progress has been made to delineate their key role in structuring, organizing, and protecting DNA. Histone variants are now recognized as central players in controlling chromatin integrity, functionalizing chromosome territories from large-scale domains to individual regulatory elements, and establishing and maintaining epigenetic information at the time scale of development.

Here we sought to review these advances in the study of histone variants. We considered the multiple dimensions of exploration in which they have investigated: (1) variant nucleosomes and their biochemical and biophysical properties; (2) histone variants out of the nucleosomes and their handling by histone chaperones—escort factors that handle specific variants throughout their cellular life; (3) their dynamic regulation throughout the cell cycle and at specific genomic locations; (4) their evolution and importance to functionally organize chromosome territories and cellular functions; (5) their developmental dynamics and roles; and (6) their impact in genome instability and disease.

A range of methodological approaches have been specifically developed to probe the variants. We wished to cover how they have impacted the fields of biochemistry, biophysics, molecular biology, epigenomics, genetics, bioinformatics and mathematical modeling, cellular and developmental biology, evolution, and pathology. We thus hope that this book will constitute not only a laboratory guide for the implementation of specific methods but also a bridge to link multidisciplinary approaches, a cross-fertilization mechanism that we believe will be crucial in moving forward in this exciting field as a community. We also hope that this book will benefit readers not only as a detailed guide to particular protocols but also as an inspiration to explore the experimental possibilities away from their comfort zone, catalyzing new ideas and potentially triggering collaborations.

This book became a reality thanks to a fantastic group of authors who have generously given their time and energy to share their authoritative collective knowledge. We are above all grateful for their invaluable contributions.

Paris, France

Guillermo A. Orsi
Geneviève Almouzni

Contents

Part III Microscopy and Cell Biology

Part IV Evolution, Development, and Diseases

Contributors

GENEVIÈVE ALMOUZNI • *Institut Curie, PSL Research University, CNRS, UMR3664, Equipe Labellisée Ligue contre le Cancer, Paris, France; Sorbonne Universités, UPMC Univ Paris 06, CNRS, UMR3664, Paris, France*

ALICIA ALONSO • *Division of Hematology/Oncology, Department of Medicine, Epigenomics Core Facility, Weill Cornell Medical College, New York, NY, USA*

YASUHIRO ARIMURA • *Laboratory of Chromatin Structure and Function, Institute for Quantitative Biosciences, The University of Tokyo, Tokyo, Japan; Graduate School of Advanced Science and Engineering, Waseda University, Tokyo, Japan*

SOPHIE BARRAL • *Institut Albert Bonniot, CNRS UMR 5309, INSERM U1209, Université Grenoble Alpes, Grenoble, France*

TERESA K. BARTH • *Research Unit Protein Science, Helmholtz Zentrum München, Munich, Germany*

EMILY BERNSTEIN • *Departments of Oncological Sciences, Icahn School of Medicine at Mount Sinai, New York, NY, USA; Departments of Dermatology, Icahn School of Medicine at Mount Sinai, New York, NY, USA*

THIERRY BUCHOU • *Institut Albert Bonniot, CNRS UMR 5309, INSERM U1209, Université Grenoble Alpes, Grenoble, France*

CAITRÍONA M. COLLINS • *Centre for Chromosome Biology, Biomedical Sciences, National University of Ireland Galway, Galway, Ireland*

IVAN R. CORRÊA JR • *New England Biolabs, Ipswich, MA, USA*

JULIETTE DABIN • *Epigenetics and Cell Fate Centre, UMR7216 CNRS/Paris Diderot University, Paris, France*

INES A. DRINNENBERG • *Institut Curie, Paris Sciences et Lettres Research University, Centre National de la Recherche Scientifique UMR 3664, Paris, France*

ELAINE M. DUNLEAVY • *Centre for Chromosome Biology, Biomedical Sciences, National University of Ireland Galway, Galway, Ireland*

ROBERT J. DURONIO • *Integrative Program in Biological and Genome Sciences, Curriculum in Genetics and Molecular Biology, Department of Biology, Department of Genetics, Lineberger Comprehensive Cancer Center, University of North Carolina, Chapel Hill, NC, USA*

LARA E. EMERY • *Department of Molecular Biosciences, The University of Texas at Austin, Austin, TX, USA; Institute for Cellular and Molecular Biology, The University of Texas at Austin, Austin, TX, USA*

DANIELE FACHINETTI • *Institut Curie, PSL Research University, CNRS, UMR 144, Paris, France*

DONG FANG • *Department of Pediatrics, Institute for Cancer Genetics, Columbia University, New York, NY, USA; Department of Genetics and Development, Institute for Cancer Genetics, Columbia University, New York, NY, USA*

IGNASI FORNE • *Protein Analysis Unit, Biomedical Center, Ludwig Maximilian University of Munich, Munich, Germany*

ANNA FORTUNY • *Epigenetics and Cell Fate Centre, UMR7216 CNRS/Paris Diderot University, Paris, France*

RISA FUJITA • *Laboratory of Chromatin Structure and Function, Institute for Quantitative Biosciences, The University of Tokyo, Tokyo, Japan; Graduate School of Advanced Science and Engineering, Waseda University, Tokyo, Japan*

RICHARD J. GIBBONS • *Medical Research Council (MRC) Molecular Haematology Unit, Weatherall Institute of Molecular Medicine, University of Oxford, Oxford, UK*

MARILUZ GÓMEZ-RODRÍGUEZ • *Instituto Gulbenkian de Ciência, Oeiras, Portugal; Departamento de Ciencias Naturales and Matemáticas, Pontificia Universidad Javeriana, Cali, Colombia*

JÉRÔME GOVIN • *Large Scale Biology Laboratory, Biosciences and Biotechnologies Institute of Grenoble, INSERM U1038, CEA, Université Grenoble Alpes, Grenoble, France*

DAN HASSON • *Departments of Oncological Sciences, Icahn School of Medicine at Mount Sinai, New York, NY, USA; Departments of Dermatology, Icahn School of Medicine at Mount Sinai, New York, NY, USA*

STEVEN HENIKOFF • *Basic Sciences Division, Fred Hutchinson Cancer Research Center, Seattle, WA, USA; Howard Hughes Medical Institute, Seattle, WA, USA*

DOUGLAS R. HIGGS • *Medical Research Council (MRC) Molecular Haematology Unit, Weatherall Institute of Molecular Medicine, University of Oxford, Oxford, UK*

SEBASTIAN HOFFMANN • *Institut Curie, PSL Research University, CNRS, UMR 144, Paris, France*

NAGHMEH HOGHOUGHI • *Institut Albert Bonniot, CNRS UMR 5309, INSERM U1209, Université Grenoble Alpes, Grenoble, France*

NAOKI HORIKOSHI • *Graduate School of Advanced Science and Engineering, Waseda University, Tokyo, Japan*

AXEL IMHOF • *Chromatin Proteomics, Biomedical Center, Ludwig Maximilian University of Munich, Munich, Germany; Protein Analysis Unit, Biomedical Center, Ludwig Maximilian University of Munich, Munich, Germany*

LARS E. T. JANSEN • *Instituto Gulbenkian de Ciência, Oeiras, Portugal*

SAADI KHOCHBIN • *Institut Albert Bonniot, CNRS UMR 5309, INSERM U1209, Université Grenoble Alpes, Grenoble, France*

TOMOYA KUJIRAI • *Laboratory of Chromatin Structure and Function, Institute for Quantitative Biosciences, The University of Tokyo, Tokyo, Japan; Graduate School of Advanced Science and Engineering, Waseda University, Tokyo, Japan*

HITOSHI KURUMIZAKA • *Laboratory of Chromatin Structure and Function, Institute for Quantitative Biosciences, The University of Tokyo, Tokyo, Japan; Graduate School of Advanced Science and Engineering, Waseda University, Tokyo, Japan*

MARY LEATHAM-JENSEN • *Integrative Program in Biological and Genome Sciences, Curriculum in Genetics and Molecular Biology, Department of Biology, Department of Genetics, Lineberger Comprehensive Cancer Center, University of North Carolina, Chapel Hill, NC, USA*

JUSTIN W. C. LEUNG • *Department of Molecular Biosciences, The University of Texas at Austin, Austin, TX, USA; Institute for Cellular and Molecular Biology, The University of Texas at Austin, Austin, TX, USA; Department of Radiation Oncology, Department of Biochemistry and Molecular Biology, University of Arkansas for Medical Sciences, Little Rock, AR, USA*

ALEJANDRA LOYOLA • *Fundación Ciencia & Vida, Santiago, Chile*

SHINICHI MACHIDA • *Graduate School of Advanced Science and Engineering, Waseda University, Tokyo, Japan*
SEBASTIÁN MARTY-LOMBARDI • *Fundación Ciencia & Vida, Santiago, Chile*
A. GREGORY MATERA • *Integrative Program in Biological and Genome Sciences, Curriculum in Genetics and Molecular Biology, Department of Biology, Department of Genetics, Lineberger Comprehensive Cancer Center, University of North Carolina, Chapel Hill, NC, USA*
DANIEL J. MCKAY • *Integrative Program in Biological and Genome Sciences, Curriculum in Genetics and Molecular Biology, Department of Biology, Department of Genetics, Lineberger Comprehensive Cancer Center, University of North Carolina, Chapel Hill, NC, USA*
MICHAEL P. MEERS • *Integrative Program in Biological and Genome Sciences, Curriculum in Genetics and Molecular Biology, Department of Biology, Department of Genetics, Lineberger Comprehensive Cancer Center, University of North Carolina, Chapel Hill, NC, USA*
JAKUB MIECZKOWSKI • *Neurobiology Center, Nencki Institute of Experimental Biology, Warsaw, Poland*
KYLE M. MILLER • *Department of Molecular Biosciences, The University of Texas at Austin, Austin, TX, USA; Institute for Cellular and Molecular Biology, The University of Texas at Austin, Austin, TX, USA*
ANTOINE MOLARO • *Division of Basic Sciences, Fred Hutchinson Cancer Research Center, Seattle, WA, USA*
YUICHI MOROZUMI • *Institut Albert Bonniot, CNRS UMR 5309, INSERM U1209, Université Grenoble Alpes, Grenoble, France*
GUILLERMO A. ORSI • *Institut Curie, PSL Research University, CNRS, UMR3664, Equipe Labellisée Ligue contre le Cancer, Paris, France; Sorbonne Universités, UPMC Univ Paris 06, CNRS, UMR3664, Paris, France*
TAYLOR J. R. PENKE • *Integrative Program in Biological and Genome Sciences, Curriculum in Genetics and Molecular Biology, Department of Biology, Department of Genetics, Lineberger Comprehensive Cancer Center, University of North Carolina, Chapel Hill, NC, USA*
SANDRA PIQUET • *Epigenetics and Cell Fate Centre, UMR7216 CNRS/Paris Diderot University, Paris, France*
SOPHIE E. POLO • *Epigenetics and Cell Fate Centre, UMR7216 CNRS/Paris Diderot University, Paris, France*
SRINIVAS RAMACHANDRAN • *Basic Sciences Division, Fred Hutchinson Cancer Research Center, Seattle, WA, USA; Howard Hughes Medical Institute, Seattle, WA, USA*
DOMINIQUE RAY-GALLET • *Institut Curie, PSL Research University, CNRS, UMR3664, Equipe Labellisée Ligue contre le Cancer, Paris, France; Sorbonne Universités, UPMC Univ Paris 06, CNRS, UMR3664, Paris, France*
SOPHIE ROUSSEAUX • *Institut Albert Bonniot, CNRS UMR 5309, INSERM U1209, Université Grenoble Alpes, Grenoble, France*
FRANCISCO SAAVEDRA • *Fundación Ciencia & Vida, Santiago, Chile*
ANDREAS SCHMIDT • *Protein Analysis Unit, Biomedical Center, Ludwig Maximilian University of Munich, Munich, Germany*
WOJCIECH SIWEK • *Instituto Gulbenkian de Ciência, Oeiras, Portugal*

TATIANA A. SOBOLEVA • *The John Curtin School of Medical Research, The Australian National University, Canberra, ACT, Australia*
DANIEL SOBRAL • *Instituto Gulbenkian de Ciência, Oeiras, Portugal*
HIDEAKI TAGAMI • *Graduate School of Natural Sciences, Nagoya City University, Nagoya, Aichi, Japan*
JELENA TELENIUS • *Medical Research Council (MRC) Molecular Haematology Unit, Weatherall Institute of Molecular Medicine, University of Oxford, Oxford, UK*
MICHAEL Y. TOLSTORUKOV • *Department of Molecular Biology, Massachusetts General Hospital and Harvard Medical School, Boston, MA, USA*
JÚLIA TORNÉ • *Institut Curie, PSL Research University, CNRS, UMR3664, Equipe Labellisée Ligue contre le Cancer, Paris, France; Sorbonne Universités, UPMC Univ Paris 06, CNRS, UMR3664, Paris, France*
DAVID J. TREMETHICK • *The John Curtin School of Medical Research, The Australian National University, Canberra, ACT, Australia*
JULIA TRUCH • *Medical Research Council (MRC) Molecular Haematology Unit, Weatherall Institute of Molecular Medicine, University of Oxford, Oxford, UK*
MORITZ CARL VÖLKER-ALBERT • *Chromatin Proteomics, Biomedical Center, Ludwig Maximilian University of Munich, Munich, Germany*
HEPING WANG • *Department of Neurosurgery, Tongji Hospital of Tongji Medical College, Huazhong University of Science and Technology, Wuhan, P. R. China*
ZHIGUO ZHANG • *Department of Pediatrics, Institute for Cancer Genetics, Columbia University, New York, NY, USA; Department of Genetics and Development, Institute for Cancer Genetics, Columbia University, New York, NY, USA*

Part I

Biochemistry

Chapter 1

Methods for Preparing Nucleosomes Containing Histone Variants

Tomoya Kujirai, Yasuhiro Arimura, Risa Fujita, Naoki Horikoshi, Shinichi Machida, and Hitoshi Kurumizaka

Abstract

Histone variants are key epigenetic players that regulate transcription, repair, replication, and recombination of genomic DNA. Histone variant incorporation into nucleosomes induces structural diversity of nucleosomes, consequently leading to the structural versatility of chromatin. Such chromatin diversity created by histone variants may play a central role in the epigenetic regulation of genes. Each histone variant possesses specific biochemical and physical characteristics, and thus the preparation methods are complicated. Here, we introduce the methods for the purification of human histone variants as recombinant proteins, and describe the preparation methods for histone complexes and nucleosomes containing various histone variants. We also describe the detailed method for the preparation of heterotypic nucleosomes, which may function in certain biological phenomena. These methods are useful for biochemical, structural, and biophysical studies.

Key words Histone variants, Recombinant histones, Nucleosome reconstitution, Chromatin, Histone

1 Introduction

In the nucleus of the eukaryotic cell, the genomic DNA is highly compacted and organized into chromatin, in which a basic repeating unit, the nucleosome, is connected by various lengths of linker DNAs [1]. Chromatin formation generally renders the genomic DNA inaccessible to DNA-binding proteins, which regulate genomic DNA functions, such as replication, repair, recombination, and transcription [2, 3]. Therefore, the structures and dynamics of chromatin may regulate the genomic DNA functions by restricting the accessibility of transacting DNA-binding factors [2, 3].

This chromatin-mediated genomic DNA regulation is considered to be a central mechanism for epigenetics, by which the functional status of genomic DNA is inherited by daughter cells [4]. However, the mechanism underlying epigenetic genomic DNA

Guillermo A. Orsi and Geneviève Almouzni (eds.), *Histone Variants: Methods and Protocols*, Methods in Molecular Biology, vol. 1832, https://doi.org/10.1007/978-1-4939-8663-7_1, © Springer Science+Business Media, LLC, part of Springer Nature 2018

regulation remains poorly understood, because information about the higher-order chromatin structures and chromatin dynamics has remained elusive.

The protein components of the nucleosome are histones H2A, H2B, H3, and H4 [5]. In the nucleosome, two copies each of histones H2A, H2B, H3, and H4 form a histone octamer, and about 150 base-pairs of DNA are left-handedly twined around it [5]. The overall nucleosome shape resembles a disk [5]. Various posttranslational modifications (PTMs) of histones, such as methylation, acetylation, crotonylation, hydroxyisobutyrylation, succinylation, malonylation, glutarylation, O-GlcNAcylation, phosphorylation, and ubiquitination, have been identified [6–10]. In addition, non-allelic isoforms of histones, called histone variants, have been found in many living organisms [11–25]. These histone modifications and variants are thought to affect the higher order chromatin configuration, providing versatility for the chromatin structure and dynamics [2, 3, 6].

The method for nucleosome reconstitution with recombinant histones was established by Luger et al. [5] and Dyer et al. [26]. Using this technology, the biochemical, biophysical, and structural aspects of various histone variants have been studied in nucleosomes and chromatin [27–44]. Here, we describe methods for the preparation of histone variants, histone complexes (H2A-H2B dimer, H3-H4 tetramer, and H2A-H2B-H3-H4 octamer), and nucleosomes containing histone variants. We also describe the reconstitution method for heterotypic nucleosomes, in which histone variants are heterotypically incorporated into the nucleosome. These methods are useful for further biochemical, biophysical, and structural studies of the chromatin architecture.

The methods for the purification and reconstitution of nucleosomes containing histone variants are summarized in Table 1.

2 Materials

2.1 Instruments and Equipment

1. Centrifugal concentrator (pore size: 10 kDa).
2. Centrifugal concentrator (pore size: 30 kDa).
3. Dialysis membrane (MWCO: 6–8 kDa).
4. Syringe filter unit (pore size: 0.22 μm).
5. HPLC.
6. Model 491 Prep Cell (Bio-Rad) with accessory equipment (power supply, peristaltic pump, UV detector, chart recorder, and fraction collector) or equivalent.

2.2 Matrices and Columns

1. Ni-NTA agarose.
2. Econo-Column (Bio-Rad) or equivalent.

Table 1
Corresponding section numbers for preparation methods of histone variants

Histones	Histone purification	H2A-H2B dimer reconstitution	H3-H4 tetramer reconstitution	Octamer reconstitution	Nucleosome reconstitution
H2A	3.1	3.2		3.4	3.5
H2A.Z.1	3.1	3.2		3.4	3.5
H2A.Z.2	3.1	3.2		3.4	3.5
H2A.B	–	3.3		–	3.6
H2B	3.1	3.2		3.4	3.5
H3.1	3.1		3.2	3.4	3.5
H3.2	3.1		3.2	3.4	3.5
H3.3	3.1		3.2	3.4	3.5
H3T	3.1		3.2	–	3.6
H3.5	3.1		3.2	3.4	3.5
H3.Y	3.1		3.2	3.4	3.5
H3.6	3.1		3.2	3.4	3.5
CENP-A	3.1		3.2	3.4	3.5
H4	3.1		3.2	3.4	3.5

3. MonoS 10/100 GL column (GE Healthcare) or equivalent.
4. HiLoad 16/60 Superdex 200 prep grade column (GE Healthcare) or equivalent.
5. Superdex 200 gel filtration column (1.5 cm diameter × 20 cm height) or equivalent.

2.3 Reagents and Buffers

1. LB medium: tryptone 25 g, NaCl 25 g, yeast extract 12.5 g, dissolved in 2.5 L of water. After autoclaving, add 1.25 mL of ampicillin (100 mg/mL) to a 50 μg/mL final concentration.
2. Thrombin protease.
3. PreScission protease (GE Healthcare) or equivalent.
4. Buffer 1: 50 mM Tris-HCl, pH 8.0, 500 mM NaCl, 1 mM PMSF, 5% glycerol.
5. Buffer 2: 50 mM Tris-HCl, pH 8.0, 500 mM NaCl, 7 M guanidine-HCl, 5% glycerol (*see* **Note 1**).
6. Buffer 3: 50 mM Tris-HCl, pH 8.0, 500 mM NaCl, 6 M urea, 5% glycerol, 5 mM imidazole (*see* **Note 1**).
7. Buffer 4: 50 mM Tris-HCl, pH 8.0, 500 mM NaCl, 6 M urea, 5% glycerol, 500 mM imidazole (*see* **Note 1**).

8. Buffer 5: 10 mM Tris-HCl, pH 8.0, 2 mM 2-mercaptoethanol.
9. Buffer 6: 20 mM NaOAc, pH 5.2, 200 mM NaCl, 6 M urea, 5 mM 2-mercaptoethanol, 1 mM EDTA (*see* **Note 1**).
10. Buffer 7: 20 mM NaOAc, pH 5.2, 900 mM NaCl, 6 M urea, 5 mM 2-mercaptoethanol, 1 mM EDTA (*see* **Note 1**).
11. Histone powder dissolving buffer: 20 mM Tris-HCl, pH 7.5, 7 M guanidine-HCl, 20 mM 2-mercaptoethanol.
12. 2 M Refolding buffer (2MRB): 10 mM Tris-HCl, pH 7.5, 2 M NaCl, 1 mM EDTA, 5 mM 2-mercaptoethanol.
13. 1 M Refolding buffer (1MRB): 10 mM Tris-HCl, pH 7.5, 1 M NaCl, 1 mM EDTA, 5 mM 2-mercaptoethanol.
14. 0.5 M Refolding buffer (0.5MRB): 10 mM Tris-HCl, pH 7.5, 0.5 M NaCl, 1 mM EDTA, 5 mM 2-mercaptoethanol.
15. 0.1 M Refolding buffer (0.1MRB): 10 mM Tris-HCl, pH 7.5, 0.1 M NaCl, 1 mM EDTA, 5 mM 2-mercaptoethanol.
16. Reconstitution buffer-high (RB-high): 10 mM Tris-HCl, pH 7.5, 2 M KCl, 1 mM EDTA, 1 mM DTT.
17. Reconstitution buffer-low (RB-low): 10 mM Tris-HCl, pH 7.5, 250 mM KCl, 1 mM EDTA, 1 mM DTT.
18. 1× TBE buffer: 90 mM Tris, 90 mM boric acid, 2 mM EDTA.
19. Elution buffer: 20 mM Tris-HCl, pH 7.5, 1 mM DTT.

3 Methods

3.1 Purification of Human Histone Variants

3.1.1 Plasmid Construction

The DNA fragments encoding human histones are each inserted into the pET15b vector using the *Nde*I and *Bam*HI sites, except for H2A.B and H3T [37, 40]. Recombinant histones are produced as the N-terminally hexa-histidine (His_6)-tagged proteins. In the histones produced with this system, the thrombin protease recognition sequence (L-V-P-R-G-S) is located just before the histone portion. Therefore, the His_6-tag peptide can be removed by thrombin protease treatment. After the removal of the His_6-tag peptide, four amino acid residues (G-S-H-M) remain at the N-terminus of the histones. For the production of histone H4, the codons were optimized for *Escherichia coli*, as described previously [45].

For the production of the histone H3 variant, human CENP-A, the codons were optimized for *E. coli*. The sequence from the initiating codon (ATG) to the stop codon (TAA) is as follows:

ATG GGC CCG CGT CGT CGT AGC CGT AAA CCG GAA GCA CCG CGT CGT CGT AGT CCG AGT CCG ACC CCG ACC CCG GGT CCG TCA CGT CGT GGT CCG TCT CTG GGT GCA TCT TCT CAT CAG CAT AGC CGT CGT CGC CAG GGT TGG CTG AAA GAA ATT CGC AAA CTG CAG

AAA AGC ACC CAT CTG CTG ATT CGC AAA CTG CCG TTT AGC CGC CTG GCG CGC GAA ATT TGC GTG AAA TTT ACC CGC GGC GTG GAT TTT AAC TGG CAG GCG CAG GCA CTG CTG GCA CTG CAA GAA GCG GCG GAA GCA TTT CTG GTG CAT CTG TTT GAA GAT GCG TAT CTG CTG ACC CTG CAT GCG GGC CGC GTG ACC CTG TTT CCG AAA GAT GTG CAG CTG GCG CGT CGC ATT CGT GGC CTG GAA GAA GGT CTG GGC TAA.

3.1.2 Histone Production and Purification

The purification method described here is applicable for the core histones H2A, H2B, H3, H4, and their variants, except for H2A.B.

1. Transform the *E. coli* BL21 (DE3) cells with a histone expression plasmid. For H4 and CENP-A, transform the *E. coli* JM109 (DE3) and BL21 (DE3) codon plus RIL strains, respectively. Plate the cells onto an LB agar plate containing 100 μg/mL ampicillin, and incubate the plate at 37 °C overnight.
2. Pick colonies from the plate, and inoculate 5 L of liquid LB media containing 50 μg/mL ampicillin.
3. Culture at 37 °C overnight with shaking.
4. Harvest the cells from the overnight culture by centrifugation at 7000 × *g* at 4 °C for 10 min.
5. Suspend the cells in 50 mL of buffer 1.
6. Disrupt the cells by 5–20 rounds of sonication (1 round is 200 s with cooling intervals). Avoid foaming and keep the sample on ice.
7. Centrifuge the whole cell lysate at 39,191 × *g* at 4 °C, and discard the supernatant. Histones are recovered in the insoluble fraction.
8. Resuspend the pellet by sonication in 50 mL of buffer 1.
9. After two rounds of sonication and centrifugation, resuspend the pellet using a spatula, and disrupt the pellet by sonication in 50 mL of buffer 2 (*see* **Note 2**).
10. Rotate the sample at 4 °C overnight.
11. Centrifuge the sample at 39,191 × *g* at 4 °C, and collect the supernatant.
12. Add 2 mL of prewashed Ni-NTA agarose beads to the supernatant, and rotate for 1 h at 4 °C.
13. Centrifuge the sample at 120 × *g* for 10 min, and discard the supernatant.
14. Pack the Ni-NTA beads into an Econo-Column.
15. Wash the beads with 50 column volumes (100 mL) of buffer 3, using a peristaltic pump (1 mL/min).

16. Elute the histones by a linear gradient of buffer 3 (50 mL) and buffer 4 (50 mL). After elution, analyze the fractions by 16% SDS-PAGE.
17. Collect the fractions containing the histones, and dialyze the histone fractions against 3 L of buffer 5 at 4 °C overnight, using the dialysis membrane tube.
18. Add the appropriate amount of thrombin protease for the His_6-tag peptide removal (see below). Gently mix the protein sample in thrombin protease buffer, containing 20 mM Tris-HCl (pH 7.5), 150 mM NaCl, and 2.5 mM $CaCl_2$ by inverting the tube 2–3 times, and then add the thrombin protease. Incubate the sample at room temperature for 3 h with rotation. For CENP-A preparation, the samples are incubated at room temperature overnight, without rotation (*see* **Note 3**). For H3.Y, the cleavage reaction is conducted without the thrombin buffer. To determine the proper amount of thrombin protease, a small-scale reaction is required, as described later.

    Small-scale reaction:

    Incubate the purified His_6-tagged histones (200 μL) with various amounts of thrombin protease at room temperature for 3 h (or overnight for CENP-A) (*see* **Note 4**). After the cleavage reaction, place the samples on ice and check them by 16% SDS-PAGE (*see* **Note 5**).
19. After the His_6-tag peptide removal, filter the sample with a 0.22 μm filter and apply the sample to a Mono S 10/100 GL column connected to an HPLC or an equivalent instrument.
20. Wash the Mono S 10/100 GL column with 4 column volumes of buffer 6, and elute the purified histones by a linear gradient of 200–900 mM NaCl with 20 column volumes of buffer 7 by the HPLC program (*see* **Note 6**).
21. Analyze the peak fractions by 16% SDS-PAGE, and collect the fractions containing the purified histones (*see* **Note 7**).
22. Dialyze the sample 4 times for 4 h each against 3 L of water containing 2 mM 2-mercaptoethanol for desalting.
23. Transfer the sample to a 50 mL tube, freeze the sample in liquid nitrogen, and lyophilize the frozen sample.
24. Store the lyophilized histone powder at 4 °C.

The concentration of purified histones can be estimated by measuring the absorbance at 280 nm. We usually determined the extinction coefficient value of each histone with the absorbance of 1 mg of a lyophilized histone dissolved in 1 mL water. These extinction coefficient values are summarized in Table 2. The experimental extinction coefficient value of H2A.Z.1 is employed as the H2A.Z.2 value. Similarly, the experimental extinction coefficient

Table 2
Extinction coefficients for histones

Histones	Experimental value [M^{-1} cm^{-1}]	Calculated value [M^{-1} cm^{-1}]
H2A	4215	4470
H2A.Z.1	5527	2980
H2A.Z.2	–	2980
H2A.B	–	4470
H2B	3101	7450
H3.1	1782	4470
H3.2	–	4470
H3.3	–	4470
H3T	–	4470
H3.5	2924	4470
H3.Y	9140	9970
H3.6	–	4470
CENP-A	12,860	12,490
H4	4030	5960
H2A^{peptide}	9678	15,470
H3.3^{SUMO}	–	5960

value of H3.1 is used for those of H3.2, H3.3, H3T, and H3.6. For H2A.B and His$_6$-SUMO-H3.3, we used the calculated extinction coefficient values that were estimated by the ProtParam program (http://web.expasy.org/protparam/). The approximate extinction coefficient of histone complexes can be estimated from the sum of the coefficient of each histone. Since the final nucleosome concentration is determined as the nucleosomal DNA concentration, the experimental and calculated extinction coefficient values are both provided.

3.2 Reconstitution of H2A-H2B Dimer and H3-H4 Tetramer

1. Measure the weight of the lyophilized histone powders, and mix them in 1: 1 stoichiometry. For the H2A-H2B dimer, 1.2 mg of H2A and 1.2 mg of H2B are adequate. For the H3-H4 tetramer, 1.4 mg of H3 and 1.0 mg of H4 are adequate.
2. Fix the histone concentration to 1.0 mg/mL, by adding histone powder dissolving buffer to the histone mixture.

3. Rotate the sample for 1.5 h at 4 °C.
4. Dialyze the sample against 500 mL of 2 M refolding buffer (2MRB) at 4 °C for 4 h. After dialysis in 2MRB, change the dialysis buffer to 500 mL of 1MRB, 0.5MRB, and 0.1MRB, in a stepwise manner, and in each step, dialyze the sample for 4 h at 4 °C. If the sample contains a precipitate, remove it by centrifugation (*see* **Note 8**).
5. After four rounds of dialysis, filter the sample with a 0.22 μm filter and concentrate the volume to 200 μL with a centrifugal concentrator (pore size: 10 kDa). If the sample contains a precipitate, remove it by centrifugation.
6. Load the sample onto a Superdex 200 gel filtration column (1.5 cm diameter × 20 cm height) or a HiLoad 16/60 Superdex 200 prep grade column. Equilibrate the gel filtration column with 1.2 column volumes of 0.1MRB, just before loading the sample.
7. Analyze the eluted fractions by 16% SDS-PAGE, and collect the fractions containing H2A and H2B, or H3 and H4 in equal stoichiometry.
8. Concentrate the histone complexes with a centrifugal concentrator (pore size: 10 kDa).
9. Measure the absorbance at 280 nm, and determine the concentration with the extinction coefficient of the histone complexes.
10. Freeze the histone complex in liquid nitrogen, and store at −80 °C.

3.3 Reconstitution of H2A.B-H2B Dimer

1. Culture BL21(DE3) cells producing H2A.B in 5 L of liquid LB media containing 50 μg/mL ampicillin, at 37 °C overnight.
2. The His_6-tagged H2A.B is purified by Ni-NTA chromatography by the same method described in Subheading 3.1.2.
3. After elution from the Ni-NTA beads as described in Subheading 3.1.2, **step 16**, concentrate the fractions containing H2A.B with a centrifugal concentrator (pore size: 10 kDa) to 0.5 mg/mL.
4. After concentration, mix 2.4 mL of 0.5 mg/mL of His_6-tagged H2A.B with an equal molar amount of lyophilized H2B powder (1.2 mg of H2B powder), and form the His-H2A.B-H2B dimer.
5. Rotate the sample for 1.5 h at 4 °C.
6. Dialyze the sample against 100 mL of histone powder dissolving buffer at 4 °C for 4 h.

7. After dialysis, change the dialysis buffer to 500 mL of 2MRB, 1MRB, 0.5MRB, and 0.1MRB, sequentially, and dialyze for more than 4 h against each buffer.
8. Centrifuge the sample to remove any precipitate, and collect the supernatant (*see* **Note 8**).
9. To remove the His_6-tag peptide, add thrombin protease (approximately 1 unit to 1 mg of the H2A.B molecule), and rotate the sample for 3 h at room temperature.
10. After checking the His_6-tag peptide removal by SDS-PAGE, filter the sample with a 0.22 μm filter and apply the sample to a HiLoad 16/60 Superdex 200 prep grade column equilibrated with 2MRB.
11. Analyze the eluted fractions by 16% SDS-PAGE, and collect the fractions containing H2A.B and H2B in equal stoichiometry.
12. Concentrate the sample with a centrifugal concentrator (pore size: 10 kDa).
13. Measure the absorbance at 280 nm, and determine the sample concentration with the extinction coefficient of the histone complexes.
14. Freeze the H2A.B-H2B dimer in liquid nitrogen, and store at −80 °C.

3.4 Reconstitution of Histone Octamer

1. Measure the weight of each lyophilized histone powder, and mix them in about 1: 1: 1: 1 molar stoichiometry. For the formation of a canonical histone octamer, 1.2 mg of H2A, 1.2 mg of H2B, 1.4 mg of H3, and 1.0 mg of H4 are appropriate (*see* **Note 9**).
2. Dissolve these histone powders (1.5 mg/mL concentration) in histone powder dissolving buffer, and rotate the sample for 1.5 h at 4 °C.
3. Dialyze the sample for 4 h against 500 mL of 2MRB, 4 times. After dialysis, if the sample contains a precipitate, remove it by centrifugation (*see* **Note 8**).
4. Filter the sample with a 0.22 μm filter and purify the sample on a HiLoad 16/60 Superdex 200 prep grade column. Equilibrate the gel filtration column with 1.2 column volumes of 2MRB, just before loading the sample.
5. Analyze the peak fractions by 16% SDS-PAGE, and collect the fractions containing H2A, H2B, H3, and H4 in equal stoichiometry (Fig. 1).
6. Concentrate the peak fractions with a centrifugal concentrator (pore size: 30 kDa).

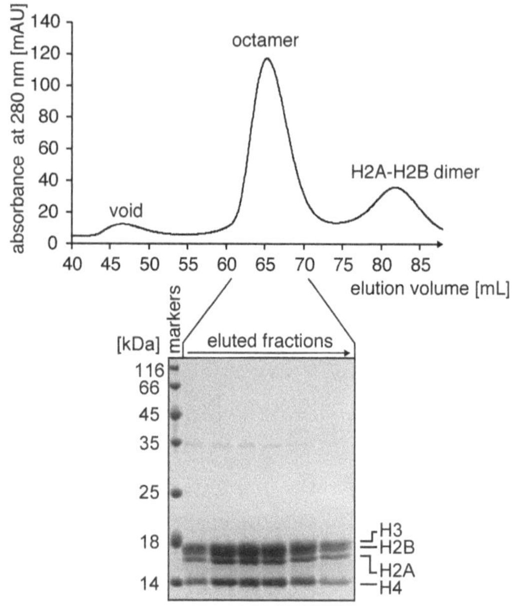

Fig. 1 A chromatogram of the histone octamer eluted from a HiLoad 16/60 Superdex 200 prep grade gel filtration column. Peaks corresponding to void, octamer, and H2A-H2B dimer are shown. The octamer peak fractions were analyzed by 16% SDS-PAGE with Coomassie Brilliant Blue staining

7. Measure the absorbance at 280 nm, and determine the sample concentration with the extinction coefficient of the histone complexes.
8. Freeze the histone octamer in liquid nitrogen, and store at −80 °C.

3.5 Reconstitution of Nucleosome with Histone Octamer

We perform the nucleosome reconstitution with a histone octamer by the salt dialysis method. Since some histone variants, such as human histones H2A.B and H3T, are defective in histone octamer formation, the nucleosomes containing these histone variants are reconstituted with the H2A-H2B dimer and the H3-H4 tetramer by the method described following Subheading 3.6 (see below).

1. Mix the histone octamer and DNA (145-200 base pairs of DNA) in a histone octamer: DNA molar ratio of 1.0–2.0: 1.0 for a 146 base-pair α-satellite DNA [5] or a 145 base-pair 601 sequence DNA [46]. To determine the appropriate histone octamer: DNA ratio, a small-scale reconstitution is required before the large-scale reconstitution. Titrate a histone octamer to a small amount of DNA (*c.a.* 50 μg), and reconstitute the nucleosomes according to the following procedures (*see* **Note 10**).
2. Fix the final concentration of DNA to 0.8 mg/mL, under conditions with 2 M KCl. In this step, the water, 4 M KCl, DNA,

and histone octamer should be mixed in this order. The histone octamer should not be exposed to a KCl concentration lower than 2 M.

3. Dialyze the sample mixture against 400 mL of RB-high.
4. Decrease the KCl concentration by exchanging the dialysis solution to that containing 250 mM KCl with RB-low (1.6 L) using a peristaltic pump (0.8 mL/min).
5. After the salt dialysis step, dialyze the sample against 400 mL of RB-low for 4 h.
6. Check the sample by native-PAGE. If the reconstituted nucleosome exhibits multiple bands on the native polyacrylamide gel, the sample may be incubated at 55 °C for 2 h to induce nucleosome repositioning (*see* **Note 11**).

3.6 Reconstitution of Nucleosome with H2A-H2B Dimer and H3-H4 Tetramer

1. Mix the H2A-H2B dimer: H3-H4 tetramer: DNA in an approximately 3: 1.5: 1 molar ratio under 2 M KCl conditions. The final concentration of DNA is 0.65 mg/mL.
2. After mixing the DNA and the histone complexes, dialyze the sample mixture against RB-high, and reconstitute the nucleosome by the salt dialysis method, as described in Subheading 3.5, **steps 3–5**.

3.7 Purification of Nucleosome by the Prep Cell Apparatus

The reconstituted nucleosomes are purified by preparative native-PAGE with the Prep Cell apparatus, according to the following method. In this step, the hexasome (a complex composed of one H2A-H2B dimer, one H3-H4 tetramer, and DNA), the free DNA, and the free histones are removed. Dialysis Membrane is used as the filtering membrane for the Prep Cell.

1. Prepare a native-PAGE gel (0.5× TBE (or 0.2× TBE), 6% polyacrylamide) (*see* **Note 12**) using a Model 491 Prep Cell apparatus (Bio-Rad), according to the manufacturer's instructions. For nucleosome purification, a 6.5 cm height of the gel is adequate.
2. Prepare 20 mM Tris-HCl (pH 7.5) buffer containing 1 mM DTT for the elution solution.
3. Add 30% sucrose to the nucleosome sample to a 5% final concentration.
4. After pre-running at 10 W for 1 h at a flow rate of 1 mL/min, apply the nucleosome sample with 5% sucrose to the Prep Cell apparatus, and run at 10 W for approximately 2 h. The flow rate and fraction size are set to 2.0 mL/min and 2.0 mL/fraction, respectively.
5. Monitor the eluted fractions by measuring the absorbance at 260 nm. A representative elution profile of the nucleosomes from the Prep Cell apparatus is shown in Fig. 2.

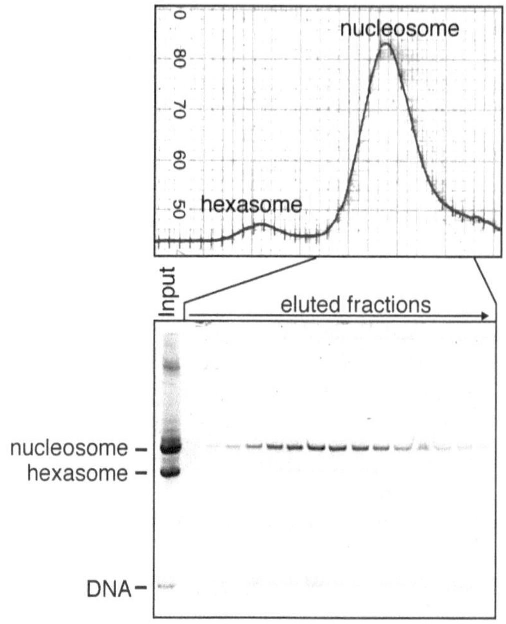

Fig. 2 A Prep Cell elution profile of a nucleosome containing 193 base-pairs of the 601 positioning sequence DNA. Nucleosomes reconstituted by the salt dialysis method were fractionated by a Prep Cell (6.5 cm height, 6% acrylamide, 0.5× TBE). The nucleosome peak fractions were analyzed by 6% 0.5× TBE native-PAGE with ethidium bromide staining

6. Analyze the peak fractions by 6% native-PAGE in 0.5× TBE (or 0.2× TBE) buffer.
7. Collect the fractions containing the nucleosomes, and concentrate with a centrifugal concentrator (pore size: 30 kDa).
8. Determine the concentration of the purified nucleosomes by measuring the absorbance at 260 nm, and store at 4 °C.

3.8 Reconstitution of the Heterotypic Nucleosome (H2A.Z-H2A)

The heterotypic nucleosome containing H2A.Z and H2A is reportedly formed in cells [47, 48]. We previously established the purification method for the H2A.Z/H2A heterotypic nucleosome [32].

1. Prepare the peptide-fused H2A. A peptide containing 120 amino acid residues is fused to the C-terminus of human histone H2A ($H2A^{peptide}$) (*see* **Note 13**). The purification method for $H2A^{peptide}$ is the same as that for the canonical histones, as described in Subheading 3.1.2, **steps 1–16**.
2. After purification by Ni-NTA agarose column chromatography, dialyze the sample against 3 L of sterile water at 4 °C for 4 h.
3. After four rounds of dialysis, lyophilize the sample.

4. Mix the lyophilized H2A^{peptide}, H2A.Z, H2B, H3, and H4 (0.5: 0.5: 1: 1: 1 molar ratio) in histone powder dissolving buffer (1.0 mg/mL concentration).
5. Rotate the sample mixture for 1.5 h at 4 °C.
6. Prepare the histone octamers containing H2A^{peptide} with the protocol described in Subheading 3.4. In this step, the resulting histone octamers contain the H2A^{peptide}/H2A^{peptide} homotypic, H2A.Z/H2A^{peptide} heterotypic, and H2A.Z/H2A.Z homotypic octamers.
7. Reconstitute the nucleosomes by the protocol described in Subheading 3.5. The reconstituted nucleosome sample contains the H2A^{peptide}/H2A^{peptide} homotypic, H2A.Z/H2A^{peptide} heterotypic, and H2A.Z/H2A.Z homotypic nucleosomes. The nucleosome formation can be monitored by 6% native-PAGE in 0.2x TBE (or 0.5× TBE) buffer. The H2A.Z/H2A^{peptide} heterotypic nucleosome corresponds to the middle band among the three nucleosomal bands.
8. Prepare a native-PAGE gel (0.5× TBE (or 0.2× TBE), 6% polyacrylamide) (*see* **Note 12**), using a Model 491 Prep Cell apparatus (Bio-Rad), according to the manufacturer's instructions. For nucleosome purification, a 6.5 cm height of the gel is adequate.
9. Prepare 20 mM Tris-HCl (pH 7.5) buffer containing 1 mM DTT for the elution solution.
10. Add 30% sucrose to the nucleosome sample to a 5% final concentration.
11. After pre-running at 10 W for 1 h, apply the nucleosome sample with 5% sucrose to the Prep Cell apparatus, and run at 10 W for approximately 2 h. The flow rate and fraction size are set to 2.5 mL/min and 2.5 mL/fraction, respectively.
12. Collect the fractions containing the H2A.Z/H2A^{peptide} heterotypic nucleosome.
13. Add thrombin protease (5–10 units) to 1 mg of the H2A^{peptide} molecule in the H2A.Z/H2A^{peptide} heterotypic nucleosome, and incubate at 16 °C overnight to remove the additional peptide.
14. Confirm the removal of the peptide by native-PAGE and SDS-PAGE.
15. Purify the resulting H2A.Z/H2A heterotypic nucleosome using a Model 491 Prep Cell with the same settings as described above.
16. Collect the fractions containing the H2A.Z/H2A heterotypic nucleosome.

17. Concentrate the purified H2A.Z/H2A heterotypic nucleosome to a concentration greater than 0.1 mg/mL with a centrifugal concentrator (pore size: 30 kDa).
18. Store the sample at 4 °C.

3.9 Reconstitution of Heterotypic Nucleosome (CENP-A-H3)

The heterotypic nucleosome containing CENP-A and H3.3 has been found in aggressive cancer cells [49]. We previously reported the method for the heterotypic nucleosome preparation with CENP-A and H3.3 [44]. In this section, we describe the details of the method.

1. Prepare the H3.3-SUMO fusion protein. A His_6-SUMO peptide is fused to the N-terminus of human histone H3.3 ($H3.3^{SUMO}$) (*see* **Note 13**). The purification method for $H3.3^{SUMO}$ is the same as that for canonical histones, as described in Subheading 3.1.2, without His-SUMO tag cleavage.
2. Mix the lyophilized powders of H2A, H2B, $H3.3^{SUMO}$, CENP-A, and H4 in a 1: 1: 0.6: 0.4: 1 ratio in histone powder dissolving buffer (1.0 mg/mL concentration).
3. After rotation at 4 °C for 1.5 h, dialyze the sample mixture against 2MRB for 4 h.
4. After four rounds of dialysis, remove the precipitate by centrifugation (*see* **Note 8**).
5. After filtering the sample with a 0.22 μm filter, load the sample onto a HiLoad 16/60 Superdex 200 prep grade column previously equilibrated with 1.2 column volumes of 2MRB by the same method as described in Subheading 3.4, **step 4**.
6. After elution of the histone complexes with 1 column volume of 2MRB, analyze the eluted fractions by SDS-PAGE, and collect the fractions containing the H2A, H2B, $H3.3^{SUMO}$, CENP-A and H4.
7. Reconstitute the nucleosome by the method described in Subheading 3.5. Three nucleosomes are formed: (1) the homotypic $H3.3^{SUMO}$/$H3.3^{SUMO}$ nucleosome, (2) the homotypic CENP-A/CENP-A nucleosome, and (3) the heterotypic CENP-A/$H3.3^{SUMO}$ nucleosome.
8. Purify the heterotypic CENP-A/$H3.3^{SUMO}$ nucleosome using a Model 491 Prep Cell (Bio-Rad), as described in Subheading 3.8, **steps 8–11**.
9. Collect the fractions containing the CENP-A/$H3.3^{SUMO}$ nucleosome.
10. Concentrate the purified CENP-A/$H3.3^{SUMO}$ heterotypic nucleosome with a centrifugal concentrator (pore size: 30 kDa) to a concentration greater than 0.1 mg/mL.

11. Add PreScission protease (approximately 50 units) to 1 mg of the $H3.3^{SUMO}$ molecule in the CENP-A/$H3.3^{SUMO}$ heterotypic nucleosome, and incubate at 4 °C overnight to remove the His_6-SUMO portion of H3.3.
12. Confirm the removal of the His_6-SUMO peptide by native-PAGE and SDS-PAGE.
13. Purify the heterotypic nucleosome containing CENP-A and H3.3 with a Model 491 Prep Cell with the same settings as described above.
14. Collect the fractions containing the CENP-A/H3.3 heterotypic nucleosome.
15. Concentrate the purified CENP-A/H3.3 heterotypic nucleosome with a centrifugal concentrator (pore size: 30 kDa) to concentrate greater than 0.1 mg/mL.
16. Store the sample at 4 °C.

4 Notes

1. Solutions containing urea or guanidine-HCl should be prepared immediately before use. Urea should be deionized by a treatment with Amberlite IRN150 resin.
2. Histone recovery is improved, if the pellet is dissolved by sonication as much as possible.
3. To avoid protein precipitation, the protein solution should be gradually mixed with the thrombin protease buffer. Freezing and thawing should also be avoided.
4. The titration range of thrombin protease is from 0.2 to 1.6 U/mg, in increments of 0.2 U/mg.
5. His_6-tagged histones migrated slowly as compared to the untagged histones on the gel. Therefore, the removal of the His_6-tag can be monitored by SDS-PAGE.
6. The sample should be washed until the absorption at 280 nm reaches the baseline. Four column volumes may be sufficient. To elute the sample by a linear gradient, the protocols provided with an HPLC are useful.
7. Collect fractions without degraded histone fragments or His_6-tagged histones.
8. The precipitate is observed as a white material, and can be pelleted by centrifugation at 15,000 × g for 10 min at 4 °C. Recover the resulting supernatant.
9. For the reconstitution of the histone octamers containing histone variants, the molar ratio of H2A, H2B, H3, and H4 is approximately 1:1:1:1.

10. The titration range of a histone octamer to DNA in molar ratio is from 1.0–2.0, in increments of 0.2.
11. Multiple bands suggest that the sample contains sub-nucleosomes (e.g., hexasomes), or differently positioned nucleosomes. Hexasomes will disappear by increasing the amount of the histone octamer or H2A-H2B dimer.
12. For the separation of nucleosomes and hexasomes containing the 146 base-pair α-satellite derivative, the nucleosome migrates faster than the hexasome under electrophoresis conditions with 0.2× TBE.
13. Nucleosomes containing a SUMO-tagged histone or a peptide-fused histone migrate slowly in native PAGE. The heterotypic nucleosomes can be separated on the gel because of their different migrations.

Acknowledgments

We thank Ms. Yukari Iikura (Waseda University) for her assistance. This work was supported in part by JSPS KAKENHI Grant Numbers JP25116002 and JP17H01408, by Platform Project for Supporting Drug Discovery and Life Science Research (Basis for Supporting Innovative Drug Discovery and Life Science Research (BINDS)) from AMED under Grant Number JP18am0101076, and by JST CREST Grant Number PMJCR16G1, Japan [to H.K.]. H.K. and N.H. were supported by the Waseda Research Institute for Science and Engineering. T.K. and R.F. were supported by Research Fellowships from JSPS for Young Scientists [JP15J06807 and JP16J10043, respectively]. Y.A. and S.M. were supported in part by JSPS KAKENHI Grant Numbers JP17K15043 [to Y.A.] and JP16K18473 [to S.M.].

References

1. Wolffe A (1998) Chromatin: structure and function, 3rd edn. Academic Press, San Diego
2. Luger K, Dechassa ML, Tremethick DJ et al (2012) New insights into nucleosome and chromatin structure: an ordered state or a disordered affair? Nat Rev Mol Cell Biol 13:436–447
3. Venkatesh S, Workman JL (2015) Histone exchange, chromatin structure and the regulation of transcription. Nat Rev Mol Cell Biol 16:178–189
4. Campos EI, Stafford JM, Reinberg D (2014) Epigenetic inheritance: histone bookmarks across generations. Trends Cell Biol 24:664–674
5. Luger K, Mäder AW, Richmond RK et al (1997) Crystal structure of the nucleosome core particle at 2.8 Å resolution. Nature 389:251–260
6. Strahl BD, Allis CD (2000) The language of covalent histone modifications. Nature 403:41–45
7. Kouzarides T (2007) Chromatin modifications and their function. Cell 128:693–705
8. Bhaumik SR, Smith E, Shilatifard A (2007) Covalent modifications of histones during development and disease pathogenesis. Nat Struct Mol Biol 14:1008–1016

9. Bannister AJ, Kouzarides T (2011) Regulation of chromatin by histone modifications. Cell Res 21:381–395
10. Rousseaux S, Khochbin S (2015) Histone acylation beyond acetylation: terra incognita in chromatin biology. Cell J 17:1–6
11. Szenker E, Ray-Gallet D, Almouzni G (2011) The double face of the histone variant H3.3. Cell Res 21:421–434
12. Franklin SG, Zweidler A (1977) Non-allelic variants of histones 2a, 2b and 3 in mammals. Nature 266:273–275
13. Talbert PB, Henikoff S (2010) Histone variants - ancient wrap artists of the epigenome. Nat Rev Mol Cell Biol 11:264–275
14. Palmer DK, O'Day K, Wener MH et al (1987) A 17-kD centromere protein (CENP-A) copurifies with nucleosome core particles and with histones. J Cell Biol 104:805–815
15. Albig W, Ebentheuer J, Klobeck G et al (1996) A solitary human H3 histone gene on chromosome 1. Hum Genet 97:486–491
16. Witt O, Albig W, Doenecke D (1996) Testis-specific expression of a novel human H3 histone gene. Exp Cell Res 229:301–306
17. Brush D, Dodgson JB, Choi OR et al (1985) Replacement variant histone genes contain intervening sequences. Mol Cell Biol 5:1307–1317
18. Schenk R, Jenke A, Zilbauer M et al (2011) H3.5 is a novel hominid-specific histone H3 variant that is specifically expressed in the seminiferous tubules of human testes. Chromosoma 120:275–285
19. Wiedemann SM, Mildner SN, Bönisch C et al (2010) Identification and characterization of two novel primate-specific histone H3 variants, H3.X and H3.Y. J Cell Biol 190:777–791
20. Taguchi H, Xie Y, Horikoshi N et al (2017) Crystal structure and characterization of novel human histone H3 variants, H3.6, H3.7, and H3.8. Biochemistry 56:2184–2196
21. West MH, Bonner WM (1980) Histone 2A, a heteromorphous family of eight protein species. Biochemistry 19:3238–3245
22. Iouzalen N, Moreau J, Méchali M (1996) H2A.ZI, a new variant histone expressed during Xenopus early development exhibits several distinct features from the core histone H2A. Nucleic Acids Res 24:3947–3952
23. Eirín-López JM, González-Romero R, Dryhurst D et al (2009) The evolutionary differentiation of two histone H2A.Z variants in chordates (H2A.Z-1 and H2A.Z-2) is mediated by a stepwise mutation process that affects three amino acid residues. BMC Evol Biol 9:31
24. Chadwick BP, Willard HF (2001) A novel chromatin protein, distantly related to histone H2A, is largely excluded from the inactive X chromosome. J Cell Biol 152:375–384
25. Pehrson JR, Fried VA (1992) MacroH2A, a core histone containing a large nonhistone region. Science 257:1398–1400
26. Dyer PN, Edayathumangalam RS, White CL et al (2004) Reconstitution of nucleosome core particles from recombinant histones and DNA. Methods Enzymol 375:23–44
27. Suto RK, Clarkson MJ, Tremethick DJ et al (2000) Crystal structure of a nucleosome core particle containing the variant histone H2A.Z. Nat Struct Biol 7:1121–1124
28. Fan JY, Gordon F, Luger K et al (2002) The essential histone variant H2A.Z regulates the equilibrium between different chromatin conformational states. Nat Struct Biol 9:172–176
29. Park YJ, Dyer PN, Tremethick DJ et al (2004) A new fluorescence resonance energy transfer approach demonstrates that the histone variant H2AZ stabilizes the histone octamer within the nucleosome. J Biol Chem 279:24274–24282
30. Horikoshi N, Sato K, Shimada K et al (2013) Structural polymorphism in the L1 loop regions of human H2A.Z.1 and H2A.Z.2. Acta Crystallogr D Biol Crystallogr 69:2431–2439
31. Sugiyama M, Horikoshi N, Suzuki Y et al (2015) Solution structure of variant H2A.Z.1 nucleosome investigated by small-angle X-ray and neutron scatterings. Biochem Biophys Rep 4:28–32
32. Horikoshi N, Arimura Y, Taguchi H et al (2016) Crystal structures of heterotypic nucleosomes containing histones H2A.Z and H2A. Open Biol 6:160127
33. Bao Y, Konesky K, Park YJ et al (2004) Nucleosomes containing the histone variant H2A.Bbd organize only 118 base pairs of DNA. EMBO J 23:3314–3324
34. Gautier T, Abbott DW, Molla A et al (2004) Histone variant H2ABbd confers lower stability to the nucleosome. EMBO Rep 5:715–720
35. Angelov D, Verdel A, An W et al (2004) SWI/SNF remodeling and p300-dependent transcription of histone variant H2ABbd nucleosomal arrays. EMBO J 23:3815–3824
36. Sugiyama M, Arimura Y, Shirayama K et al (2014) Distinct features of the histone core structure in nucleosomes containing the histone H2A.B variant. Biophys J 106:2206–2213

37. Arimura Y, Kimura H, Oda T et al (2013) Structural basis of a nucleosome containing histone H2A.B/H2A.Bbd that transiently associates with reorganized chromatin. Sci Rep 3:3510
38. Chakravarthy S, Gundimella SK, Caron C et al (2005) Structural characterization of the histone variant macroH2A. Mol Cell Biol 25:7616–7624
39. Tachiwana H, Osakabe A, Shiga T et al (2011) Structures of human nucleosomes containing major histone H3 variants. Acta Crystallogr D Biol Crystallogr 67:578–583
40. Tachiwana H, Kagawa W, Osakabe A et al (2010) Structural basis of instability of the nucleosome containing a testis-specific histone variant, human H3T. Proc Natl Acad Sci U S A 107:10454–10459
41. Urahama T, Harada A, Maehara K et al (2016) Histone H3.5 forms an unstable nucleosome and accumulates around transcription start sites in human testis. Epigenetics Chromatin 9:2
42. Kujirai T, Horikoshi N, Sato K et al (2016) Structure and function of human histone H3.Y nucleosome. Nucleic Acids Res 44:6127–6141
43. Tachiwana H, Kagawa W, Shiga T et al (2011) Crystal structure of the human centromeric nucleosome containing CENP-A. Nature 476:232–235
44. Arimura Y, Shirayama K, Horikoshi N et al (2014) Crystal structure and stable property of the cancer-associated heterotypic nucleosome containing CENP-A and H3.3. Sci Rep 4:7115
45. Tanaka Y, Tawaramoto-Sasanuma M, Kawaguchi S et al (2004) Expression and purification of recombinant human histones. Methods 33:3–11
46. Lowary PT, Widom J (1998) New DNA sequence rules for high affinity binding to histone octamer and sequence-directed nucleosome positioning. J Mol Biol 276:19–42
47. Viens A, Mechold U, Brouillard F et al (2006) Analysis of human histone H2AZ deposition in vivo argues against its direct role in epigenetic templating mechanisms. Mol Cell Biol 26:5325–5335
48. Nekrasov M, Amrichova J, Parker BJ et al (2012) Histone H2A.Z inheritance during the cell cycle and its impact on promoter organization and dynamics. Nat Struct Mol Biol 19:1076–1083
49. Lacoste N, Woolfe A, Tachiwana H et al (2014) Mislocalization of the centromeric histone variant CenH3/CENP-A in human cells depends on the chaperone DAXX. Mol Cell 53:631–644

Chapter 2

Characterization of Posttranslational Modifications on Histone Variants

Francisco Saavedra, Sebastián Marty-Lombardi, and Alejandra Loyola

Abstract

The study of histone variants and histone posttranslational modifications (PTMs) is a trending topic in different fields including developmental biology, neurobiology, and immunology; as well as in the understanding of molecular mechanisms leading to diverse diseases, such as cancer. Since the establishment of histone PTMs starts immediately after their synthesis and it continues once they are assembled into chromatin, here we describe a classic protocol of subcellular fractionation aiming to study histones at different stages of maturation. This includes newly synthesized histones enriched in cytosolic fractions; a pool of newly synthesized, evicted, and stored histones enriched in nuclear soluble fractions; and chromatin-associated histones enriched in chromatin pellet. To study specific histone variants and the establishment of their PTMs, we describe a protocol for obtaining histone variants expressed in bacteria. In addition, we describe a Triton-Acetic acid-Urea (TAU) gel electrophoresis protocol adapted to work on mini-gels, which can be coupled to Western blot to analyze PTMs on histone variants. Finally, we describe a Chromatin immunoprecipitation (ChIP) assay for studying histone PTMs, or tagged histone variants, on specific DNA sequences.

Key words Histones, Histone variants, Recombinant histones, Histone purification, PTMs, TAU gel electrophoresis, Acid extraction, ChIP

1 Introduction

Histones are proteins that, together with DNA, structure the chromatin inside the nucleus of eukaryotic cells. They have low molecular weights, between 10 and 16 kDa, high proportion of lysines and arginines in their amino acid sequences and isoelectric points of around pH 10 [1, 2]. Consequently, histones have a positive net charge under physiological conditions, which makes them highly affine to DNA. Although the main function of his-

Francisco Saavedra and Sebastián Marty-Lombardi contributed equally to this work.

Guillermo A. Orsi and Geneviève Almouzni (eds.), *Histone Variants: Methods and Protocols*, Methods in Molecular Biology, vol. 1832, https://doi.org/10.1007/978-1-4939-8663-7_2, © Springer Science+Business Media, LLC, part of Springer Nature 2018

tones is to compact the DNA, they are far from being just DNA packaging proteins. A growing variety of histone Posttranslational Modifications (PTMs) have been described, which are capable of directly impacting into chromatin structure by altering histone-DNA interactions or by acting as an assembly point for different protein machineries. Thus, histone PTMs influence all the functions where DNA is involved, including gene expression, some being correlated with gene activation and others with silencing [3, 4]. Histone variants add another level of complexity to the chromatin structure. They are defined as histones whose expression is not linked to the cell cycle progression, unlike canonical histones H3, H4, H2A, and H2B, which are synthesized coupled to DNA replication [5–7]. Histone variants have diverse roles; for instance, H3.3 regulates gene transcription and silencing [8]; phosphorylated H2A.X participates in the response to DNA double-strand break [9].

Here, we describe a variety of techniques for the study of the different pools of histones that are found inside the cell. These pools include histones bound to chromatin, called nucleosomal histones, and diverse non-nucleosomal histones, called soluble histones. Among the soluble pool, nuclear histones are composed of evicted, stored, and newly synthesized histones, while the cytosolic population is enriched on newly synthesized histones [10, 11]. We describe a subcellular fractionation method, based in Dignam et al. [12] that isolates cytosolic extract, nuclear extract, and nuclear pellet from HeLa cells. In addition, this fractionation method can be coupled to acid extraction to obtain histone-enriched samples by exploiting their basic character [1]. These isolation methodologies allow recovering histone pools for the analyses of PTMs and histone variants in different cellular reservoirs by classic and simple assays, like Western blot, taking advantage of the extensive variety of commercially available antibodies.

Furthermore, we describe a method to obtain high amounts of recombinant histones by using bacterial expression vectors [13]. This gives the possibility of expressing the desired histone variant without PTMs, and therefore, being suitable for different assays such as the evaluation of enzymatic activities and specificities related to the establishment of PTMs. Moreover, we describe a technique to separate histone variants based on the changes in the amino acidic composition between variants and canonical histones by using TAU (Triton X-100, Acetic acid, Urea) gel electrophoresis [14]. When this protocol is coupled to Western blot assays, PTM analyses on histones variants are greatly facilitated. Finally, we describe a ChIP (Chromatin Immunoprecipitation) methodology to study PTMs and histone variants on specific DNA sequences.

2 Materials

Prepare all solutions using distilled water and analytical grade reagents. Prepare and store all stock solutions at room temperature, unless otherwise indicated, up to a month. When handling irritant or volatile compounds, like glacial Acetic acid, work under fume hood. Prepare all working solutions the day before use and keep them at 4 °C, unless otherwise indicated. Add PMSF and DTT to the working solutions at the moment of use (*see* **Note 1**). DTT can be replaced by β-Mercaptoethanol (*see* **Note 2**). Cell culture solutions must be handled in sterility conditions and stored at 4 °C.

2.1 Stock Solutions

1. 2.5 M KCl. Dissolve 186.38 g of Potassium chloride in distilled water to a final volume of 1000 mL.
2. 1 M TRIS pH 7.9. Dissolve 121.14 g of Tris(hydroxymethyl) aminomethane (TRIS) in distilled water to a final volume of 1000 mL, adjusting pH to 7.9 with concentrated HCl.
3. 5 M NaCl. Dissolve 292.20 g of Sodium chloride in distilled water to a final volume of 1000 mL.
4. 1 M $MgCl_2$. Dissolve 4.76 g of Magnesium chloride in distilled water to a final volume of 50 mL.
5. 0.5 M EDTA pH 8.0. Dissolve 73.06 g of Ethylenediaminetetraacetic acid (EDTA) in distilled water to a final volume of 500 mL, adjusting pH to 8.0 with NaOH pellets (*see* **Note 3**).
6. 0.1 M PMSF. Dissolve 0.87 g of Phenylmethanesulfonyl fluoride (PMSF) in isopropanol to a final volume of 50 mL (*see* **Note 4**).
7. 1 M DTT. Dissolve 1.54 g of Dithiothreitol (DTT) in distilled water to a final volume of 10 mL. Store at −20 °C.
8. Antibiotic 1000× (*see* **Note 5**).
9. 1 M IPTG 1000×. Dissolve 2.38 g of Isopropyl β-D-1-thiogalactopyranoside (IPTG) in distilled water to a final volume of 10 mL.
10. Running Buffer 10×: 250 mM TRIS, 2 M Glycine, 1% Sodium Dodecyl Sulfate (SDS). Dissolve 30.3 g of TRIS, 150.1 g of Glycine and 10 g of SDS in distilled water, to a final volume of 1000 mL.
11. T-TBS 10×: 200 mM TRIS pH 7.9, 1.5 M NaCl, 1% Tween 20. Mix 200 mL of 1 M TRIS pH 7.9, 300 mL of 5 M NaCl and 10 mL of 100% Tween 20, adding distilled water to a final volume of 1000 mL.
12. 30% Acrylamide-0.8% Bis-acrylamide. Dissolve 150 g of Acrylamide and 4 g of N, N′-Methylenebis(acrylamide) (Bis-

acrylamide) in distilled water, to a final volume of 500 mL. Keep at 4 °C, protected from light (*see* **Note 6**).

13. Upper Buffer: 0.5 M TRIS pH 6.8, 0.4% SDS. Dissolve 30.3 g of TRIS and 2 g of SDS in distilled water to a final volume of 500 mL, adjusting pH with concentrated HCl. Keep at 4 °C.
14. Lower Buffer: 1.5 M TRIS pH 8.8, 0.4% SDS. Dissolve 90.9 g of TRIS and 2 g of SDS in distilled water to a final volume of 500 mL, adjusting pH with concentrated HCl. Keep at 4 °C.
15. 10% APS. Dissolve 1 g of Ammonium persulfate (APS) with distilled water to a final volume of 10 mL. Make 250 μL aliquots and keep at −20 °C until use. After thawing an aliquot, keep it at 4 °C up to a week.
16. Laemmli Buffer 5×: 312.5 mM TRIS pH 7.9, 10% SDS, 25% glycerol, 5% β-mercaptoethanol, 0.005% Bromophenol Blue [15]. Mix 3.125 mL of 1 M TRIS pH 7.9, 1 g of SDS, 2.5 mL of 100% Glycerol and 0.5 mg of Bromophenol Blue. Dissolve in distilled water to a final volume of 10 mL. Store this buffer at −20 °C in 1.8 mL aliquots. Before use, thaw one aliquot and add 200 μL of 100% β-mercaptoethanol. When β-mercaptoethanol is added, store this buffer at room temperature.
17. 40% Acrylamide-0.67% Bis-acrylamide. Dissolve 100 g of Acrylamide and 1.675 g of Bis-acrylamide in distilled water to a final volume of 250 mL. Keep at 4 °C, protected from light (*see* **Note 6**).
18. 1% Methyl green. Dissolve 0.1 g of Methyl green in distilled water to a final volume of 10 mL. Make 1 mL aliquots and keep at −20 °C.
19. 25 mg/mL Protamine sulfate. Dissolve 0.5 g of Protamine sulfate in distilled water to a final volume of 20 mL. Make 1 mL aliquots and keep at −20 °C.
20. TAU sample Buffer 2×: 6 M Urea, 0.02% Methyl green, 5% Acetic acid, 12.5 mg/mL, 12.5 mg/mL Protamine Sulfate, 10% glycerol. Mix 0.36 g of Urea, 20 μL of 1% Methyl green, 50 μL of glacial Acetic acid, 500 μL of 25 mg/mL Protamine sulfate, 100 μL of 100% glycerol. Add distilled water to a final volume of 1 mL. Keep at −20 °C.
21. 2.5 M LiCl. Dissolve 5.3 g of Lithium chloride in distilled water to a final volume of 50 mL.
22. 2.5 M Glycine. Dissolve 9.4 g of Glycine in distilled water to a final volume of 50 mL.
23. 1 M $NaHCO_3$. Dissolve 4.2 g of Sodium bicarbonate in distilled water to a final volume of 50 mL.

24. 10% Triton X-100. Mix 1 mL of Triton X-100 and 9 mL of distilled water.
25. 10% SDS. Dissolve 1 g of SDS in distilled water to a final volume of 10 mL.
26. 10% Sodium deoxycholate. Dissolve 1 g of Sodium deoxycholate in distilled water to a final volume of 10 mL.
27. 10% NP40. Mix 1 mL of NP40 and 9 mL of distilled water.
28. 0.5 M HEPES pH 7.9. Dissolve 59.6 g of 4-(2-hydroxyethyl)-1-piperazineethanesulfonic acid (HEPES) in distilled water to a final volume of 500 mL, adjusting pH to 7.9 with concentrated HCl.

2.2 Working Solutions and Buffers

1. Buffer A. Mix 5 mL of 1 M TRIS pH 7.9, 750 μL of 1 M $MgCl_2$ and 2 mL of 2.5 M KCl. Add distilled water to a final volume of 500 mL and keep at 4 °C. At the moment of use, prepare an aliquot of 10 mL and add 5 μL of 1 M DTT and 20 μL of 0.1 M PMSF.
2. Buffer B 10×. Mix 30 mL of 1 M TRIS pH 7.9, 56 mL of 2.5 M KCl and 3 mL of 1 M $MgCl_2$. Add distilled water to a final volume of 100 mL. Keep at 4 °C.
3. Buffer C. Mix 10 mL of 1 M TRIS pH 7.9, 750 μL of 1 M $MgCl_2$, 42 mL of 5 M NaCl, 200 μL of 0.5 M EDTA and 125 mL of 100% glycerol. Add distilled water to a final volume of 500 mL. At the moment of use, prepare an aliquot of 10 mL and add 5 μL of 1 M DTT and 50 μL of 0.1 M PMSF. Keep at 4 °C.
4. BC50 Buffer. Mix 20 mL of 1 M TRIS pH 7.9, 20 mL of 2.5 M KCl, 400 μL of 0.5 M EDTA and 200 mL of 100% glycerol. Add distilled water to a final volume of 1000 mL. At the moment of use, add 500 μL of 1 M DTT (or 700 μL of 100% β-mercaptoethanol) and 2 mL of 0.1 M PMSF. Keep at 4 °C.
5. Buffer E. Mix 2.5 mL of 1 M TRIS pH 7.9, 250 μL of 1 M $MgCl_2$, 50 μL of 0.5 M EDTA and 12.5 mL of 100% glycerol. Add distilled water to a final volume of 50 mL. At the moment of use, add 25 μL of 1 M DTT and 100 μL of 0.1 M PMSF. Keep at 4 °C.
6. LB (Luria-Bertani) Medium. Mix 10 g of NaCl, 5 g of yeast extract and 10 g of Tryptone. Add distilled water to a final volume of 1000 mL and sterilize by autoclaving. Keep at 4 °C.
7. 0.5 M HCl-10% Glycerol. Mix 420 μL of 37% Hydrochloric acid and 1 mL of 100% glycerol. Add distilled water to a final volume of 10 mL.

8. Inclusion Bodies Wash Buffer. Mix 35 mL of 1 M TRIS pH 7.9, 10 mL of 5 M NaCl and 1 mL of 0.5 M EDTA. Add distilled water to a final volume of 500 mL.
9. Unfolding Buffer. Mix 66.8 g of Guanidium chloride, 2 mL of 1 M TRIS pH 7.9 and 1 mL of 1 M DTT. Add distilled water to a final volume of 100 mL. Keep at 4 °C.
10. T-TBS 1×. Mix 100 mL of T-TBS 10× and 900 mL of distilled water.
11. Running Buffer 1×. Mix 100 mL of Running Buffer 10× and 900 mL of distilled water.
12. Transfer Buffer. Mix 200 mL of 100% methanol, 100 mL of Running Buffer 10× and 700 mL of distilled water.
13. Red Ponceau staining solution: 0.1% Ponceau salt, 5% Acetic acid. Mix 0.5 g of Ponceau salt and 25 mL of glacial Acetic acid. Add distilled water to a final volume of 500 mL.
14. Blocking Buffers. To prepare T-TBS 5% Milk, dissolve 2.5 g of skim milk in T-TBS 1× to a final volume of 50 mL. To prepare T-TBS 1% BSA, dissolve 0.5 g of lyophilized Bovine Serum Albumin in T-TBS 1× to a final volume of 50 mL (*see* **Note 7**).
15. Laemmli Buffer 1×. Mix 200 μL of Laemmli Buffer 5× and 800 μL of distilled water.
16. TAU Running Buffer: 5% Acetic acid. Dilute 50 mL of glacial Acetic acid in distilled water to a final volume of 1000 mL.
17. TAU gel Wash Buffer 1: 0.5% SDS, 50 mM Acetic acid. Dissolve 2.5 g of SDS in 1.45 mL of glacial Acetic acid and add distilled water to a final volume of 250 mL.
18. TAU gel Wash Buffer 2: 62.5 mM TRIS pH 6.8, 2% SDS, 5% β-mercaptoethanol. Dissolve 1.9 g of TRIS and 5 g of SDS in distilled water adding 12.5 mL of β-mercaptoethanol, and adjusting pH to 6.8 with concentrated HCl, to a final volume of 250 mL.
19. Cell Lysis Buffer. Mix 500 μL of 1 M TRIS pH 7.9, 100 μL of 5 M NaCl and 1 mL of 10% NP40. Add distilled water to a final volume of 50 mL. Filter with a 0.2 μm filter. Add protease inhibitors cocktail 10× to final concentration of 1× at the moment of use (*see* **Note 8**). Keep at 4 °C.
20. Sonication Buffer. Mix 5 mL of 0.5 M HEPES pH 7.9, 1.4 mL of 5 M NaCl, 100 μL of 0.5 M EDTA, 5 mL of 10% Triton X-100, 500 μL of 10% Sodium deoxycholate and 500 μL of 10% SDS. Add distilled water to a final volume of 50 mL. Filter with a 0.2 μm filter.
21. Elution Buffer. Mix 500 μL of 1 M $NaHCO_3$ and 1 mL of 10% SDS. Add distilled water to a final volume of 10 mL. Filter

with a 0.2 μm filter. This solution must be prepared at the moment of use.

22. IP Wash Buffer. Mix 10 mL of 2.5 M LiCl, 5 mL of 1 M TRIS pH 7.9, 5 mL of 10% NP40 and 500 μL of 10% Sodium deoxycholate. Add distilled water to a final volume of 50 mL. Filter with a 0.2 μm filter.
23. Nuclei Buffer. Mix 2.5 mL of 1 M TRIS pH 7.9, 1 mL of 0.5 M EDTA and 5 mL of 10% SDS. Add distilled water to a final volume of 50 mL. Filter with a 0.2 μm filter. Add protease inhibitors cocktail 10× to final concentration of 1× at the moment of use.
24. TE Buffer. Mix 2.5 mL of 1 M TRIS pH 7.9 and 200 μL of 0.5 M EDTA. Add distilled water to a final volume of 50 mL. Filter with a 0.2 μm filter.

2.3 Cell Culture

1. Culture plates.
2. HeLa cells.
3. PBS 1×: Dilute 1 volume of sterile PBS 10× to 1× by adding 9 volumes of sterile water.
4. Culture medium, 10% FBS: DMEM medium high-glucose supplemented with 10% Fetal Bovine Serum (FBS) and 1% Penicillin-Streptomycin.
5. 0.01% Trypsin-EDTA: Dilute 1 volume of commercially available 0.05% Trypsin-EDTA by adding 4 volumes of sterile PBS 1×.

2.4 Other Reagents, Materials, and Equipment

1. Laboratory consumables/plastic material (centrifuge tubes, serological pipettes, etc.).
2. CO_2 incubator for cell culture.
3. Refrigerated microcentrifuge (e.g., Heraeus™ Fresco™ 17 Microcentrifuge from Thermo Fisher Scientific).
4. Dounce tissue grinder with a loose type pestle.
5. Probe type sonicator (E.g., Sonic Dismembrator, Model 120 from *Thermo Fisher Scientific*).
6. Cold acetone (−20 °C).
7. Trichloroacetic acid (TCA).
8. 37% Formaldehyde solution.
9. Bacterial strains for protein expression (E.g., BL21), previously transformed with a plasmid coding for histones (untagged Histone H3 sequence cloned on the vector pET3a [16]).
10. Mini-PROTEAN® Tetra Cell system from *Bio-Rad* or equivalent.
11. Tetra Blotting Module system from *Bio-Rad* or equivalent.

12. Nitrocellulose or PVDF membrane.
13. Electrophoresis power supply.
14. 100% methanol.
15. 100% Tetramethylethylenediamine (TEMED).
16. Lab filter paper, Whatman® type, grade 3 MM.
17. Prestained protein ladder (e.g., PageRuler™ Prestained Protein Ladder, 10–180 kDa from Thermo Fisher Scientific).
18. Primary antibodies against histones or histone PTMs.
19. Secondary antibodies conjugated to Horseradish peroxidase (HRP), directed against the primary antibodies Ig type.
20. Chemiluminescence substrate for the detection of HRP conjugated secondary antibodies (e.g., Pierce™ Enhanced Chemiluminescence (ECL) Western Blotting Substrate from *Thermo Fisher Scientific*).
21. Autoradiography cassettes and films.
22. Urea.
23. 50 μg/μL Cytochrome-C, diluted in distilled water.
24. Glacial Acetic acid.
25. Ethylene Glycol bis(succinimidyl succinate) (EGS).
26. Protease inhibitors (Cocktail).
27. Bovine Serum Albumin.
28. Orbital incubator shaker.
29. Microbiological incubator.
30. SOC medium for *E. coli*.
31. Dialysis tubing of 1 K MWCO (e.g., Spectra/Por®7 Dialysis Membrane, Thomas Scientific).
32. Bradford reagent (e.g., Coomassie Protein Assay Reagent, Thermo Scientific).

3 Methods

3.1 Isolation of Histones

3.1.1 Histones from Cytosolic, Nuclear and Chromatin Extracts Derived from HeLa Cells

As mentioned previously, there are different populations of histones inside the cell and each population is at a different state of maturation. The cellular fractionation protocol described here is based on Dignam et al. [12]. Although we described the protocol for HeLa cells, it can be used with a variety of cells, including human and non-human cell lines, as well as cells from primary culture. For a schematic view of the procedure, *see* Fig. 1.

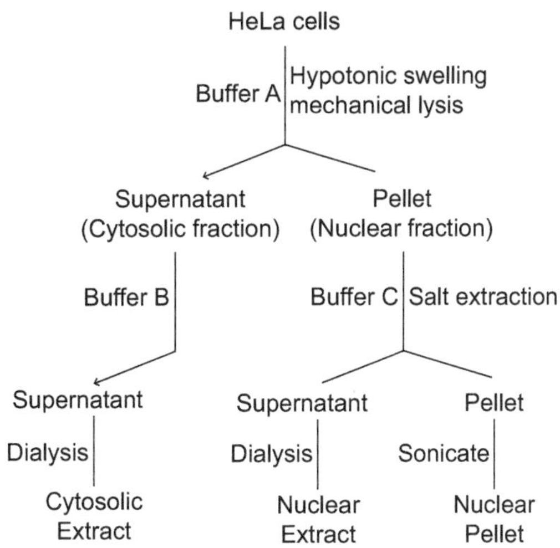

Fig. 1 Subcellular fractionation scheme. After hypotonic swelling and mechanical lysis of the cells, the homogenate is centrifuged to recover the Cytosolic Extract in the supernatant, and Nuclear and Chromatin Extracts from the pellet. This pellet is salt extracted to obtain Nuclear Extract, while the unsoluble material corresponds to the Nuclear Pellet

HeLa cell culture

1. Seed 3 × 10^6 cells per 150 mm plates, 24 h before harvesting. To harvest, remove the culture medium by aspiration and wash the cells twice with warm PBS 1× (37 °C).
2. Add 5 mL of warm 0.01% Trypsin-EDTA. Incubate for 7 min, or until the cells detach from the plates, at 37 °C in a cell culture CO_2 incubator.
3. Add 5 mL of culture medium supplemented with 10% FBS (*see* **Note 9**). Recover the cells in a centrifuge tube.
4. Centrifuge the cells 8 min at 500 × *g*, at room temperature, and wash them twice with warm PBS 1×.
5. Keep the cell pellet, approximately 50 μL, on ice. Proceed quickly with the subcellular fractionation protocol.

Subcellular fractionation to obtain cytosolic, nuclear, and chromatin extracts

1. Add 250 μL of Buffer A to 50 μL of cell pellet, corresponding to five volumes of the cell pellet. Resuspend by a gentle up-and-down pipetting (*see* **Note 10**). Incubate the cells for 10 min on ice.
2. Centrifuge the cells at 10,000 × *g* 10 min, at 4 °C. Remove the supernatant carefully. Add 100 μL of Buffer A, corresponding to two volumes of the initial cell pellet (*see* **Note 11**). Resuspend the cells by gentle up-and-down pipetting. Transfer the cell suspension to the Dounce tissue grinder (*see* **Note 12**).

3. Homogenize the cells 10 times with the Dounce tissue grinder, using the "loose" type pestle (*see* **Note 13**). Transfer the homogenate to a new tube and centrifuge at 10,000 × *g* for 10 min, at 4 °C.
4. Take the supernatant carefully and leave it in a new tube. Keep the remaining pellet for nuclear soluble extraction (**step** 7).
5. Measure the volume of the supernatant recovered from **step 4**. Dilute Buffer B 10× in this supernatant to 1× concentration. In our example, for 200 μL of the recovered supernatant, add 22 μL of Buffer B 10×. Mix well and centrifuge at 13,800 × *g* 1 h, at 4 °C (*see* **Note 14**).
6. Take the supernatant from **step 5** and dialyze against 1000 mL of BC50 Buffer 2 h at 4 °C, using dialysis tubing of 1 K MWCO. Measure protein concentration (Bradford method [17]) and store the *Cytosolic Extract* at −80 °C.
7. To obtain the *Nuclear Extract*, take the pellet from **step 4** and add a volume of Buffer C equal to the volume obtained of *Cytosolic Extract*, in our case, 200 μL. Resuspend the pellet thoroughly by pipetting up-and-down, and incubate for 30 min at 4 °C in continuous agitation.
8. Centrifuge at 13,800 × *g* 30 min at 4 °C. Recover the supernatant carefully and transfer into a new tube, keeping the new pellet for nuclear pellet solubilization (*see* **step 10**).
9. Measure the volume of the supernatant recovered from **step 8** and dialyze against 1000 mL of BC50 Buffer 2 h at 4 °C. Measure protein concentration and store the *Nuclear Extract* at −80 °C.
10. To obtain soluble *Nuclear Pellet*, resuspend the pellet obtained from **step 8** in 200 μL of Buffer E (equal to the volume obtained from Cytosolic and Nuclear Extracts). Sonicate the pellet by three pulses of 30 s each time, 60% amplitude, or until complete solubilization. Store the *Nuclear Pellet* at −80 °C.

Western blot analyses using samples derived from cytosolic, nuclear, and chromatin extracts are shown in Fig. 2.

Acid extraction of histones

1. Measure the sample volume to use for the acid extraction (e.g., 200 μL of *Nuclear Pellet* obtained from the subcellular fractionation, **step 10**). Add one volume (200 μL) of 0.5 M HCl-10% Glycerol and resuspend thoroughly by pipetting up and down.
2. Incubate the sample with rotation for 1.5 h at 4 °C.

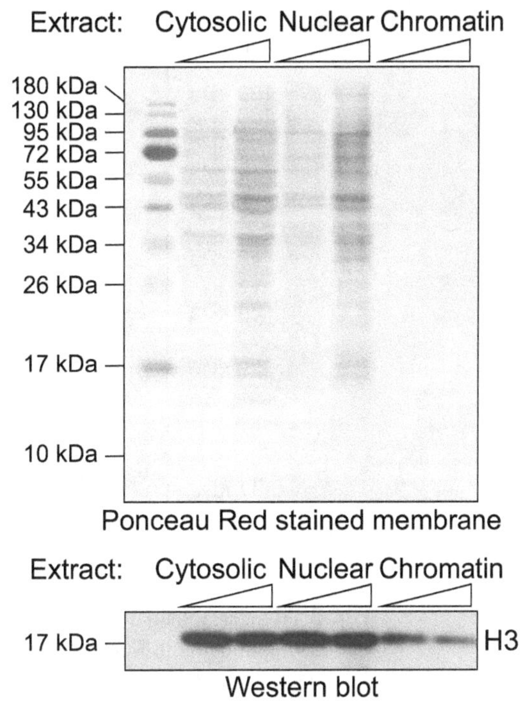

Fig. 2 Subcellular fractionation analyzed by Western blot. After subcellular fractionation of HeLa cells aliquots derived from Cytosolic Extract (1 and 3% of the total Cytosolic Extract), Nuclear Extract (1 and 3% of the total Nuclear Extract), and Nuclear Pellet (0.005% and 0.015% of the total Nuclear Pellet) were loaded onto a 15% SDS-PAGE. First lane, molecular weight standards. Upper panel: Ponceau Red stained membrane after electrotransfer the gel. Lower panel: Western blot analysis against histone H3

3. Centrifuge at 13,800 × *g* for 10 min, at 4 °C. Take the supernatant and transfer into a new tube.
4. Add TCA to a final concentration of 25%. In our example, 130 μL of 100% TCA.
5. Incubate the sample with rotation for 1 h at 4 °C.
6. Centrifuge at 13,800 × *g* for 10 min, at 4 °C. Discard the supernatant.
7. Wash the pellet with 200 μL cold acetone and vortex 5 s. Centrifuge at 13,800 × *g* for 5 min and discard the supernatant. Repeat this step twice.
8. Let the pellet dry at room temperature for about 20 min. Resuspend the pellet in 50 μL of Buffer E and measure protein concentration (*see* **Note 15**). To analyze the purity of extracted histones, run 1 μg of protein in a 15% SDS-PAGE and stain the gel with Coomassie blue staining (*see* Subheading 3.2 and Fig. 3). Store histones at −80 °C.

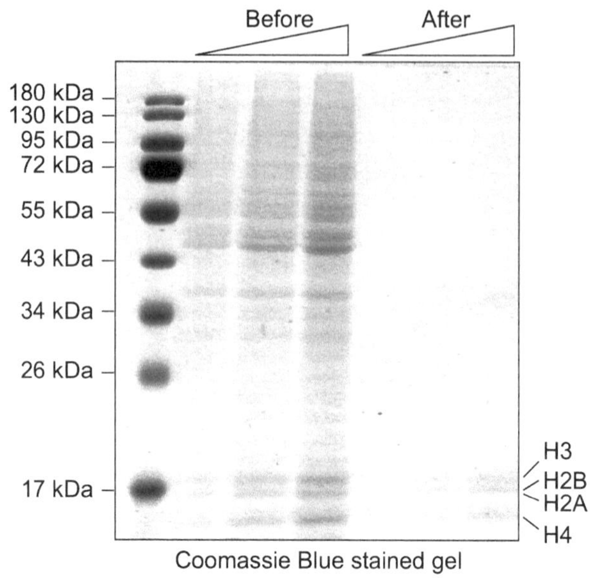

Fig. 3 Proteins obtained from Nuclear Pellet before and after acid extraction. Increasing amounts of Nuclear Pellet aliquots before (0.4%, 0.8%, and 1.2% of the total Nuclear Pellet) and after (2.5%, 5%, and 7.5% of the total acid extraction recovered material) acid extraction were loaded on 15% SDS-PAGE and stained with Coomassie blue for protein visualization. First lane, molecular weight standards

3.1.2 Recombinant Histones

This protocol, based on Luger et al. [13], is designed to purify high amounts of histone proteins by using bacterial expression vectors. The purification does not rely on "tags," given that histones are purified from inclusion bodies. Since the proteins are expressed in bacteria, they lack PTMs, which makes them the ideal substrates for the evaluation of enzymes involved in the "de novo" establishment of histone PTMs [18]. Another advantage of this system is that histones (canonical and variant) can be mutated at any residue, to analyze, for example, the effect of PTMs over the nucleosome structure [19].

Bacterial culture and histone expression

1. Transform BL21 bacterial strain with the histone expression vector (e.g., pET-histone plasmid). For this, thaw 50 μL of chemically competent bacteria on ice, and incubate with 25 ng of the pET-histone plasmid for 30 min on ice. After this, incubate for 30 s at 42 °C and then place the tube on ice for 2 min. Add 250 μL of warm SOC medium and incubate in agitation for 1 h. Plate the transformed bacteria on LB-agar containing the appropriated antibiotic and incubate at 37 °C, overnight.
2. Inoculate 50 mL of LB-containing antibiotic with a single colony. Then, incubate at 37 °C, 250 r.p.m., overnight.

3. The next day, inoculate 1000 mL of LB-containing antibiotic with the 50 mL saturated culture obtained from **step 2**. Incubate in a shaker at 37 °C, 250 r.p.m., until 600 nm optical density (OD600) reaches a value between 0.6 and 0.8.
4. Take 1 mL sample from the culture and store it on ice for further analysis (*see* **Note 16**). Then, induce histone expression by adding 1 mL of 1 M IPTG, incubating in the shaker at 37 °C, 250 r.p.m., 2 h.
5. After induction, take 1 mL sample from the culture and store it on ice for further analysis. Then, harvest bacteria by centrifuging the culture at 15,000 × *g*, 10 min. The obtained pellet can be stored at −80 °C until performing the *Histone purification step.*
6. Using the samples from **steps 4** and **5**, verify histone induction. To do this, centrifuge the samples at 13,800 × *g*, eliminate the supernatant, and resuspend the pellet in 1/10 of the original volume using distilled water. Sonicate the samples by 3 pulses of 30 s each, 60% amplitude, with resting periods of 45 s between the pulses, and add *Laemmli Buffer* 5× to a final concentration of 1×. Run 15 μL of the samples in a 15% SDS-PAGE and stain gel with Coomassie Blue to visualize the induced histone band.

Histone purification from inclusion bodies

1. Thaw the stored bacteria pellet in a warm bath (37 °C) and resuspend it in 50 mL of Inclusion Bodies Wash Buffer by pipetting up and down.
2. Perform 3 cycles of "freeze and thaw." To do this, keep the resuspended bacteria 20 min at −80 °C and then 5 min at 37 °C. Repeat twice.
3. Put the bacterial pellet on ice and sonicate by three pulses of 45 s each, 60% amplitude, with 45 s of rest between pulses.
4. Centrifuge the sonicated bacteria at 15,000 × *g*, 4 °C, 20 min and discard the supernatant.
5. Resuspend the pellet in 50 mL of Inclusion Bodies Wash Buffer plus 1% Triton X-100 and centrifuge at 12,000 × *g*, 10 min, 4 °C. Repeat this step once. It is difficult to resuspend the pellet at this step; however, the quality of histone purification is dependent on the efficiency of this step.
6. Resuspend the pellet in 50 mL of Inclusion Bodies Wash Buffer (without Triton X-100) and centrifuge at 12,000 × *g*, 10 min, 4 °C. Repeat this step once.
7. Resuspend the pellet, which contains the inclusion bodies, in 15 mL of Unfolding Buffer and incubate at 37 °C, with agitation, 1 h.

8. Eliminate undissolved material by 20 min centrifugation at 15,000 × *g*, 4 °C. Recover the supernatant and discard the pellet.
9. Dialyze the supernatant against 2 L of distilled water at 4 °C, using dialysis tubing of 1 K MWCO. Replace water every 2 h at least twice and then perform a final overnight dialysis step.
10. Recover the content inside the dialysis tubing and centrifuge at 12,000 × *g*, 4 °C, for 15 min.
11. Recover the supernatant and evaluate histone purification by performing 15% SDS-PAGE and stain the gel with Coomassie blue for visualization of the histone bands (*see* **Note 17**) (Fig. 4). Store the purified histones at −80 °C until use.

3.2 Analysis of PTMs

3.2.1 Western Blot

Western blot is one of the most utilized techniques in the field of cell biology, molecular biology, and biochemistry. This method, originated in the late 1970s and early 1980s [20–22], can be summarized in three steps: protein separation from a complex sample by gel electrophoresis, protein transfer from the gel to a solid membrane, and detection of the protein of interest (or PTMs), typically by antibodies. Although there are many protocols available in the literature, here we describe the conditions optimized for the analysis of histones and histone PTMs. For a more general

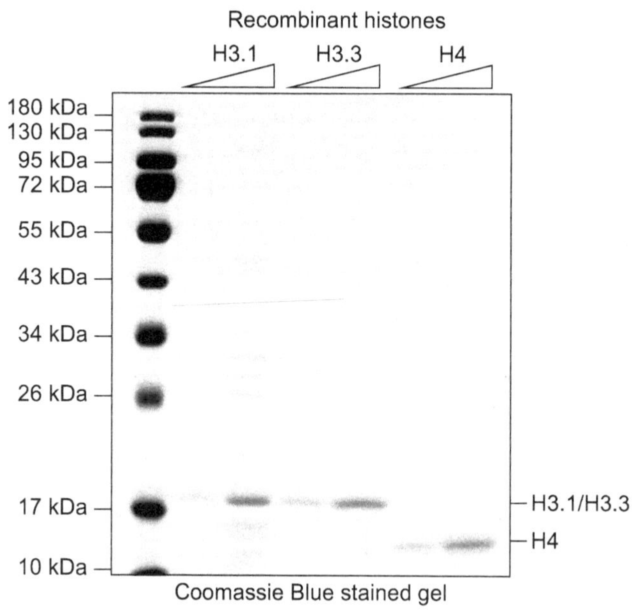

Fig. 4 Recombinant histones purified from bacteria. After histone purification, 15% SDS-PAGE gel was loaded with 0.5 and 2 μg of H3.1, H3.3, and H4 proteins. Then, the gel was Coomassie blue stained for visualization of histones. H3.1 and H3.3 have the same electrophoretic migration. First lane, molecular weight standards

Western blot protocol, we recommend consulting the work by Mahmood and Yang [23].

15% SDS-PAGE gel preparation

1. Assemble clean glass plates into a casting frame to prepare a 1.5 mm thick mini-gel.
2. To prepare the resolving gel, mix 2 mL of distilled water, 2 mL of Lower Buffer, and 4 mL of 30% Acrylamide-0.8% Bis-acrylamide. Then, add 20 μL of 10% APS and 20 μL of TEMED. Mix well avoiding the formation of bubbles, and promptly proceed to the next step.
3. Place the resolving gel mix into the assembled glass plates, leaving space for the stacking gel. Then, gently overlay 1 mL of isopropanol on the surface of the gel mix. Allow the polymerization reaction to occur for at least 40 min.
4. Remove the isopropanol, wash the surface of the gel with distilled water and eliminate the excess of water by draining over a paper towel.
5. To prepare the stacking gel, mix 2.4 mL of distilled water, 1 mL of Upper Buffer and 600 μL of 30% Acrylamide-0.8% Bis-acrylamide. Then, add 20 μL of 10% APS and 20 μL of TEMED. Mix well, avoiding the formation of bubbles, and promptly place the mix on the top of the resolving gel. Immediately introduce the comb to form the loading wells.
6. Let the gel to polymerize for at least 20 min and do not remove the comb until loading the samples.

Sample preparation for SDS-PAGE gel electrophoresis

1. Measure the protein concentration of your samples.
2. Take the amount of sample needed and transfer into a new tube (*see* **Note 18**). Add *Laemmli* Buffer 5× to a final concentration of 1×.
3. Heat the samples 5 min at 100 °C.
4. Spin the samples 10 s to precipitate condensed water from the top. Maintain the tubes at room temperature until loading the gel.

SDS-PAGE running

1. Assemble the glass plates into the electrode gasket.
2. Place the assembled electrode gasket into the electrophoresis chamber. Fill the gasket and the electrophoresis chamber with Running Buffer 1× and gently remove the comb (*see* **Note 19**).
3. Load the samples into the gel's wells by using Hamilton syringes or 200 μL long pipette tips (*see* **Notes 20** and **21**),

saving one well for prestained protein ladder. Fill any non-used well with an equivalent Laemmli Buffer 1× volume (*see* **Note 22**).

4. Load prestained protein ladder, close the electrophoresis chamber, and run the gel at 110 V, room temperature, until the Bromophenol blue dye marker reaches the end of the gel. Then, proceed with the protein transfer to nitrocellulose or PVDF membrane.

Electrophoretic protein transfer (Electroblotting)

1. After finishing the SDS-PAGE running, recover the gel and remove the stacking gel by using a spatula. Place it into Transfer Buffer.
2. Cut two 7.5 × 10 cm pieces of 3 MM grade Whatmann® type filter paper and place into Transfer Buffer.
3. Cut one 7.5 × 10 cm piece of nitrocellulose or PDVF membrane (*see* **Note 23**). When using PVDF, activate the membrane with 100% methanol. Nitrocellulose membranes do not require methanol activation. Keep membranes in Transfer Buffer.
4. Assemble the protein-transferring sandwich by disposing the following components, in the indicated order, into a gel holder cassette:
   (a) A flat foam pad.
   (b) A piece of filter paper.
   (c) The nitrocellulose or PDVF membrane.
   (d) The acrylamide gel.
   (e) A piece of filter paper.
   (f) A flat foam pad.

   Eliminate any bubble between the layers of the sandwich (*see* **Note 24**).
5. Introduce the protein-transferring sandwich into the electrode module, with the gel facing toward the cathode and the membrane to the anode. Then, put the electrode module into the transferring chamber and fill it with Transfer Buffer. Place a cooling tray (−20 °C) inside the chamber and cap it, connecting the electrodes to a power supply.
6. Perform the electrotransfer 1.5 h at 400 mA, 4 °C (*see* **Note 25**).
7. Disassemble the transferring sandwich and stain the membrane with Ponceau to verify the transfer. To do this, incubate the membrane with 50 mL of Ponceau staining solution about 5 min. Remove the staining solution and wash the membrane with 100% methanol (PVDF) or T-TBS 1× (nitrocellulose).

Membrane blocking and antibody incubation

1. Block the membrane by incubating for 1 h in blocking solution (T-TBS 1% BSA or 5% Milk, depending on the antibody to use (*see* **Note 7**)).
2. Incubate the membrane with the primary antibody overnight at 4 °C and rotation. Dilute the antibody in T-TBS 1×, according to the manufacturer's instructions. Always use validated antibodies (*see* **Note 26**).
3. Wash the membrane 3 times with T-TBS 1×, 5 min each.
4. Incubate the membrane with the secondary antibody conjugated to HRP 1 h at room temperature and rotation. Dilute the antibody in T-TBS 1×, according to the manufacturer's instructions.
5. Wash the membrane 3 times with T-TBS 1×, 5 min each. Perform all the subsequent steps in the **dark room**.
6. Develop the assay by incubating the membrane with a chemiluminescence substrate mix (e.g., ECL from *Thermo Scientific*) 5 min. Prepare the mix according to the manufacturer's instructions.
7. Place the membrane between two transparent plastic sheets, removing the excess of the chemiluminescence substrate mix. Then, place the membrane in an autoradiographic cassette.
8. Place an autoradiographic film on top of the covered membrane for 1 min, or the required time, closing the autoradiographic cassette.
9. Remove and develop the film.

3.2.2 TAU Gel Electrophoresis Coupled to Western Blot

Three are the fundamental principles of TAU gels: First, with the exception of the linker histones, core histones associate to non-ionic detergents, such as Triton, increasing their "effective mass." Second, histones associate with non-ionic detergents at different degrees based on the amino acidic sequence composition. This differential binding is enhanced by urea, because this agent disrupts the interaction between detergent and histones. And third, histones acquire a positive net charge under acidic conditions, which is achieved by acetic acid. The present protocol can differentiate between the histone variants H3.1 and H3.3; and H2A.1, H2A.2, H2A.3, and H2A.4 [24]. By adapting the procedure described by Lennox and Cohen in 1989 [25], the present protocol utilizes mini-gels, facilitating the implementation of the technique as well as the electrotransfer to PVDF membranes for PTM analyses by Western blot.

TAU acrylamide gel preparation (12% Acrylamide-0.2% Bis-acrylamide, 6 M Urea, 6 mM Triton X-100, 5% Acetic Acid)

1. Assemble the clean glass plates into a casting frame to prepare a 1.5 mm thick mini-gel.

2. To prepare the resolving TAU gel, mix 3 mL of 40% Acrylamide-0.67% Bis-acrylamide, 3.6 g of Urea and 0.5 mL of glacial Acetic acid. Add distilled water to a final volume of 9.6 mL and mix until the urea is completely dissolved.
3. Add 400 μL of 10% Triton X-100 and gently mix to avoid formation of bubbles (*see* **Note 27**).
4. Add 190 μL of 10% APS and 190 μL of TEMED and mix gently. Promptly proceed to the next step.
5. Place the resolving TAU gel mix into the assembled glass plates, leaving space for stacking gel. Then, gently overlay 1 mL of isopropanol on the surface of the gel mix. Allow the polymerization reaction to occur at least 40 min.
6. Remove the isopropanol, wash the surface of the gel with distilled water and eliminate the excess of water by draining over a paper towel.
7. To prepare the stacking TAU gel (5% Acrylamide-0.08% Bis-acrylamide, 3 M Urea, 6 mM Triton X-100, 5% Acetic Acid), mix 625 μL of 40% Acrylamide-0.67% Bis-acrylamide, 0.9 g of Urea and 250 μL of glacial Acetic acid. Add distilled water to a final volume of 4.8 mL and mix until the urea is completely dissolved.
8. Add 400 μL of 10% Triton X-100 and gently mix to avoid formation of bubbles (*see* **Note 27**).
9. Add 200 μL of 10% APS, 200 μL of TEMED and mix gently. Promptly place the mix on top of the resolving gel. Immediately place the appropriated comb to form the loading wells. Let it polymerize at least 40 min.
10. Assemble the glass plates into an electrode gasket and place the gasket into the electrophoresis chamber. Fill the gasket and chamber with TAU Running Buffer and gently remove the comb (*see* **Note 28**).
11. Close the chamber, connect the electrodes to the power supply in an **inverse manner**, that is the positive terminal of the gasket into the negative terminal of the power supply and vice versa (*see* **Note 29**). Prerun the gel overnight at 90 V (*see* **Note 30**).
12. The next morning, replace the TAU Running Buffer with fresh buffer, clean up the wells by gently pipetting TAU Running Buffer, and proceed to load the samples.

Sample preparation for TAU gel electrophoresis

1. Mix each sample with one volume of TAU Sample Buffer 2×. It is recommended to prepare the same volume for all the samples, up to 25 μL.

2. Keep the samples at room temperature until loading the gels. Do not heat the samples (*see* **Note 31**).

TAU gel running

1. Load 5 μL of 50 μg/μL of Cytochrome-C into the prerun TAU gel. Cytochrome-C is an H4 migration marker that can be visualized while running the gel by its brown color. Then, load the samples and fill any remaining well with TAU Sample Buffer 1×.
2. Close the electrophoresis chamber. Connect the electrodes to the power supply in the **inverse manner**, as detailed in **TAU acrylamide gel preparation, step 11**, and run the gel at 120 V, at room temperature, until the Cytochrome-C band migrates ¾ of the total gel length. This will take approximately 2 h. Once the running has finished, promptly proceed with the protein transfer to the PVDF membrane. Alternatively, stain the TAU gel with Coomassie blue without any preparation step (Fig. 5).

Preparing TAU gels for electrotransfer

1. Disassemble the electrophoresis chamber and gently take the TAU gel, removing the stacking gel.
2. Wash the gel twice 30 min with the TAU gel Wash Buffer 1.
3. Wash the gel once 30 min with TAU gel Wash Buffer 2.
4. Leave the gel in Transfer Buffer and proceed with Electroblotting, blocking of the membrane and incubation with primary and secondary antibodies as described previously (*see* Subheadings 3.2 and 3.2.1) (Fig. 6).

3.2.3 Chromatin Immunoprecipitation (ChIP)

Chromatin immunoprecipitation relays on the recognition of chromatin fragments by specific antibodies to study proteins that are in contact with DNA. In a first step, cells are fixed using a cross-linker reagent to stabilize the interaction between proteins and DNA. Then, the fixed chromatin is disrupted by sonication to obtain chromatin fragments up to 500 bp. Chromatin fragments are then immunoprecipitated, and the DNA fragments purified and analyzed. The present protocol, adapted from Soutoglou and Talianidis, 2002 [26] describes a method optimized for the study of histone PTMs located on specific DNA sequences, like promoters. A scheme of the fundamental steps of this technique is shown on Fig. 7. Alternatively, when samples are obtained from cells expressing histone variants fused to a tag sequence, such as Flag-tag, antibodies directed against the tag sequence are utilized to explore the binding of variants to specific DNA sequences.

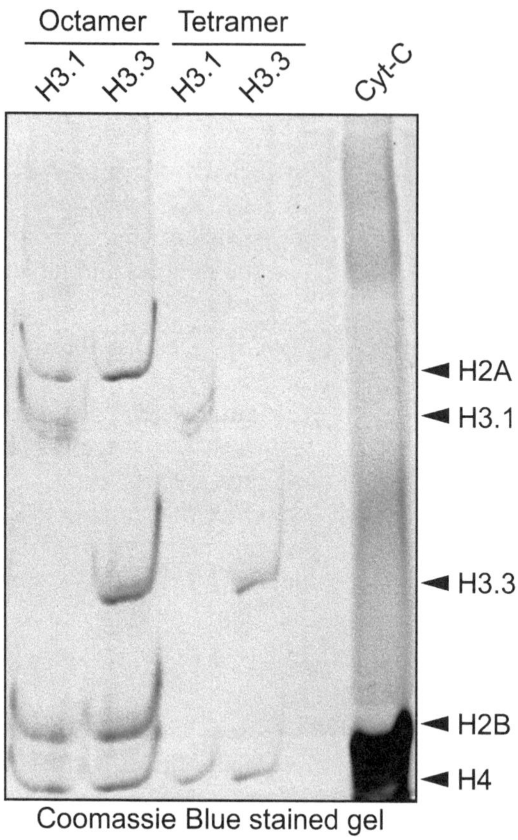

Fig. 5 Analysis of recombinant histone variants by TAU gel electrophoresis. 5 μg of either octamer or tetramers H3-H4 proteins were loaded into the TAU gel and then Coomassie blue stained. The migration of the different histones is pointed on the right

Cell fixation

1. Seed 3×10^6 cells into a 150 mm plate, 24 h before harvest.
2. Collect the cells in a 15 mL conical tube and leave it on ice.
3. Centrifuge the tube at 800 × *g* 5 min and wash the cells with cold PBS 1×. Repeat this wash step once and centrifuge the cells at 800 × *g* 5 min.
4. Mix 100 μL of 37% formaldehyde solution and 3.6 mL of PBS 1× to freshly prepare 3.7 mL of 1% formaldehyde in PBS.
5. Add 500 μL of 1% formaldehyde to the cells to start the cross-linking. Then, transfer the cells to a 1.5 mL eppendorf tube and incubate 10 min at room temperature, with gentle agitation.
6. Quench the reaction by adding 25 μL of 2.5 M Glycine. Incubate with rotation 5 min at room temperature.

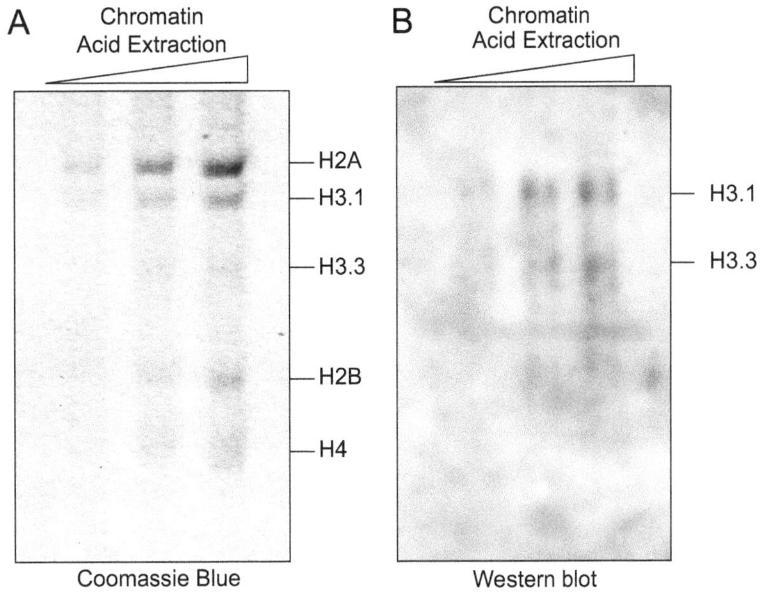

Fig. 6 Analysis of histone variants derived from Nuclear Pellet by TAU gel electrophoresis analysis coupled to Western blot. (**a**) 2.5%, 5%, and 7.5% of the total extracted Nuclear Pellet sample were loaded into the TAU gel and then Coomassie blue stained. (**b**) 0.1%, 0.2%, and 0.3% of the total extracted Nuclear Pellet sample were loaded into the TAU gel and then electrotransferred. Then, the membrane was Western blotted against histone H3. The migration of the different histones is pointed on the right

7. Centrifuge the cells 5 min at 400 × *g*, 4 °C. Discard the supernatant.
8. Mix 450 μL of cold PBS 1× with 50 μL of protease inhibitors cocktail and wash the cells with this solution. Repeat this step once. If EGS fixation is required, continue to **EGS fixation.** Otherwise, proceed with **Cell lysis and chromatin fragmentation.**

EGS fixation (Optional).

EGS is a crosslinker that contains amine-reactive ends around a 12-atom spacer arm, having a longer cross-linking radius in comparison to formaldehyde. Therefore, EGS fixation is utilized to analyze protein complexes associated with DNA.

1. Freshly prepare a solution of 50 mg/mL of EGS (EGS 50×) in 50% Acetic acid, diluted in water. To facilitate the dissolution of EGS, pre-warm the 50% Acetic acid solution between 60 and 70 °C.
2. Dilute *EGS 50×* to a final concentration of 1× in PBS 1×, at room temperature. Keep at 37 °C.

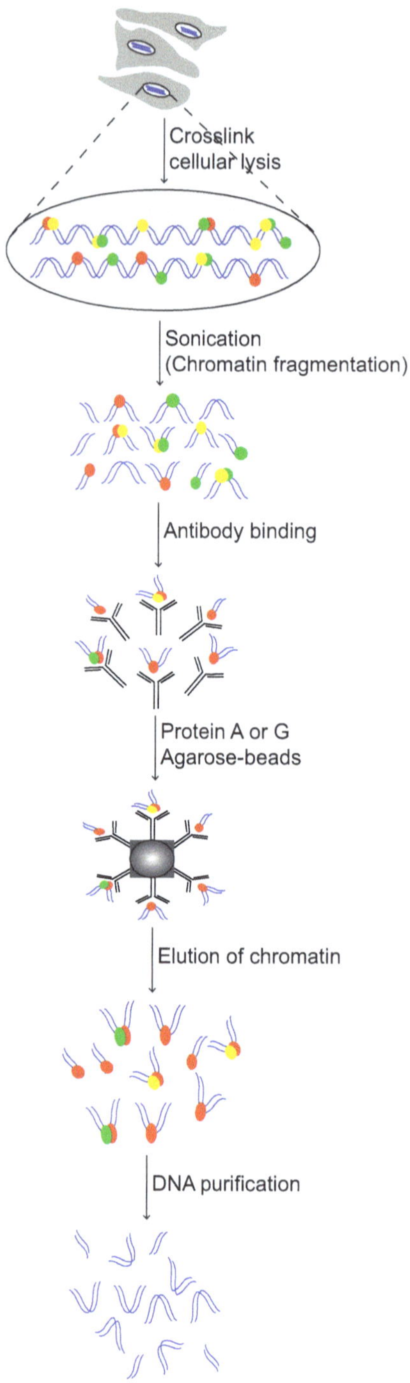

Fig. 7 Steps of Chromatin Immunoprecipitation. Fixed cells are lysed and sonicated to obtain chromatin fragments up to 500 bp. Fragments are then immunoprecipitated by using specific antibodies and then agarose beads conjugated to protein A or G. Chromatin fragments are eluted, proteins degraded and DNA purified and quantified for further analyses. Adapted from [27]

3. Add 500 μL of EGS 1× to the cell pellet. Incubate for 1 h at room temperature with rotation.
4. Wash the cells 3 times with 500 μL of PBS 1× at room temperature, centrifuging 5 min the cells at 400 × *g*.

Cell lysis and chromatin fragmentation

1. Add 50 μL of protease inhibitors cocktail 10× to 450 μL of Cell Lysis Buffer (*see* **Note 8**). Use this solution to resuspend the cell pellet. Incubate for 10 min on ice.
2. Homogenize the cells with a Dounce tissue grinder, 10 times, using the loose pestle.
3. Centrifuge the lysate for 1 min at 5400 × *g*, at 4 °C. The nuclei will remain in the pellet.
4. Mix 25 μL of protease inhibitors cocktail 10× and 225 μL of Nuclei Buffer. Use this solution to resuspend the nuclei pellet. Incubate for 10 min on ice. Add 225 μL of Sonication Buffer and 25 μL of protease inhibitors cocktail 10×.
5. Perform chromatin fragmentation by sonicating the resuspended nuclei by 80 pulses of 15 s each time, 60% amplitude, with pauses of 15 s between each pulse of sonication.
6. Centrifuge the sonicated samples for 10 min at 17,000 × *g*, at 4 °C. Transfer the supernatant into a clean tube and discard the pellet. Repeat this step once. Store this *Chromatin sample* at −80 °C or proceed with the immunoprecipitation step (*see* **Note 32**).

Preparation of protein A (or G) agarose beads

1. Take twice the volume of a 50% slurry agarose beads required for the immunoprecipitation (*see* **Note 33**). In our example, take 50 μL of protein A-agarose beads and centrifuge 5 min at 1500 × *g*, 4 °C. Discard the supernatant.
2. Wash the beads by adding 100 μL of PBS 1× and centrifuge 5 min at 1500 × *g*, 4 °C. Discard the supernatant and repeat this step once.
3. Add one volume of PBS 1× to the beads. In our example, add 25 μL of PBS 1×. This is the *Beads solution*. Keep at 4 °C or use immediately.

Immunoprecipitation

1. Measure the DNA concentration of the *Chromatin sample*, using a mixture of Nuclei Buffer and Sonication Buffer (1:1) as blank. Aliquot the *Chromatin sample* in fractions of 1 μg of DNA per tube and add Sonication Buffer, plus protease inhibitors, to a final volume of 500 μL. Keep one of these aliquots as *Input*.

2. Perform a "pre-clearing step" to each *Chromatin sample.* To do this, add 50 μL of the *Beads solution* and 2 μg of immunoglobulin (e.g., rabbit IgG (*see* **Note 34**)) to each *Chromatin sample.* Incubate with rotation for 2 h at 4 °C.
3. Centrifuge 5 min at 1500 × *g*, 4 °C, and transfer the supernatant into a clean tube. Discard the pellet.
4. Add 1 μg of antibody to the precleared *chromatin sample.* Incubate with rotation overnight at 4 °C (*see* **Note 35**).
5. Add 10 μL of *Beads solution* and incubate in rotation 1 h at 4 °C. Then, centrifuge 2 minutes at 1500 × *g*, 4 °C, and discard the supernatant.
6. Add 500 μL of Sonication Buffer and incubate 5 min, with rotation, at 4 °C. Centrifuge 5 min at 1500 × *g*, 4 °C, and discard the supernatant (*see* **Note 36**).
7. Add 500 μL of IP Wash Buffer and incubate for 5 min, with rotation, at 4 °C. Centrifuge for 5 min at 1500 × *g*, 4 °C, and discard the supernatant. Repeat this step once (*see* **Note 36**).
8. Wash the beads with 500 μL of TE Buffer (pH 8.0). Incubate the beads for 5 min, with rotation, at 4 °C. Centrifuge 5 min at 1500 × *g* and discard the supernatant (*see* **Note 36**).

DNA recovery from the immunoprecipitation

1. Add 100 μL of Elution Buffer to the beads obtained from the last step of the immunoprecipitation (*see* **Note 37**).
2. From now on, include the Input sample. Vortex gently 15 s and incubate for 15 min at 65 °C. Vortex gently 15 s once more.
3. Centrifuge 1 minute at 17,000 × *g*. Transfer the supernatant into a new tube and discard the pellet. Add 4 μL of 5 M NaCl.
4. Add 1 μL of 20 mg/mL of RNase A and incubate overnight at 65 °C.
5. Add 2.5 μL of Proteinase K to a final concentration of 1 U/mL. Then, add Proteinase K Buffer 10× to a final concentration of 1×. Incubate for 2 h at 50 °C and then add 95 μL of Buffer TE (pH 8.0) (*see* **Note 38**).
6. Purify the DNA by phenol-chloroform extraction or by using the purification kit of your choice (*see* **Note 39**).
7. Use the purified DNA for real-time qPCR, using primers directed against the sequence of interest.

4 Notes

1. PMSF is not stable in aqueous solution, having a half-life of 35 min in this medium [28]. For this reason, this compound is added to the buffers shortly before use.
2. Some investigators prefer DTT over β-mercaptoethanol because it is less irritant, but both are used for the same purpose: preventing protein oxidation. When replacing one by another, 0.5 mM DTT is equivalent to 10 mM β-Mercaptoethanol.
3. Dissolve EDTA in 400 mL of distilled water, add NaOH pellets until reaching the desired pH and then complete volume to 500 mL. EDTA will not dissolve until the pH is above 8.0.
4. When prepared in anhydrous isopropanol, PMSF is stable for months [28].
5. The antibiotic needed will depend on the resistance marker associated with the plasmid in which the gene coding for the recombinant protein is cloned. Always consult the plasmid datasheet.
6. If not protected from UV light, acrylamide will enter into polymerization reactions, forming clusters that precipitate and reduce the actual concentration of the stock acrylamide solution. To avoid this, use amber bottles to store acrylamide solution or wrap conventional bottles with foil.
7. The use of T-TBS 5% Milk or T-TBS 1% BSA as blocking solutions is dependent on the antibody employed. Some antibodies give stronger signals when the membrane is blocked with BSA. This is common for the analysis of histone PTMs. In contrast, blocking with milk might help to reduce the recognition of unspecific bands. Therefore, we highly recommend testing the best blocking solution for each antibody.
8. Commercially available protease inhibitors cocktails can be replaced by homemade cocktails. Prepare 1000× aprotinin (10 mg/mL), 1000× leupeptin (5 mg/mL) and 1000× pepstatin (7 mg/mL) and dilute them to 1× concentration in the desired buffer.
9. Inactivation of the trypsin is achieved thanks to the presence of fetal bovine serum in the culture medium. Thus, it is important to wash the cells with PBS 1× before adding trypsin.
10. Handle carefully the cells to minimize their premature disruption and leaking the nuclear content.
11. The cell pellet increases twice its initial size due to the swelling of the cells in Buffer A.

12. It is recommended to precool the Dounce tissue grinder on ice before the homogenization.
13. The "loose" type pestle allows the disruption of the plasma membrane without affecting the nuclear membrane integrity. Ten up-and-down cycles are sufficient for the mechanical disruption of the cells.
14. In the original protocol, cytosolic extract is the soluble fraction recovered after centrifugation at 100,000 × *g*. However, the present protocol is adapted to work by using refrigerated microcentrifuges available in almost any laboratory. When possible, centrifuge the extract at 100,000 × *g*.
15. When starting from 50 μL of cell pellet, 100 μL of a final concentration of at least 1 μg/μL is obtained.
16. This sample reveals the proteins synthesized by the bacteria before IPTG mediated protein induction.
17. About 2 mg of histone proteins can be obtained from 1000 mL of bacterial culture. 20 μL of the final 15 mL histone solution should be enough to visualize the purified histones.
18. To perform Western blot analyses, the optimal amount of protein to load into the gel depends of several things. This includes the sample source (e.g., Cytosolic Extracts or Nuclear Pellet) and the sensitivity of the antibody employed for detection (e.g., Antibodies against histones or histone PTMs). As a reference, 5–20 μg of proteins derived from cytosolic, nuclear and whole cell extracts, and 50–200 ng of proteins derived from solubilized chromatin or acid extractions are sufficient for the analysis of histones H3 and H4, and their PTMs.
19. Although it is possible to remove the comb from the glass plates on dry, removing the comb while immersed into Running Buffer 1× facilitates the process.
20. Based on our experience, avoid loading protein samples on the wells located at the gel side edges. Protein bands from those lanes usually run with a "smile" shape.
21. Samples can be loaded using common 200 μL yellow pipette tips, but it is recommended to use long pipette tips or Hamilton syringes to facilitate the loading process.
22. Filling any non-used well with Laemmli Buffer 1× helps to keep the proteins in one lane, without expanding to the other lanes.
23. PVDF is more resistant than nitrocellulose membrane. In our experience, we prefer PVDF when analyzing histone proteins.
24. To ensure that bubbles are completely eliminated, assemble the sandwich in a deep tray filled with sufficient Transfer Buffer

as to cover all the sandwich layers. Use a 15 mL conical tube as a roller to flatten the layers and remove bubbles.

25. To keep the temperature low, place the transferring chamber into a box filled with ice.
26. The most critical aspect in validating results obtained from Western blot analyses (especially against histone PTMs) is the quality of the antibody utilized. Rothbart and collaborators [29] have created a database which summarizes the specificity of over 100 commercially available antibodies against histones.
27. Since Triton X-100 is a detergent, abrupt agitation will produce a persistent foam. If that happens, let the mixture rest until bubbles disappear.
28. The TAU stacking gel is more "gooey" than the SDS-PAGE gel, therefore, remove the comb carefully.
29. Under TAU conditions, proteins are positively charged thanks to the acetic acid. Therefore, the proteins migrate toward the negative electrode.
30. The prerunning step is necessary to eliminate all the remaining APS from the gel to avoid protein oxidation. Furthermore, this prerunning step "ages" the gel to eliminate residual free radicals from the acrylamide polymerization reaction [25].
31. Since TAU Sample Buffer has high concentration of urea to induce protein denaturation, no heat is needed. Moreover, heating proteins in the presence of high concentrations of urea induces protein carbamylation, a modification that affects protein charge and, consequently, affects its electrophoretic migration on TAU gels [30].
32. After sonication, the expected size of chromatin fragments is shorter than 500 bp. Confirm the effectiveness of the sonication step by running the DNA on an agarose gel. This sonication protocol has been standardized for human cell lines. When chromatin samples are derived from other sources the correct sonication protocol requires some adjustments.
33. The use of either protein A or G agarose beads will depend on the isotype of the antibody used. As a guide, see Bonifacino et al., 2001 [31].
34. The immunoglobulin used will depend on the antibody used to immunoprecipitate.
35. When evaluating histone PTMs, 1 μg of antibody is enough for immunoprecipitation. When evaluating non-histone proteins bound to DNA, the amount of antibody must be standardized.

36. Pour buffer on the tube's wall and shake it by taping with your fingers. Do not pipet the beads at this step.
37. Use not more than 7 days old Elution Buffer, to avoid changes on the pH.
38. Remove the tubes using a warm water recipient, in order to avoid SDS precipitation. When adding TE Buffer, the final volume in the tube should be 200 μL.
39. Do not use TE Buffer or DNA Elution Buffer from commercial kits. These buffers often come with EDTA that reduces the PCR efficiency.

Acknowledgments

We thank Francisca Alvarez for valuable advice on the TAU gel electrophoresis and ChIP protocols, Karla Veas for providing purified H3.1 and H3.3. F.S. is supported by a grant from Comisión Nacional de Ciencia y Tecnología (PCHA/Doctorado Nacional/2014-21140346). AL's team is supported by grants from Comisión Nacional de Ciencia y Tecnología (FONDECYT 1160480 and Basal Project AFB 170004).

References

1. Mirsky AE, Pollister AW (1946) Chromosin, a desoxyribose nucleoprotein complex of the cell nucleus. J Gen Physiol 30:117–148
2. Daly MM, Mirsky AE, Ris H (1951) The amino acid composition and some properties of histones. J Gen Physiol 34:439–450
3. Allis CD, Jenuwein T (2016) The molecular hallmarks of epigenetic control. Nat Rev Genet 17:487–500
4. Kouzarides T (2007) Chromatin modifications and their function. Cell 128:693–705
5. Weber CM, Henikoff S (2014) Histone variants: dynamic punctuation in transcription. Genes Dev 28:672–682
6. Gurard-Levin ZA, Quivy J-P, Almouzni G (2014) Histone chaperones: assisting histone traffic and nucleosome dynamics. Annu Rev Biochem 83:487–517
7. Alvarez F, Loyola A (2017) Histone variants: structure, function, and implication in diseases. In: Gene regulation, epigenetics and hormone signaling. Wiley-VCH Verlag GmbH & Co. KGaA, Weinheim, Germany, pp 209–226
8. Szenker E, Ray-Gallet D, Almouzni G (2011) The double face of the histone variant H3.3. Cell Res 21:421–434
9. Bao Y (2011) Chromatin response to DNA double-strand break damage. Epigenomics 3:307–321
10. Polo SE, Roche D, Almouzni G (2006) New histone incorporation marks sites of UV repair in human cells. Cell 127:481–493
11. Loyola A, Bonaldi T, Roche D et al (2006) PTMs on H3 variants before chromatin assembly potentiate their final epigenetic state. Mol Cell 24:309–316
12. Dignam JD, Lebovitz RM, Roeder RG (1983) Accurate transcription initiation by RNA polymerase D in a soluble extract from isolated mammalian nuclei. Nucleic Acids Res 11:2561–2568
13. Luger K, Rechsteiner TJ, Richmond T. (1999) Methods in Enzymology 304:3–19 Expression and purification of recombinant histones and nucleosome reconstitution. In: Chromatin protocols. Humana Press, New Jersey, pp 1–16

14. Zweidler A (1978) Resolution of histones by polyacrylamide gel electrophoresis in presence of nonionic detergents. Methods Cell Biol 17:223–233
15. Laemmli UK (1970) Cleavage of structural proteins during the assembly of the head of bacteriophage T4. Nature 227:680–685
16. Luger K, Rechsteiner TJ, Flaus AJ et al (1997) Characterization of nucleosome core particles containing histone proteins made in bacteria. J Mol Biol 272:301–311
17. Bradford MM (1976) A rapid and sensitive method for the quantitation of microgram quantities of protein utilizing the principle of protein-dye binding. Anal Biochem 72:248–254
18. Himpel S, Joost H-G, Becker W (1999) Preparation of recombinant histone H3 as a substrate for protein kinase assays. Anal Biochem 274:138–141
19. Manohar M, Mooney AM, North JA et al (2009) Acetylation of histone H3 at the nucleosome dyad alters DNA-histone binding. J Biol Chem 284:23312–23321
20. Renart J, Reiser J, Stark GR (1979) Transfer of proteins from gels to diazobenzyloxymethyl-paper and detection with antisera: a method for studying antibody specificity and antigen structure. Proc Natl Acad Sci U S A 76:3116–3120
21. Towbin H, Staehelin T, Gordon J (1979) Electrophoretic transfer of proteins from polyacrylamide gels to nitrocellulose sheets: procedure and some applications. Proc Natl Acad Sci U S A 76:4350–4354
22. Burnette WN (1981) "Western blotting": electrophoretic transfer of proteins from sodium dodecyl sulfate--polyacrylamide gels to unmodified nitrocellulose and radiographic detection with antibody and radioiodinated protein A. Anal Biochem 112:195–203
23. Yang P-C, Mahmood T (2012) Western blot: technique, theory, and trouble shooting. N Am J Med Sci 4:429
24. Zweidler A (1984) Core histone variants of the mouse: primary structure and differential expression. In: Stein JL, Stein GS, Marzluff WF (eds) Histone genes: structure, organization, and regulation. John Wiley and sons, New York, pp 339–371
25. Lennox RW, Cohen LH (1989) Analysis of histone subtypes and their modified forms by polyacrylamide gel electrophoresis. Methods Enzymol 170:532–549
26. Soutoglou E, Talianidis I (2002) Coordination of PIC assembly and chromatin remodeling during differentiation-induced gene activation. Science 295:1901–1904
27. Massie CE, Mills IG (2008) ChIPping away at gene regulation. EMBO Rep 9:337–343
28. James GT (1978) Inactivation of the protease inhibitor phenylmethylsulfonyl fluoride in buffers. Anal Biochem 86:574–579
29. Rothbart S, Dickson B, Raab J et al (2015) An interactive database for the assessment of histone antibody specificity. Mol Cell 59:502–511
30. Kollipara L, Zahedi RP (2013) Protein carbamylation: in vivo modification or in vitro artefact? Proteomics 13:941–944
31. Bonifacino JS, Dell'Angelica EC, Springer TA (2001) Immunoprecipitation. In: Current protocols in molecular biology. John Wiley & Sons, Inc., Hoboken, p Unit 10.16

Chapter 3

Purification of Histone Variant-Interacting Chaperone Complexes

Hideaki Tagami

Abstract

Identification of the interaction partners of a protein is one of useful straightforward methods to gain insight into the molecular functions of the protein in cells. The pre-deposited forms of histones are associated with the specific histone chaperones to assemble into chromatin. Here, I describe an affinity purification method using the FLAG/HA double epitope-tagging technique and its application to purify particular histone variant-interacting chaperone complexes from soluble fraction to study dynamic chromatin functions. The purification is performed under low salt condition to obtain native histone variant complexes, and it would be useful to identify the specific chaperone proteins involved in the specific chromatin functions via histone variants.

Key words Protein complex, Pre-deposited histone variants, Soluble fraction, FLAG, HA, Epitope tag, Immunoprecipitation

1 Introduction

While DNA carries genetic information, the organization of chromatin structure can provide epigenetic information by being maintained after cell division or during differentiation. The chromatin-encoded epigenetic information that includes DNA methylation and various histone modifications organizes specification of cell types. Covalent modifications of histones, such as acetylation, phosphorylation, and methylation, have been shown to contribute to the formation and maintenance of transcriptionally active and inactive chromatin. For instance, histone H3 acetylated at K14 is found in the active chromatin, whereas histone H3 methylated at K9 is predominantly found in the inactive chromatin. The "histone code" hypothesis proposed that different combinations of histone modifications provide distinct recognition by chromatin-binding proteins, resulting in a variety of chromatin states [1]. In addition to the modification of histones, histone variants that mark specific chromatin loci could play important roles in gene expression and

Guillermo A. Orsi and Geneviève Almouzni (eds.), *Histone Variants: Methods and Protocols*, Methods in Molecular Biology, vol. 1832, https://doi.org/10.1007/978-1-4939-8663-7_3, © Springer Science+Business Media, LLC, part of Springer Nature 2018

genome maintenance [2, 3]. For instance, H3.3, which differs from the major H3.1 at only 4–5 amino acid positions, is predominantly deposited into active chromatin throughout the cell cycle. The centromere-specific H3 variant (CENP-A in vertebrates) is deposited selectively at centromeres and is required for centromeric chromatin maintenance and kinetochore formation. A key subject concerning chromatin regulation is to understand the functional difference of specialized chromatin, the formation of those structural functions, and the molecular mechanisms of epigenetic information maintenance. Here, I describe how to take advantage of the FLAG/HA double epitope-tagging technique to identify and characterize specific histone variant-interacting chaperone proteins as multimeric protein complexes from mammalian cells [4] and yeast cells.

2 Materials

1. cOmplete Mini, EDTA-free (Roche 11,836,170,001).
2. Spectra/Por Dialysis Membrane MWCO:8000 (SpectrumLabs 132,115).
3. ANTI-FLAG M2 Affinity Gel (SIGMA-ALDRICH A2220).
4. 3× FLAG Peptide (SIGMA-ALDRICH F4799).
5. Anti-HA (12CA5) antibody-conjugated Protein A Sepharose: Immobilize Anti-HA antibody to nProtein A Sepharose 4 Fast Flow (GE healthcare 17,528,001) and crosslink using a standard protocol [5].
6. HA Peptide.
7. Poly-Prep Chromatography Columns (Bio-Rad 7,311,550).
8. Wizard Minicolumns (Promega A7211).
9. PMSF: 1 M PMSF solution in DMSO. Store at −20 °C.
10. DTT: 1 M DTT solution in H_2O, ~100 μL aliquots. Store at −20 °C.
11. 2-Mercaptoethanol.
12. Tween20.
13. Buffer A: 100 mM HEPES-KOH (pH 7.9), 245 mM KCl, 1 mM EDTA, 5 mM EGTA. Add 2.5 mM DTT and cOmplete mini, just prior to use.
14. Buffer B: 20 mM Tris-HCl (pH 8.0), 100 mM KCl, 0.5 mM EDTA, 20% glycerol, 0.1% Tween20. Add 0.5 mM DTT and cOmplete mini, just prior to use.
15. 0.3B: 20 mM Tris-HCl (pH 8.0), 300 mM KCl, 5 mM $MgCl_2$, 0.2 mM EDTA, 10% glycerol, 0.1% Tween20. Add 10 mM 2-mercaptorthanol and 0.25 mM PMSF, just prior to use.

16. 0.1B: 20 mM Tris-HCl (pH 8.0), 100 mM KCl, 5 mM $MgCl_2$, 0.2 mM EDTA, 10% glycerol, 0.1% Tween20. Add 10 mM 2-mercaptorthanol and 0.25 mM PMSF, just prior to use.
17. Frozen-EZ Yeast Transfornation II kit (ZYMO RESEARCH T2001).
18. GenePulser Xcell (Bio-Rad).

3 Methods

3.1 Generation of FLAG/HA-Tagged Histone in its C-Terminus-Expressing Cells

A method of the generation of recombinant mammalian cells was described in detail elsewhere [6] (*see* **Note 1**). Here, I describe FLAG/HA-tagging methods in yeast (*see* **Note 2**).

1. Design primers to make the PCR products containing the homologous region (~50 nt for budding yeast and ~500 nt for fission yeast) of the just before and after the stop codon of histone gene, and the FLAG/HA, stop codon, terminator and the selection marker of the vectors. We use 2-step PCR module methods to make long PCR products for fission yeast.
2. Transform the PCR products into yeast cells using Frozen-EZ Yeast Transfornation II kit or electroporation using GenePulser Xcell according to the manufacturer's manuals. We can construct FLAG/HA-tagged histone gene in its C-terminus expressed from the native promoter.
3. Confirm the DNA sequence of tagged histone gene in the recombinant yeast cells. Determine the expression of FLAG/HA-tagged histone by immunoblotting and immunofluorescence microscopy with anti-FLAG and/or anti-HA antibodies.

3.2 Preparation of Nuclear Extracts or Whole Cell Extracts.

3.2.1 Mammalian System

1. Large-scale HeLa S3 cell culture and cell fractionation are described in detail in Ref. 6. HeLa S3 cells expressing FLAG/HA-tagged histone (canonical histone and variants) can be grown in Joklik's medium as a suspension culture at the 5–20-L scale, and the density of cells kept between 5–8 × 10^5/mL at 37 °C.
2. Harvest cells by centrifugation (700 × *g* for 8 min at 4 °C).
3. Prepare cytoplasmic and nuclear extracts as well as nuclear pellet fractions which contain most chromatin from fresh HeLa cells. We use Dignam's method [7] (*see* **Note 3**). These fractions can be frozen by liquid nitrogen and can be stored at −80 °C.
4. Use nuclear extract for the purification of soluble histone interacting protein complexes as described in Subheading 3.3.

The preparation of whole cell extract from small-scale cell culture is described later (*see* **Note 4**).

5. Harvest cells by scraping and centrifugation (500 × *g* for 1 min).
6. Wash cells with 1 mL PBS and transfer to 1.5 mL tube.
7. Centrifuge (800 × *g* for 1 min).
8. Repeat PBS wash twice.
9. Add 5 vol. of 0.3B to the cell pellet.
10. Freeze in liquid nitrogen.
11. Thaw at 37 °C just until the ice melts.
12. Repeat the freeze-thaw step again (Subheading 3.2.1, **steps 10** and **11**).
13. Rotate the tubes for 30 min at 4 °C.
14. Centrifuge (1600 × *g* for 30 min at 4 °C).
15. Transfer supernatant (Whole cell extracts: WCE) to a new 1.5 mL tube (*see* **Note 5**). Keep the pellet as chromatin fractions (*see* **Note 6**).
16. Dialyze the WCE against 0.1B (*see* **Note 7**).
17. Use WCE for the purification of soluble histone interacting protein complexes as described in Subheading 3.3.

3.2.2 Yeast System

1. Yeast cells expressing FLAG/HA-tagged histone can be grown in an appropriate medium such as YPD for budding yeast or YEA for fission yeast at 1–4-L scale until OD_{600} = 1.5~2.0 at 30 °C.
2. Harvest cells by centrifugation (2000 × *g* for 5 min at 4 °C).
3. Wash cells with ice-cold H_2O and transfer to 50 mL tube.
4. Centrifuge (2000 × *g* for 5 min at 4 °C).
5. Discard the supernatant and weigh the cell pellet.
6. Freeze the cell pellet by liquid nitrogen (*see* **Note 8**).
7. Add same volume of glass beads (0.5 mm) to the cell pellet.
8. Seal the cap with parafilm and set on FastPrep-24 (MP Bio) using BigPrep Adaptor (50 mL × 2).
9. Disrupt cells at 6.5 M/S for 20 s.
10. Leave on ice for 5 min.
11. Disrupt cells at 6.5 M/S for 20 s.
12. Leave on ice for 5 min and check the cells using a microspore (*see* **Note 9**).
13. Add half volume of ice cold Buffer A to the cell pellet.
14. Disrupt cells at 6.5 M/S for 20 s.
15. Centrifuge (3600 × *g* for 10 min at 4 °C) to precipitate glass beads and cells.

16. Transfer the supernatant to a new ultracentrifugation tube (whole cell extracts: WCE).
17. Add half volume of ice-cold Buffer A to the cell pellet and vortex well.
18. Centrifuge (3600 × *g* for 10 min at 4 °C) to precipitate glass beads and cells.
19. Transfer the supernatant to WCE tube (Subheading 3.2.2, **step 16**). Keep the pellet at −80 °C.
20. Spin WCE in an ultracentrifuge (110,000 × *g* for 2 h at 4 °C).
21. Transfer the supernatant to a dialysis tube (Spectra/Por Dialysis Membrane MWCO: 8000). Keep the pellet as chromatin fraction at −80 °C.
22. Dialyze the supernatant against 1~2 L Buffer B for 2 h at 4 °C (*see* **Note 7**).
23. Transfer the supernatant to a new ultracentrifugation tube.
24. Spin by Ultracentrifugation (110,000 × *g* for 30 min at 4 °C).
25. Transfer the supernatant to a new 15 mL tube.
26. Use WCE for the purification of soluble histone interacting protein complexes as described later.

3.3 Immuno-Affinity Purification of Soluble Histone Variant Complexes

Here, I describe our immuno-affinity purification method from nuclear extracts and whole cell extracts for soluble histone variant complexes. I recommend performing parallel mock purification using extract prepared from control cells expressing non-tagged proteins. The cell extracts from HeLa cells and yeast contain proteins that specifically bind to ANTI-FLAG M2 Affinity Gel. Most "contaminants" can be removed by the second immune-affinity purification using anti-HA (12CA5) antibody-conjugated beads. The experimental procedure is the same for both mammalian and yeast cells.

3.3.1 Anti-FLAG Antibody Immuno-Affinity Purification

1. If an extract is stored at −80 °C, thaw it rapidly in 37 °C water bath, shaking occasionally. Put the tube on ice just before thawing the extract completely.
2. Centrifuge the sample (10,000 × *g* for 30 min at 4 °C).
3. Wash ANTI-FLAG M2 Affinity Gel with 0.1 M glycine-HCl (pH 2.5) during the centrifugation of the extract.
4. Add appropriate amount of beads onto a Poly-Prep Chromatography Column. Use 100 μL beads per 10 mL extract (*see* **Note 10**).
5. Wash the beads with 10 bed vol. of 0.1 M glycine-HCl (pH 2.5) to remove uncrosslinked antibody.
6. Immediately spin the column (200 × *g* for 1 min). (Do not expose to the acidic buffer for long time).

7. Add 10 vol. of 1 M Tris-HCl (pH 8.0) and pipette well to neutralize the beads.
8. Centrifuge the column (200 × *g* for 1 min).
9. Add 10 vol. of ice-cold 0.1B and pipette well.
10. Centrifuge the column (200 × *g* for 1 min).
11. Cap the bottom and add 1 vol. of ice-cold 0.1B.
12. Transfer the bead suspension (50% slurry) to a new 15 mL or 1.5 mL tube.
13. Immediately after centrifugation of the extract (Subheading 3.3.1, **step 2**), transfer the supernatant to the tube containing M2 beads (*see* **Note 4**).
14. Close the cap tightly, and rotate gently for 2–4 h at 4 °C.
15. Pour the sample onto a Poly-Prep Chromatography Column set over a 15 mL tube or 50 mL tube on ice or at 4 °C.
16. Allow the sample to drain completely by gravity flow.
17. Transfer the flow-through to an appropriate tube and store it at −80 °C (FLAG FT).
18. Wash the beads by filling the column to the top with ~12 mL of ice-cold 0.1B, and allowing it to drain completely by gravity flow.
19. Repeat the washing step twice (Subheading 3.3.1, **step18**) for a total of three washes.
20. Set the column over a new 15 mL tube.
21. Spin the column at 200 × *g* for 1 min to remove liquid from the beads.
22. Immediately cap the bottom of the column, and seal it with parafilm.
23. Add an equal volume of FLAG-elution buffer (160 μg/mL 3× FLAG peptide in 0.1B) to the packed volume of the M2-beads (*see* **Note 11**).
24. Suspend the beads well by tapping gently, and cap the top of the column.
25. Incubate for 1 h at 4 °C, mixing gently.
26. Remove the top and bottom caps and set on a new 15 mL tube.
27. Centrifuge the column (200 × *g* for 1 min) to elute the FLAG-purified complexes.
28. Transfer the purified sample to a new 0.5 mL tube (Elution 1).
29. Keep the samples on ice when the sample is further purified by anti-HA antibody immunoprecipitation. If not, freeze it by liquid nitrogen and store at −80 °C.

30. Add an equal volume of 0.1B to the column, and suspend the beads well by tapping gently.
31. Centrifuge the column (200 × *g* for 1 min).
32. Transfer the sample to a new 0.5 mL tube (Elution 2).
33. If necessary, repeat the elution step (Subheading 3.3.1, **steps 30–32**) for Elution 3.

3.3.2 Anti-HA Antibody Immuno-Affinity Purification

1. Wash anti-HA 12CA5 antibody-conjugated beads with 0.1 M glycine-HCl (pH 2.5).
2. Take appropriate amount of beads into a 1.5 mL tube. Use 10~20 μL beads per 100 μL of anti-FLAG purified sample.
3. Centrifuge the tube (2500 × *g* for 1 min).
4. Discard the supernatant by careful micropipetting.
5. Add 0.5~1 mL of 0.1 M glycine-HCl (pH 2.5), and pipetting gently as is the case of M2 beads (*see* Subheading 3.3.1, **step 5**).
6. Immediately centrifuge the tube (2500 × *g* for 1 min). (Do not expose to the acidic buffer for long time.)
7. Add 1 mL of 1 M Tris-HCl (pH 8.0) and pipette well to neutralize the beads.
8. Centrifuge the tube (2500 × *g* for 1 min).
9. Discard the supernatant by careful micropipetting.
10. Add 1 mL of ice-cold 0.1B and pipette well.
11. Centrifuge the tube (2500 × *g* for 1 min).
12. Discard the supernatant completely.
13. Add ~200 μL of anti-FLAG purified sample.
14. Rotate gently the tube for 2–4 h at 4 °C (*see* **Note 12**).
15. Centrifuge the tube (6000 × *g* for 1 min).
16. Transfer the supernatant to a new 0.5 mL tube, and store it at −80 °C (HA FT).
17. Add 0.5~1 mL of ice-cold 0.1B, and mix gently by tapping.
18. Centrifuge the tube (6000 × *g* for 1 min).
19. Discard the supernatant by careful micropipetting.
20. Repeat washing step twice (Subheading 3.3.2, **steps 17–19**) for a total of three washes.
21. Discard the supernatant completely.
22. Add an equal volume of HA-elution buffer (200 μg/mL HA peptide in 0.1B) to the packed volume of the beads (*see* **Note 13**).
23. Suspend the beads well by tapping gently.
24. Incubate for 1 h at 4 °C, mixing gently.

25. Transfer the suspension to Wizard Minicolumn set on a new 1.5 mL tube.
26. Centrifuge the tube (6000 × *g* for 1 min).
27. Transfer the purified sample to a new 0.5 mL tube (Elution 1).
28. Add equal volume of 0.1B to the column, and suspend the beads well by tapping gently.
29. Centrifuge the column (6000 × *g* for 1 min).
30. Transfer the sample to a new 0.5 mL tube (Elution 2).
31. If necessary, repeat the elution step (Subheading 3.3.2, **steps 28–30**) for Elution 3.
32. Freeze the samples by liquid nitrogen and store at −80 °C.

3.4 Analysis of the Affinity-Purified Materials

We usually analyze the purified samples (~2 μL) on a 4–12% Bis-Tris NuPAGE gel (Thermo Fisher) and visualize by silver staining (Fig. 1). Protein bands are cut out from the gels, digested with trypsin, and analyzed by mass spectrometry. If necessary, the sample is further separated on glycerol gradient by ultracentrifugation or by reciprocal immunoprecipitation using a specific antibody against the component protein in the purified sample.

The pre-deposited forms of histone (both canonical histones and its variants) are intermediates of nucleosome assembly and disassembly. Thus, the purification of soluble histone complexes is a powerful tool to identify factors involved in chromatin dynamics. The purified histone variant complexes are useful not only for the identification of the histone variant-interacting chaperone proteins and other factors, but also for further analyses of the biochemical activities of histone variant in vitro.

4 Notes

1. To generate stable mammalian cells that express C-terminally FLAG/HA-tagged histone and its variant of interest, we use a retroviral vector, pOZ-FH-C. The vast majority of known covalent modification sites are located in the N-terminal tails of histones. Thus, we usually fuse FLAG- and HA- epitope tags with its C-terminus. The tagged histone should be expressed at 5–10% of the endogenous histone in the stable cell line. We found that transfection of the retroviral vector into packaging cells with FuGENE 6 (Roche) and Lipofectamine 2000 (Thermo Fisher) work as efficiently as does the calcium phosphate method described in ref. 6.
2. In yeast, the tandem affinity purification tag (TAP-tag) technique is widely used for protein complex purification [8]. However, TAP-tag that contains protein A and CBP is big compared to small histone protein, and it takes longer time for

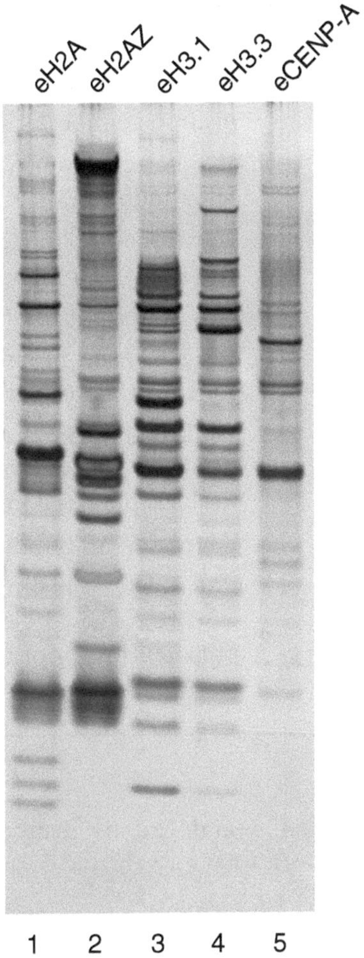

Fig. 1 Silver staining of FLAG/HA affinity purified soluble histone complexes. The soluble histone complexes of H2A (lane 1), H2A.Z (lane 2), H3.1 (lane 3), H3.3 (lane 4), and CENP-A (lane 5) were purified from nuclear extracts prepared from HeLa expressing FLAG/HA epitope-tagged each protein as described in the text

the elution. We succeeded FLAG/HA-tagging system for histone complexes both in budding and fission yeasts. We use FLAG/HA-tagging vectors: pBS1479-FHA (selection marker: TRP1) for budding yeast and pKG1810-FHA (selection marker: KanMX) for fission yeast.

3. In the Dignam's method, the nuclear proteins including most transcription factors are extracted with 0.42 M KCl, and rapidly dialyzed to 0.1 M KCl.
4. If a large-scale culture is not available, or if you wish to purify the histone variant complexes at a specific cell-cycle stage by using double thymidine treatment [9], HeLa cells can be grown in several flasks or dishes.

5. Do not take the lipid layer on the top.
6. The pellet can be solubilized by MNase, and the solubilized fraction can be used for the purification of chromatin-bound histone complexes.
7. It is better to purify the soluble histone complexes under low salt condition.
8. The frozen cell pellet can be stored at −80 °C.
9. It is enough if ~10% of cells are disrupted (They look darkly). It is better not to disrupt thoroughly in our experience.
10. I recommend preliminary experiments with each tagged histone protein to determine the optimum amount of beads.
11. We usually store 4 mg/mL of 3× FLAG peptide solutions in 0.1B at −20 °C. The FLAG-elution buffer (1/25 dilution in 0.1B) is prepared just prior to use.
12. We put an air bubble in the suspension to mix the beads.
13. We usually store 4 mg/mL of HA peptide solutions in 0.1B at −20 °C. The HA-elution buffer (1/20 dilution in 0.1B) is prepared just prior to use.

Acknowledgments

I would like to thank Drs Jun-ichi Nakayama and Guillermo A. Orsi for valuable comments.

References

1. Jenuwein T, Allis CD (2001) Translating the histone code. Science 293:1074–1080
2. Henikoff S, Smith MM (2006) Histone variants and epigenetics. In: Allis CD et al (eds) Epigenetics. CSHL Press, New York
3. Hammond CM, Stømme CB, Huang H et al (2017) Histone chaperone networks shaping chromatin function. Nat Rev Mol Cell Biol 18:141–158
4. Tagami H, Ray-Gallet D, Almouzni G, Nakatani Y (2004) Histone H3.1 and H3.3 complexes mediate nucleosome assembly pathways dependent or independent of DNA synthesis. Cell 116:51–61
5. Harlow E, Lowe D (1988) Using antibodies: a laboratory manual. CSHL Press, New York
6. Nakatani Y, Ogryzko V (2003) Immunoaffinity purification of mammalian protein complexes. Methods Enzymol 370:430–444
7. Dignam JD, Martin PL, Shastry BS, Roeder RG (1983) Eukaryotic gene transcription with purified components. Methods Enzymol 101:582–598
8. Puig O, Caspary F, Rigaut G et al (2001) The tandem affinity purification (TAP) method: a general procedure of protein complex purification. Methods 24:218–229
9. EM D, Roche D, Tagami H et al (2009) HJURP is a cell-cycle-dependent maintenance and deposition factor of CENP-A at centromeres. Cell 137:485–497

Chapter 4

Detection of Histone Modification Dynamics during the Cell Cycle by MS-Based Proteomics

Moritz Carl Völker-Albert, Andreas Schmidt, Teresa K. Barth, Ignasi Forne, and Axel Imhof

Abstract

DNA replication and subsequent deposition of nucleosomes is critical for the maintenance of the genome and epigenetic inheritance. Experiments using human tissue culture cells harvested at defined stages of the cell cycle can help to elucidate the mechanism of histone deposition and chromatin assembly in detail. Here, we describe a pulsed-SILAC approach to distinguish newly synthesized and deposited histones during S-phase of the cell cycle from parental "old" histones incorporated in previous replications and to decipher posttranslational histone modifications (PTMs).

Key words SILAC labeling, PTM-analysis, Mass spectrometry, Histone quantitation

1 Introduction

During S-phase, DNA is duplicated, which requires the disassembly and reassembly of nucleosomes that constitute chromatin. In this replication-coupled nucleosome assembly, parental ("old") histones H3-H4 as well as newly synthesized ("new") histone H3-H4 molecules are incorporated first, followed by rapid deposition of H2A-H2B dimers [1, 2]. Thereby, old and new histones are randomly distributed in a dispersive manner among the two daughter strands.

Importantly, the mixed deposition of old and new histones onto replicated DNA impacts on the propagation of histone PTMs in two basic principles: In mode 1, new histones acquire the PTM state of the parental histones within one cell cycle whereas in mode 2, the propagation relies on progressive modification of both new and parental histones (H3K9me3 and H3K27me3) [3].

The fact that canonical histone synthesis is strongly coupled to the cell cycle and becomes up-regulated during S-phase [4] can be used to selectively mark old and new histones by stable isotopic labeling of amino acids in cell culture (SILAC) [5, 6]. The pulsed-SILAC

Guillermo A. Orsi and Geneviève Almouzni (eds.), *Histone Variants: Methods and Protocols*, Methods in Molecular Biology, vol. 1832, https://doi.org/10.1007/978-1-4939-8663-7_4, © Springer Science+Business Media, LLC, part of Springer Nature 2018

method is based on cell culture media containing different isotopically labeled amino acids that are used to monitor the synthesis of proteins in a period of time. Thereby, the incorporation of labeled histones and their respective modifications can be followed in synchronized cells in a time-dependent manner via mass spectrometry.

This protocol is optimized for a comparative analysis of histone modifications on old and new histones after incorporation and their subsequent maturation. Furthermore, proteomic analysis combined with SILAC labeling can reveal histone variant dynamics together with respective histone modifications as well as protein complexes involved in DNA replication and chromatin assembly [7, 8].

2 Materials

Prepare all solutions using ultrapure water and analytical grade reagents. Prepare and store all reagents at room temperature, if not indicated differently.

2.1 Cell Culture with Human Cells

1. Use appropriate tissue culture material like sterile 10 cm dishes and 6-well plates.
2. This protocol is best suited for human cell lines like HEK293 or HeLa cells. The number of cells on a confluent plate will vary with cell type. This protocol is referred to adherent HeLa cells.
3. SILAC medium devoid of the amino acids arginine and lysine and supplemented with dialyzed FBS.
4. R^0: SILAC DMEM medium, light lysine (L-$^{12}C_6$, $^{14}N_4$-lysine) with 146 mg/L, light arginine (L-$^{12}C_6$, $^{14}N_4$-arginine) with 84 mg/L.
5. R^4: SILAC DMEM medium, medium lysine (L-lysine HCL (4,4,5,5-D4) with 146 mg/L, light arginine (L-$^{12}C_6$, $^{14}N_4$-arginine) with 84 mg/L.
6. R^{10}: SILAC DMEM medium, light lysine (L-$^{12}C_6$, $^{14}N_4$-lysine) with 146 mg/L, heavy arginine (L-$^{13}C_6$, $^{15}N_4$-arginine) with 84 mg/L.
7. R^0 T: SILAC medium, 2 mM thymidine.
8. R^4 DC: SILAC medium, 24 μM 2′-deoxycytidine-hydrochloride.
9. R^{10} DC: SILAC medium, 24 μM 2′-deoxycytidine-hydrochloride.
10. Counting device for tissue culture cells and trypan blue solution to count live and dead cell populations (this can be a basic hemocytometer).
11. Phosphate-buffered saline (PBS) buffer.

12. Trypsin: 0.05% trypsin, 0.53 mM EDTA (suitable for cell culture).
13. 0.2 μm filter device and 50 mL syringe (to filter SILAC medium after the addition of amino acids).
14. Incubator with 37 °C and 5% CO_2 concentration.

2.2 Flow Cytometry Analysis of Cell Cycle Regulated Cells

1. Ethanol 100% (ice cold).
2. PBS.
3. PBS, 1% FBS.
4. RNase A (10 mg/mL).
5. Propidium iodide (1 mg/mL).

2.3 Acid Extraction of Histones

1. 0.2 M H_2SO_4.
2. 0.1 M acetic acid: add 1 mM dithiothreitol (DTT; added fresh just before use).
3. 2 M Tris-HCl pH 6.8.
4. 100% trichloroacetic acid (TCA).
5. 1× Laemmli buffer: 10% SDS, 250 mM Tris-HCl pH 6.8, 25% (v/v) glycerol, 500 mM DTT, 0.5% (m/v) bromophenol blue.
6. Coomassie brilliant blue staining and destaining solutions (or colloidal coomassie).
7. Thin-tipped pipette (or dounce homogenizer).
8. Shaking incubator (or rotator) at 4 °C and magnetic stirrer at 4 °C.
9. SpeedVac centrifuge.

2.4 Histone Acylation and Digestion

1. 16% SDS-PAGE gel.
2. 10 mM ammonium bicarbonate.
3. Acetonitrile (HPLC grade).
4. Acetonitrile AB50: acetonitrile, 50 mM ammonium bicarbonate.
5. Propionic anhydride or acetic anhydride-d6.
6. 0.1 M ammonium bicarbonate.
7. 1 M ammonium bicarbonate.
8. 50 mM ammonium bicarbonate (ice cold).
9. 0.2 μg/μL sequencing grade trypsin (prepare according to the manufacturer's instructions).
10. Scalpel.
11. Shaking incubator, 37 °C.
12. pH-indicator paper.

2.5 Acid Extraction of Peptides After In-Gel Digestion

1. Trypsin-digested supernatant from gel and gel pieces (*see* Subheading 2.4. Histone acylation and digestion).
2. 50 mM ammonium bicarbonate.
3. Acetonitrile (HPLC grade).
4. Acetonitrile AB25: 50% (v/v) acetonitrile, 25 mM ammonium bicarbonate.
5. Formic acid 5% (v/v).
6. Acetonitrile FA2.5: 50% (v/v) acetonitrile/2.5% (v/v) formic acid.
7. TFA: 0.1% (v/v) trifluoroacetic acid, spectroscopy grade.
8. 0.5 mL low protein-binding plastic tube.
9. SpeedVac centrifuge.

3 Methods

3.1 SILAC Labeling of Cells in Tissue Culture

Work in biosafety cabinets with laminar flow when handling human cell tissues. Work as clean and sterile as possible and disinfect all materials with 70% ethanol.

1. Prepare R^0, $R,^4$ and R^{10} medium containing amino acids as specified in Subheading 2.1, **step 2** (*see* **Note 1**). Incubate the medium in 37 °C warmed water bath before usage. Otherwise, store the medium at 4 °C.
2. Cultivate HeLa cells in R^0 SILAC medium to adjust cell growth for at least eight cell divisions in 10 cm dishes. As initial seeding density, use 2.2×10^6 cells. Confluent cell density is usually reached at a concentration of 8.8×10^6 cells.
3. For cell counting, take off old medium and wash cells 2× with pre-warmed (37 °C) 5 mL PBS. Remove PBS and add 3 mL of trypsin-EDTA at room temperature. Incubate the dish for 2–4 min in 37 °C incubator and stop trypsin digestion by adding 7 mL of pre-warmed R^0 SILAC medium. Centrifuge the cells at 300 × *g* for 5 min, remove the supernatant, and resuspend cells in 2 mL of pre-warmed R^0 SILAC medium. Take an aliquot of 50 μL of cells and mix with the equal amount of trypan blue solution. Apply 10 μL of the trypan blue cell suspension to the hemocytometer and count cells according to manufacturer's instructions (*see* **Note 2**).
4. Seed 1×10^6 HeLa cells in a 10 cm dish in 10 mL pre-warmed R^0 SILAC medium and distribute equally within a 10 cm dish. Additionally, seed 0.5×10^6 cells in 4 mL in a 6-well dish for flow cytometry analysis. Adjust the number of dishes for flow cytometry analysis to the time points of observation (*see* **Note 3**). Incubate cells in CO_2 incubator at 37 °C and 5% CO_2 concentration.

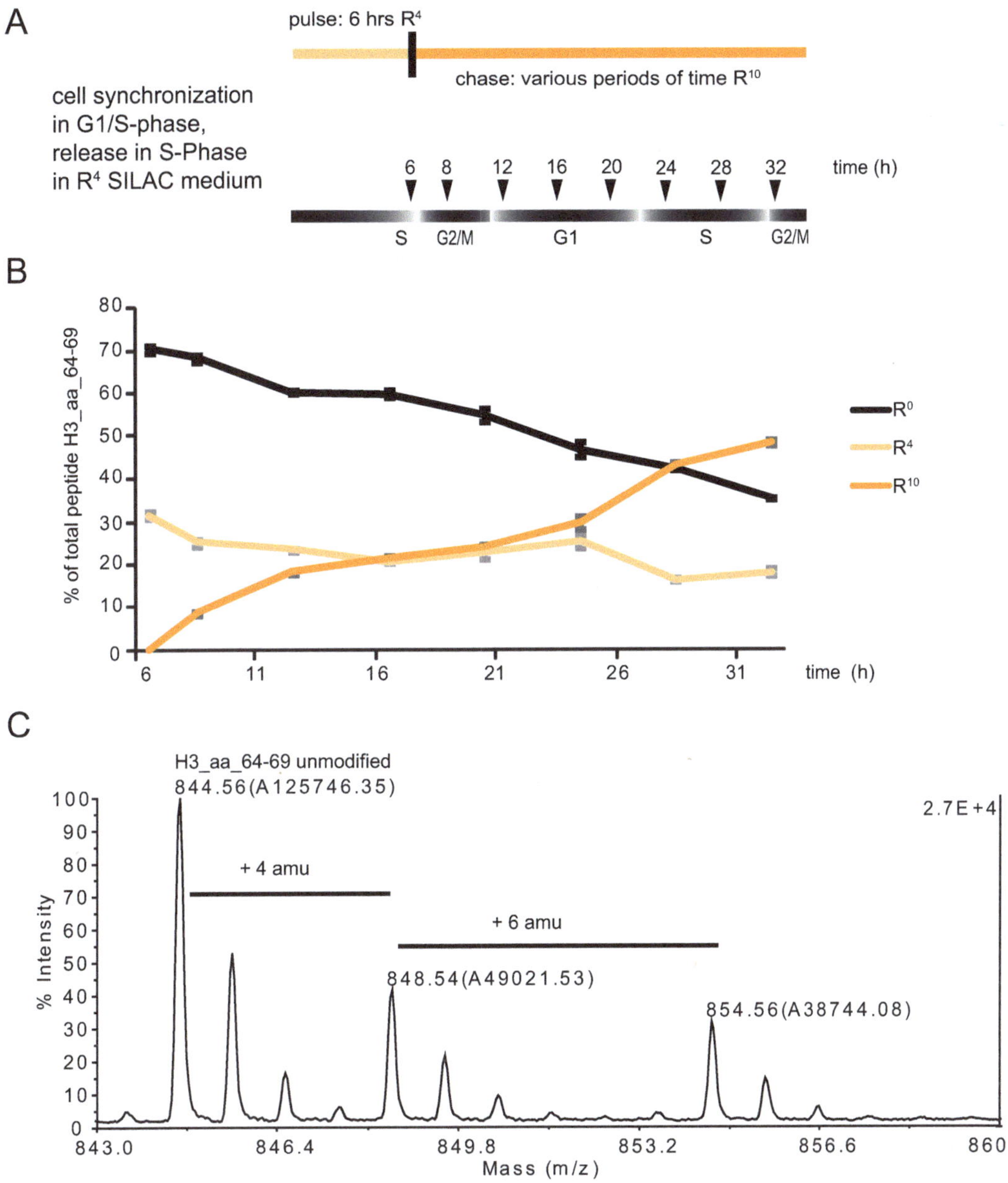

Fig. 1 Double labeling using different SILAC media allows following histones for a period longer than one cell cycle. (**a**) Experimental scheme. After synchronization, HeLa cells are released into S-phase under the R^4 medium conditions. After 6 h, the cells were transferred into a medium containing R^{10} and then harvested at indicated time points. Time (h); time after release. (**b**) Incorporation efficiency of double labeled histones for R^4 and R^{10} showing histone H3 peptide H3_aa_64–69. Error bars indicate the SEM of three independent biological replicates. (**c**) MALDI-TOF spectrum of histone H3_aa_64–69 from HeLa cells that were R^4 labeled for 6 h and afterwards R^{10} labeled for additional 6 h. Adapted from [8]

5. Between 6 and 8 h later, wash cells with pre-warmed PBS gently and apply 10 mL for 10 cm dish and 4 mL for 6-well dish of pre-warmed R^0 T SILAC medium (*see* **Note 4**) (Fig. 1).
6. Incubate cells for 17 h in 37 °C incubator with 5% CO_2 concentration.
7. For flow cytometry analysis of unsynchronized cells, take cells from one 6-well dish and wash with PBS. Treat cells with 2 mL of trypsin-EDTA to remove cells from the plastic and add 3 mL of R^0 SILAC medium. Centrifuge cells for 5 min at 300 × *g* and resuspend in 1 mL PBS. Then add 3× 900 μL 100% ice-cold EtOH to cells during slowly vortexing. Store cells at 4 °C until further procedure. Cells can be stored up to 1 week.
8. Prepare R^4 DC SILAC medium and warm up in 37 °C water bath (*see* **Note 5**).
9. For flow cytometry analysis of synchronized cells, repeat protocol steps described in **step 7**.
10. To release cells from thymidine block, wash cells 3× with pre-warmed PBS and add 10 mL for 10 cm dish and 4 mL for 6-well dish of R^4 DC SILAC medium to cells.
11. Incubate cells for up to 6 h until cells have reached late S-phase. Cells can be harvested at specific time points during S-phase progression. Each time point and cell cycle state can be defined by flow cytometry analysis.
12. Prepare R^{10} DC SILAC medium and warm up in 37 °C water bath (*see* **Note 5**).
13. For pulse-chase experiments, incubate cells with the R^4 SILAC medium for 6 h and chase with the R^{10} SILAC medium for time points of interest. Wash cells 3× with PBS between pulse and chase media.

3.2 Flow Cytometry Analysis

1. Prepare 10 mL of PBS containing 1% FBS for each flow cytometry sample.
2. Add 10 mL of PBS/1% FBS to each flow cytometry sample and centrifuge for 5 min at 300 × *g* at 4 °C.
3. Remove the supernatant and dissolve pellet in 1 mL PBS containing 10 μg/mL propidium iodide and 20 μg/mL RNAse A.
4. Mix samples well and analyze on flow cytometry machine according to manufacturer's instructions to determine cell cycle stage according to DNA content (Fig. 2).

3.3 Extraction of Soluble Proteins from Human Cell Line with 0.2 M H_2SO_4

For **step 1**, work in biosafety cabinets with laminar flow when handling human cell tissues. For all subsequent steps, work as clean and sterile as possible.

1. Collect cells from cell cultures by centrifugation at 300 × *g* for 5 min, remove the medium and wash once with 2 mL ice-cold

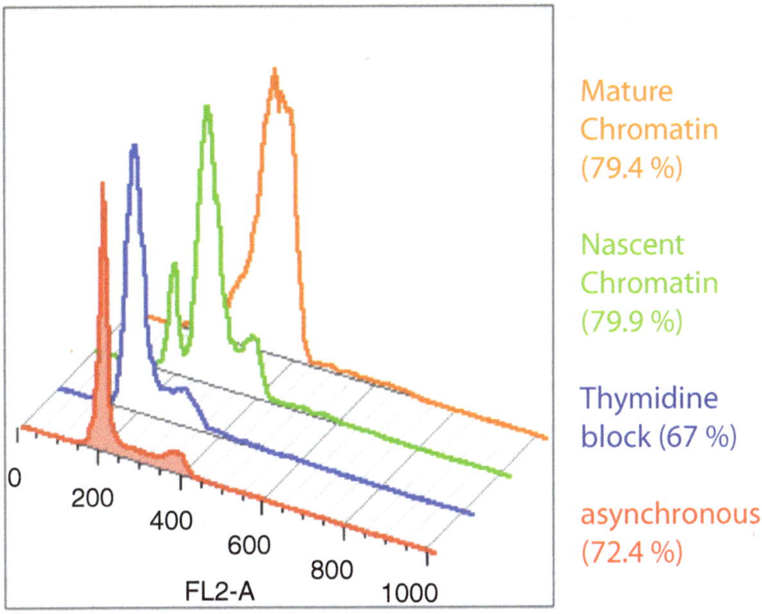

Fig. 2 Analysis of cell cycle profiles of propidium iodide-stained cells. 100,000 HeLa cells were counted. Manual selection (indicated by %) of cells was used for histogram plotting counts of cells versus FL2-A (total cell fluorescence). Nascent chromatin was collected 15 min after release from cell cycle block and mature chromatin was collected 2 h after release

PBS to reduce extracellular background, and pellet the cells again by centrifugation at 300 × *g* for 5 min. Isolated nuclei should also be pelleted by centrifugation (*see* **Note 6**).

2. Add 0.1 mL of 0.2 M H_2SO_4 per 1×10^6 cells and vortex vigorously to disperse the pellet. Place the suspension in a rotating wheel at 4 °C for 12 h (*see* **Note 7**). Subsequently, remove insoluble cell debris by centrifugation at 20,800 × *g* for 10 min at 4 °C.
3. Transfer the supernatant of the acidic extraction from **step 2** to a new vial and add TCA to a final concentration of 26%. Incubate for 30 min on ice to allow precipitation of the histones.
4. Collect precipitated proteins by centrifugation at 20,800 × *g* for 10 min at 4 °C. Remove the supernatant and wash the protein pellet 4× with 500 μL of ice-cold acetone. After the final wash, leave the pellet to dry at room temperature for 5 min.
5. Dissolve the obtained protein pellet in 1× Laemmli buffer and load the sample on a 16% polyacrylamide gel for SDS-PAGE separation (*see* **Note 8**). Run SDS-PAGE constantly at 130 V until the loading dye reaches the end of the gel. Since histones have a high electrophoretic mobility, it is important to avoid the proteins from running out of the acrylamide gel.
6. Wash the gel with water and stain with Coomassie brilliant blue to visualize protein bands (*see* **Note 9**).

3.4 Histone Acylation and Digestion

Steps 1–12 should be performed under a chemical hood to avoid contamination and the inhalation of toxic substances. Perform all steps with care and try to minimize contamination of samples: Do not lean over tubes and work as sterile as possible.

Destain gel pieces

1. Slice each purified histone sample band from the gel (*see* **Note 10**). Chop the slices into small pieces (cubes of ~1 mm^3), and place in a 0.2 mL microcentrifuge tube. Add 200 μL water to each tube and shake for 5 min at 37 °C.
2. Remove the supernatant with a pipette; take care not to aspirate the gel pieces (*see* **Note 11**).
3. Add 10 mM ammonium bicarbonate in water to each tube and shake for 5 min at 37 °C (*see* **Note 12**).
4. Remove the 10 mM ammonium bicarbonate and add 200 μL of AB50 to the gel pieces. Shake 30–60 min at 37 °C (*see* **Note 13**).
5. Remove the supernatant. If the gel pieces are still blue, repeat the ammonium bicarbonate wash; if not, proceed to **step 6**.
6. Wash with water twice as described in **step 1** (*see* **Note 14**).

Dehydrate gel pieces

7. Add 100 μL acetonitrile to the gel pieces and shake for 5 min at 37 °C.
8. Remove the supernatant (*see* **Note 15**).
9. Repeat **steps** 7 and **8**. Proceed immediately to **step 10** (*see* **Note 16**).

Perform acylation

10. Add 5 μL anhydride (propionic anhydride or acetic anhydride-d6) and 10 μL of 0.1 M ammonium bicarbonate. Open and close the tube several times (*see* **Note 17**).
11. Add 35 μL of 0.1 M ammonium bicarbonate and test the pH at 5 min after the start of the incubation by placing 1–2 μL supernatant on a piece of pH-indicator paper and, if necessary, add 1–5 μL of 1 M ammonium bicarbonate to the reaction until the pH is between 7 and 8.
12. Incubate the samples for 30–60 min at 37 °C to complete acylation (*see* **Note 18**).
13. Remove the supernatant and wash the samples three times with water as described in **steps 1** and **2**.
14. The samples can be stored at 4 °C overnight.

Digest histones

15. Dehydrate the gel pieces as described in **steps 7–9**, and place on ice.
16. Prepare a master mix containing 11 μL of 50 mM ammonium bicarbonate and 1 μL of 0.2 μg/μL trypsin for each sample

and place on ice. This amount contains a 10% excess to compensate for pipetting losses (*see* **Note 19**).

17. Immediately add 11 μL master mix to each tube with gel and incubate on ice until the gel pieces have absorbed all the supernatant (usually 5 min) (*see* **Note 20**).
18. Add 40 μL of 50 mM ammonium bicarbonate to the pieces and incubate overnight at 37 °C (*see* **Note 21**).
19. Stop the trypsin digestion reaction by placing the tubes on ice (*see* **Note 22**).

3.5 Acid Extraction of Peptides after in-Gel Digestion

1. Place the supernatant from each digestion in a clean, 0.5 mL low protein-binding plastic tube.
2. Wash the gel pieces with 50 μL (or more if necessary to completely cover the gel pieces) of the following solutions in the order indicated, shaking the sample 15 min at 37 °C or room temperature:
   50 mM ammonium bicarbonate
   AB25: 50% acetonitrile/25 mM ammonium bicarbonate
   5% formic acid
   FA2.5: 50% acetonitrile/2.5% formic acid
   100% acetonitrile
   Pool all supernatants in the same vial (*see* **Note 23**).
3. Evaporate the pooled supernatants to dryness in a SpeedVac without heating (to prevent undesired reactions).
4. Reconstitute the sample in an adequate volume of 0.1% TFA (*see* **Note 24**).
5. Samples are ready for desalting (*see* **Note 25**) and subsequent measurement on a mass spectrometer. Since the peptide aa_64–69 from histone H3 is rarely modified, it can be used as balancer for total H3 abundance. Isotopically labeled amino acids increase the mass and the m/z ratio of the measured peptide, therefore, artificially synthesized peptides can be discriminated from native H3 peptides by mass spec analysis (Fig. 1). The quantification of peptide amounts can be performed using the integrated area of the respective MS1 extracted ion chromatogram (XIC) peak [9, 10].

3.6 Histone Variant Analysis

1. For the analysis of histone modifications on histone variants, it is required to quantify unique peptides of the respective histone variants compared to canonical histones.
2. Download the FASTA files from histone variants and canonical histones of interest from public databases, e.g., http://www.uniprot.org/.
3. Copy individual histone sequences into a digestion prediction software such as GPMAW (http://www.gpmaw.com/).

Therefore, open the software and click 'Edit' → 'Edit new sequence'. Copy the sequence into the sequence field without the initial amino acid (in most cases methionine (M)), enter a name for the sequence in the field 'sequence name' and confirm on 'OK'. Then click 'Cleavage' → 'Automatic Digest' and select 'Arg-C_/R-\P' from the selection menu (*see* **Note 26**). Confirm via 'OK'.

4. The resulting list shows digested peptides according to the type of the selected digest. The column 'sequence' depicts the amino acid sequences of peptides. Compare these lists between canonical histones and histone variants to find out unique peptides for quantification. The selected peptides should be longer than three amino acids and should not contain well-described modifications to compare total histone amounts.
5. Table 1 shows an example for the comparison of histone variants H2A.Z and H2A.X to the canonical histone H2A. For MS1 quantification, colored peptides sequences were used.

4 Notes

1. Dissolve amino acids in DMEM medium. Therefore, take an aliquot of DMEM medium and dissolve amino acids in it and filter the obtained solution in a biosafety cabinet using a 50 mL syringe connected to a 0.2 μm filter and apply directly to the respective DMEM medium bottle to get the final concentration.
2. The hemocytometer is a thick crystal slide with the size of a glass slide (30 mm × 70 mm and 4 mm thickness). The glass cover is placed on the top of the hemocytometer, covering the central area. Typically, the concentration range for a cell count with hemocytometer is between 250,000 cells/mL and 2.5 million cells/mL. All further explanations and advice for counting can be found: http://www.celeromics.com/en/resources/docs/Articles/Cell-counting-Neubauer-chamber.pdf.
3. Flow cytometry analysis of cells can reveal the stage of cell cycle. Seed as many 6-well dishes as you intend to determine the exact cell cycle stage by flow cytometry analysis, e.g., before synchronization, after release from synchronization and for each time point when cells are taken for subsequent histone analysis.
4. For a single block of the cell cycle, dissolve thymidine in 40–50 mL of respective medium for 2–3 h at room temperature with a Multi-Axle Roller-Mixer. Filter the thymidine solution in a biosafety cabinet using a 50 mL syringe combined with a 0.2 μm filter and apply it to the remaining DMEM for a final concentration of 2 mM.

Table 1
GPMAW output for the canonical histone H2A and the histone variants H2A.X and H2A.Z. Each row indicates a peptide after in-silico digestion according to an Arg-C_/R-\P digestion. *Num* Number; *aa* amino acid; *MH+* positive ion mode. Colored rows indicate peptides used for MS quantification

H2A1_HUMAN (Histone H2A type 1)

Num	From aa	To aa	MH+	Sequence
1	1	3	319.34	SGR
2	4	11	801.92	GKQGGKAR
3	12	17	674.82	AKAKTR
4	18	20	349.37	SSR
5	21	29	945.11	AGLQFPVGR
6	30	32	411.48	VHR
7	33	35	401.53	LLR
8	36	42	837.91	KGNYAER
9	43	71	2917.38	VGAGAPVYLAAVLEYLTAEILELAGNAAR
10	72	77	761.86	DNKKTR
11	78	81	498.65	IIPR
12	82	88	851.04	HLQLAIR
13	89	129	4419.17	NDEELNKLLGKVTIAQGGVLPNIQAVLLPKKTESHHKAKGK

H2AZ_HUMAN Histone H2A.Z

Num	From aa	To aa	MH+	Sequence
1	1	19	1818.09	AGGKAGKDSGKAKTKAVSR
2	20	22	390.42	SQR
3	23	31	945.11	AGLQFPVGR
4	32	34	425.51	IHR
5	35	39	640.76	HLKSR
6	40	45	658.69	TTSHGR
7	46	80	3637.21	VGATAAVYSAAILEYLTAEVLELAGNASKDLKVKR
8	81	84	486.59	ITPR
9	85	91	851.04	HLQLAIR
10	92	127	3740.34	GDEELDSLIKATIAGGGVIPHIHKSLIGKKGQQKTV

H2AX_HUMAN Histone H2AX

Num	From aa	To aa	MH+	Sequence
1	!	"	319.34	SGR
2	4	11	774.9	GKTGGKAR
3	12	17	660.79	AKAKSR
4	18	20	349.37	SSR
5	21	29	945.11	AGLQFPVGR
6	30	32	411.48	VHR
7	33	35	401.53	LLR
8	36	42	860.95	KGHYAER
9	43	71	2917.38	VGAGAPVYLAAVLEYLTAEILELAGNAAR
10	72	77	761.86	DNKKTR
11	78	81	498.65	IIPR
12	82	88	851.04	HLQLAIR
13	89	142	5490.28	NDEELNKLLGGVTIAQGGVLPNIQAVLLPKKTSATVGPKAPSGGKKATQASQEY

5. To release cells from the single cell cycle block, prepare a 10× stock solution (24 mM) of 2′-deoxycytidine hydrochloride and dissolve in DMEM to have a final 1× solution (24 μM).

6. Typically, 10^6 to 10^7 cells are used as starting material, depending on organism and thus cell size. In organisms with smaller cell size, this number might have to be upscaled [11, 12].
7. For histone extraction from tissues such as dechorionated *D.melanogaster* embryos, use 1 mL of 0.2 M H_2SO_4 per gram of tissue material.
8. Due to the TCA precipitation, the sample might have a low pH indicated by a yellow color of the loading dye. In this case, adjust the pH to 6.8 using 2 M Tris-HCl, pH 6.8 until the bromophenol-blue dye of the Laemmli buffer turns blue.
9. Dependent on the species from which histones are isolated, they give rise to an abundant triplet or quadruplet of protein bands around 11–20 kDa. For further analysis by LC MS/MS, the entire cluster can be excised from the gel and processed. Keep in mind that modifications that cause a large molecular weight shift, such as ubiquitination, run with a different electrophoretic mobility and are therefore not contained in the cluster.
10. To identify the histones on a gel, recombinant histones can be loaded on a separate lane of the gel.
11. At this point, the proteins are embedded in the gel. Therefore, the same pipette tip can be used to remove all the supernatants without the risk of cross-contamination.
12. CAUTION: To prevent an explosive opening of the tube caused by the accumulated internal pressure, open and close the tubes several times after adding the ammonium bicarbonate (wear appropriate protective gear including goggles and work under a chemical hood). The ammonium bicarbonate neutralizes the leftover acetic acid employed to destain the gel. As a result, CO_2 is released and bubbles appear in the solution. The amount of formed gas depends on the amount of acetic acid left in the gel and in the amount and sequence of the protein.
13. This solution will dissolve the dye, and the supernatant will turn blue.
14. This step serves to remove the ammonium bicarbonate from the gel. Gel pieces can be stored in water at 4 °C overnight. The destaining method is modified from [13].
15. The supernatant from this step has a volume >100 μL since water is extracted from the gel pieces.
16. When the gel pieces turn opaque and small, they are dehydrated. The dehydration must be performed immediately before the subsequent steps. Otherwise, the gel portions will rehydrate from the environmental moisture.
17. Propionic anhydride facilitates analysis of methylated peptides, however, different acetylation isoforms are chromatographi-

cally separated leading to a less accurate quantitation. In order to reliably quantify acetylation isoforms for instance of histone H4_aa_4–17, D6-acetic acid anhydride should be applied in combination with a targeted MS^3 fragmentation of specific Y or B ions of the peptide isoform of interest [14].

CAUTION: The bicarbonate ions neutralize the protons released in the acylation, which causes the formation of CO_2 bubbles and pressure build-up [wear protective gear and work under a chemical hood].

18. To prevent other functional groups from being acylated, make sure that the pH of the ammonium bicarbonate solution is close to 8. At this pH, the reaction is completed upon 30 min incubation.
19. To determine which signals in the spectra originated from the degradation of trypsin, a blank reaction (trypsin in 50 mM ammonium bicarbonate) can be run in parallel.
20. Trypsin enters the gel at a temperature that prevents self-digestion of the enzyme in the solution outside the gel.
21. The peptide products of the digestion diffuse out of the gel into the supernatant. Therefore, take care from the time of the digestion reaction to prevent cross-contamination of the samples by changing the pipette tip in between samples.
22. In case the digestion is not complete (missed cleavages detected in the spectrum), add 1 μL of 0.2 μg/μL trypsin and incubate for an additional 2–4 h at 37 °C. In some cases, it is worthwhile performing acid extraction on the gel pieces (*see* Subheading 3.5) before the acquisition of the spectra. The acid extraction increases the recovery of the digestion products, which is important when the amount of sample is limited. Furthermore, some peptides may adsorb strongly to the gel or to the wall of the tube. The use of acidic solutions and acetonitrile helps to increase their recovery.
23. Some investigators use TFA instead of formic acid for extraction. Alternatively, gel pieces can be extracted by placing them in a sonication bath for 10 min per step; however, to avoid strong heating, a frozen cooling block should be placed in the sonication bath.
24. To determine the optimal volume of 0.1% TFA, the amount of protein and the sensitivity of the subsequent analytical steps must be considered. For instance, when1 μg material from human or Drosophila melanogaster nuclei is loaded on the gel, the eluted and dried peptides should be reconstituted in 10 μL 0.1% TFA. A 5 μL aliquot of this solution is sufficient to obtain good quality MALDI-TOF spectra. For a good off-line ESI-MSMS, four to eight times more protein than for the MALDI and reconstituting the peptides in 10 μL 0.1% formic acid are recommended.

25. See www.biochem.mpg.de/226863/Tutorials and [10, 15] for more details on how to prepare and use Stagetips for peptide purification. If more than eight samples need to be processed at the same time, we recommend using the tips on a 96-well plate centrifuge adapter for 200 μL tips.
26. For histone modification analysis, peptides that have been digested with trypsin after propionylation (Subheading 3.4, **step 10**) will lead to peptide sequences that end after Arginine (R) and Proline (P) ('Arg-C (R/P)').

Acknowledgments

This work was supported by a grant from the Deutsche Forschungsgemeinschaft (DFG, CRC1064-A16).

References

1. Smith PA, Jackson V, Chalkley R (1984) Two-stage maturation process for newly replicated chromatin. Biochemistry 23:1576–1581
2. Worcel A, Han S, Wong ML (1978) Assembly of newly replicated chromatin. Cell 15:969–977
3. Alabert C, Barth TK, Reveron-Gomez N et al (2015) Two distinct modes for propagation of histone PTMs across the cell cycle. Genes Dev 29:585–590
4. Osley MA (1991) The regulation of histone synthesis in the cell cycle. Annu Rev Biochem 60:827–861
5. Oda Y, Huang K, Cross FR et al (1999) Accurate quantitation of protein expression and site-specific phosphorylation. Proc Natl Acad Sci U S A 96:6591–6596
6. Ong S, Blagoev B, Kratchmarova I et al (2002) Stable isotope labeling by amino acids in cell cluture, SILAC, as a simple and accurate approach to expression proteomics. Mol Cell Proteomics 1:376–386
7. Alabert C, Bukowski-Wills JC, Lee SB et al (2014) Nascent chromatin capture proteomics determines chromatin dynamics during DNA replication and identifies unknown fork components. Nat Cell Biol 16:281–293
8. Scharf AN, Barth TK, Imhof A (2009) Establishment of histone modifications after chromatin assembly. Nucleic Acids Res 37:5032–5040
9. Egertson JD, Maclean B, Johnson R et al (2015) Multiplexed peptide analysis using data-independent acquisition and skyline. Nat Protoc 10:887–903
10. Völker-Albert MC, Schmidt A, Forne I et al (2017) Analysis of histone modifications by mass spectrometry. Curr Protoc Protein Sci Chapter 14:Unit 14.10
11. Cuomo A, Sanfilippo R, Vaccari T et al (2014) Proteomics meets genetics: SILAC labeling of Drosophila melanogaster larvae and cells for in vivo functional studies. In: Warscheid B (ed) Stable isotope labeling by amino acids in cell culture (SILAC): methods and protocols. Springer, New York, pp 293–311
12. Kruger M, Moser M, Ussar S et al (2008) SILAC mouse for quantitative proteomics uncovers kindlin-3 as an essential factor for red blood cell function. Cell 134:353–364
13. Rosenfeld J, Capdevielle J, Guillemot JC et al (1992) In-gel digestion of proteins for internal sequence analysis after one- or two dimensional gel electrophoresis. Anal Biochem 203:173–179
14. Feller C, Forne I, Imhof A et al (2015) Global and specific responses of the histone acetylome to systematic perturbation. Mol Cell 57:559–571
15. Rappsilber J, Mann M, Ishihama Y (2007) Protocol for micro-purification, enrichment, pre-fractionation and storage of peptides for proteomics using StageTips. Nat Protoc 2:1896–1906

Part II

Epigenomics

Chapter 5

Histone Native Chromatin Immunoprecipitation

Alicia Alonso, Emily Bernstein, and Dan Hasson

Abstract

Chromatin immunoprecipitation (ChIP) is becoming the standard method to study genome-wide distribution of histone variants and histone posttranslational modifications (PTMs). In this chapter, we describe a detailed native ChIP protocol and downstream procedures for the preparation of DNA libraries for next-generation sequencing. Compared to cross-linked ChIP, "native" ChIP has been shown to produce occupancy pattern data of histone PTMs and histone variants, with higher resolution and higher signal to noise ratio. We further present an adaptation of this protocol to perform native ChIP from as low as 50,000 cells.

Key words Native chromatin immunoprecipitation, Histone PTMs, Histone variants, Next-generation sequencing, ChIP-seq

1 Introduction

Chromatin is a dynamic structure that is recognized to play a central role in regulating gene expression programs important for maintaining cell type specificity and can contribute to disease when deregulated. Factors that play a role in regulating chromatin structure and output include histone post transcriptional modifications (PTMs), histone variants, transcription factors, and chromatin remodeling complexes. For example, histone variants regulate essential cellular processes such as chromosome segregation, DNA repair, and transcription initiation [1, 2]. Chromatin immunoprecipitation followed by next-generation sequencing (ChIP-seq) is a powerful method to investigate genome-wide occupancy patterns of histone PTMs, histone variants, and chromatin-binding proteins in their natural context [3–6]. Currently, most ChIP protocols use formaldehyde to cross-link proteins to the DNA, to preserve associations between macromolecules (i.e., protein-protein, and protein-DNA). This cross-linked chromatin is then fragmented by sonication, and used as the starting material for chromatin immunoprecipitation [5, 6]. However, improved resolution, elimination of nonspecific signal caused by cross-linking, and enhanced

Guillermo A. Orsi and Geneviève Almouzni (eds.), *Histone Variants: Methods and Protocols*, Methods in Molecular Biology, vol. 1832, https://doi.org/10.1007/978-1-4939-8663-7_5,

chromatin recovery due to higher epitope activity, make micrococcal nuclease (MNase)-based "native" ChIP an attractive alternative to study histone PTMs and histone variants [7, 8]. It is important to note that native ChIP is generally not applicable to non-histone components such as transcription factors, as their interactions with DNA are not stable [7]. We have optimized a native ChIP protocol [9, 10] where chromatin is digested with micrococcal nuclease (MNase) to yield mononucleosomal-sized particles, which are extracted from nuclei using 0.5 M NaCl. This is followed by their immunoprecipitation with appropriate antibodies, extraction of the associated DNA, and preparation of uniquely indexed DNA libraries for sequencing on an Illumina platform. This method was applied successfully to the identification of genomic regions enriched for an array of histone PTMs (e.g., H3K4me3 and H3K27ac) and histone variants (e.g., CENPA, H2A.Z, and macroH2A) [3, 4, 11, 12]. We foresee further adaptations to this protocol to meet more specific ChIP-seq requirements.

2 Materials

2.1 Nuclear Preparation, MNase Digestion, and Immunoprecipitation

1. Trypan Blue: prepare aliquots of 25 μL of Trypan Blue and 50 μL of 1× PBS in 1.5 mL Eppendorf tubes or a 96-well plate.
2. Trypsin: pre-warm Trypsin, 1× PBS with 1 mM $MgCl_2$.
3. Cell growth medium (use the same medium that the cells are grown in) cool to 4 °C.
4. PMSF: 100 mM phenylmethylsulfonyl fluoride (made in isopropanol every 3 weeks).
5. PI: Protease Inhibitor Cocktail (commercially available (e.g., Sigma), EDTA free in DMSO; aliquot and store in −20 °C).
6. Prechill: centrifuge adapters for 13 mL round bottom tubes (Sarstedt), low retention 1.5/2 mL Eppendorf tubes, 15 and 50 mL Falcon tubes, 13 mL Sarstedt tubes (one for each nuclear preparation).
7. Prechill: 1× PBS.
8. Dounce homogenizer: prechill (if required for nuclear preparation).
9. Prechill: Desktop centrifuge (15/50 mL tubes), Eppendorf centrifuge (1.5/2 mL tubes) and high capacity centrifuge and rotor.
10. Prepare and filter:
    (a) 100 mM $CaCl_2$.
    (b) 200 mM EGTA.

(c) 10 mM EDTA: prechill.

(d) TE buffer (10 mM Tris-HCl, 1 mM EDTA, pH to 8.2 with Acetic Acid). Keep at room temperature (RT): prechill.

(e) 5 M NaCl.

(f) 0.5 M DTT.

(g) 20% IGEPAL-CA-630 in H_2O.

(h) Buffer B (see below) with and without 0.2% Tween-20: prechill.

11. MNase (Fisher, EN0181) (*see* **Note 1**).
12. Proteinase K at 20 mg/mL.
13. RNase A at 10 mg/mL.
14. Prepare aliquots of Buffers 1, 2, and 3. For each 2–4 × 10^7 cell preparation, prepare 2 mL Buffer 1, 2 mL Buffer 2, and 8 mL Buffer 3 (see below) (*see* **Note 2**). Once aliquoted, supplement Buffers 1, 2, 3 and Buffer A with 1:1000 0.5 M DTT, 1:300 Protease inhibitor cocktail and 1:1000 100 mM PMSF. In addition, aliquot and supplement the 10 mM EDTA and TE buffer. Keep all buffers and solutions on ice.
15. For each distinct nuclear preparation, aliquot 8 mL of Buffer 3 in a Sarstedt 13 mL tube and keep on ice.
16. Thermomixer or water bath to 37 °C: preheated.
17. Agilent Technologies 2100 Bioanalyzer with an Agilent High Sensitivity DNA chip and an Agilent DNA 1000 chip.

2.2 ChIP-qPCR

1. 2× KAPA HiFi DNA Polymerase mix (Kapa Biosystems).
2. 10 mM primer mix: prepare for each positive and negative control primer pair.

2.3 Construction of Uniquely Indexed DNA Libraries for Sequencing (see Note 3)

1. 10 U/μL T4 PNK (NEB; M0201S).
2. 3 U/μL T4 DNA Polymerase (NEB; M0203S).
3. 5 U/μL DNA Polymerase Klenow Fragments (NEB; M0210S); diluted to 1U/μL when used.
4. 5 U/μL Klenow Fragment (3′-5′ exo-) (NEB; M0212S).
5. Quick T4 DNA Ligase (NEB; M2200S).
6. 2× KAPA HiF iDNA Polymerase mix (Kapa Biosystems).
7. 10 mM dNTP mix.
8. 1 mM dATP.
9. 50% glycerol in H_2O.
10. 100 bp and 25 bp DNA ladders.
11. SYBR green: 100× dilution.
12. Gel Star: 5 μL aliquots of 10,000×.

13. Sequencing adapters: prepare for each sample. Sequencing adapters are available commercially (Illumina) or can be synthesized based on Illumina guidelines. Please refer to the "Illumina Adapter Sequences" document for further details. Using barcoded adapters as opposed to indexed PCR primers, enables indexing the libraries earlier in the process to reduce possibilities of cross contamination.
14. 10 μM PCR primer mix for library amplification. PCR primers are available commercially (Illumina) or can be synthesized based on Illumina guidelines. Please refer to the "Illumina Adapter Sequences" document for further details.
15. QIAquick MinElute PCR Purification Kit (QIAGEN).
16. QIAquick PCR Purification Kit (QIAGEN).
17. QIAquick Gel Extraction Kit (QIAGEN).
18. Beckman Ampure XP beads: used for the final library purification step and size selection, to exclude adapter dimers. Alternatively, all DNA purification steps performed with columns can be performed with Ampure XP beads (*see* Subheading 3.3.7).
19. 0.5 mL Eppendorf PCR tubes: for library preparation steps.
20. 0.2 mL Eppendorf PCR tubes: for qPCR steps.
21. 1.5 mL tubes: for final elution.
22. Blue light transilluminator (*see* **Note 4**).

2.4 Buffers

All solutions should be filtered with a 0.22 μM filter and stored at 4 °C unless indicated otherwise.

1. Buffer 1: 0.32 M Sucrose, 60 mM KCl, 15 mM Tris pH 7.5, 15 mM NaCl, 5 mM $MgCl_2$, 0.1 mM EGTA.
2. Buffer 2: 0.32 M Sucrose, 60 mM KCl, 15 mM Tris pH 7.5, 15 mM NaCl, 5 mM $MgCl_2$, 0.1 mM EGTA, IGEPAL according to titration (0.03–0.2%; see below). IGEPAL is added when making aliquots.
3. Buffer 3: 1.2 M Sucrose, 60 mM KCl, 15 mM Tris pH 7.5, 15 mM NaCl, 5 mM $MgCl_2$, 0.1 mM EGTA.
4. Buffer A: 0.34 M Sucrose, 60 mM KCl, 15 mM HEPES pH 7.4, 15 mM NaCl, 4 mM $MgCl_2$.
5. Buffer B: 20 mM Tris pH 8.0, 5 mM EDTA, 500 mM NaCl.
6. Elution Buffer: 1% SDS, 100 mM $NaHCO_3$ (prepare fresh).
7. 50× TAE: 2 M Tris, 50 mM EDTA. pH to 8.3 with Acetic Acid. Keep at RT. Dilute to 1× to run gels.
8. Stop Mix Buffer: 0.6 M NaCl, 20 mM EDTA, 20 mM Tris pH 7.5, 1% SDS. Keep at −20 °C.

9. 100 bp and 25 bp DNA ladders: 20 μL DNA ladder (1 μg/μL stock), 40 μL 5× Orange G loading dye, 138 μL H_2O, 0.6 μL NaCl. Keep at 4 °C.

3 Methods

3.1 Preparation of Nuclei for MNase Digestion and Immuno precipitation

1. For general experimental considerations please refer to **Notes 5–14**.
2. Experimental flow (Fig. 1).

Before Day 1—Titrate detergent concentration (titrated once for each cell type).

Day 1—Prepare nuclei and titrate MNase digestion.

Day 2—MNase digestion followed by chromatin extraction, quantification, and visualization. Preclear the chromatin and immunoprecipitate O/N.

Day 3—Wash, elute and extract immunoprecipitated DNA followed by quantification, visualization, and ChIP enrichment analysis by ChIP-qPCR.

Before Day 1 (Performed only once for each cell type).

3.1.1 Optimization of Detergent Concentration for Preparing Nuclei

1. If starting with adherent cells, aspirate media from 1 × 10 cm plate. For suspension cells, go to **step 5**. You will need about ~500 K cells for few rounds of detergent titration (*see* **Note** 7).
2. Rinse cells with pre-warmed 1× PBS with 1 mM $MgCl_2$ at RT. Tip plate to the side and remove residual PBS (*see* **Note 15**).
3. Add 1 mL pre-warmed Trypsin (*see* **Note 16**).
4. Stop the Trypsin by adding 10 mL of 4 °C media with fetal bovine serum (FBS).
5. Gently collect cells into a 15 mL Falcon tube with 10 mL pipette.
6. Place cells on ice as soon as possible. From this point on, all reactions are done on ice.
7. Spin cells at 300 × *g* for 5 min at 4 °C.
8. Aspirate media. Flick pellet gently to disperse cells.
9. Add 10 mL of ice-cold PBS to cells and mix gently pipetting up and down.
10. Spin cells at 300 × *g* for 5 min at 4 °C.
11. Aspirate PBS. Flick pellet gently to disperse cells.
12. Use cut tip to resuspend cells in 0.5–1 mL ice-cold 1× PBS.
13. Evaluate cells under a phase-contrast microscope (*see* **Note 17**).

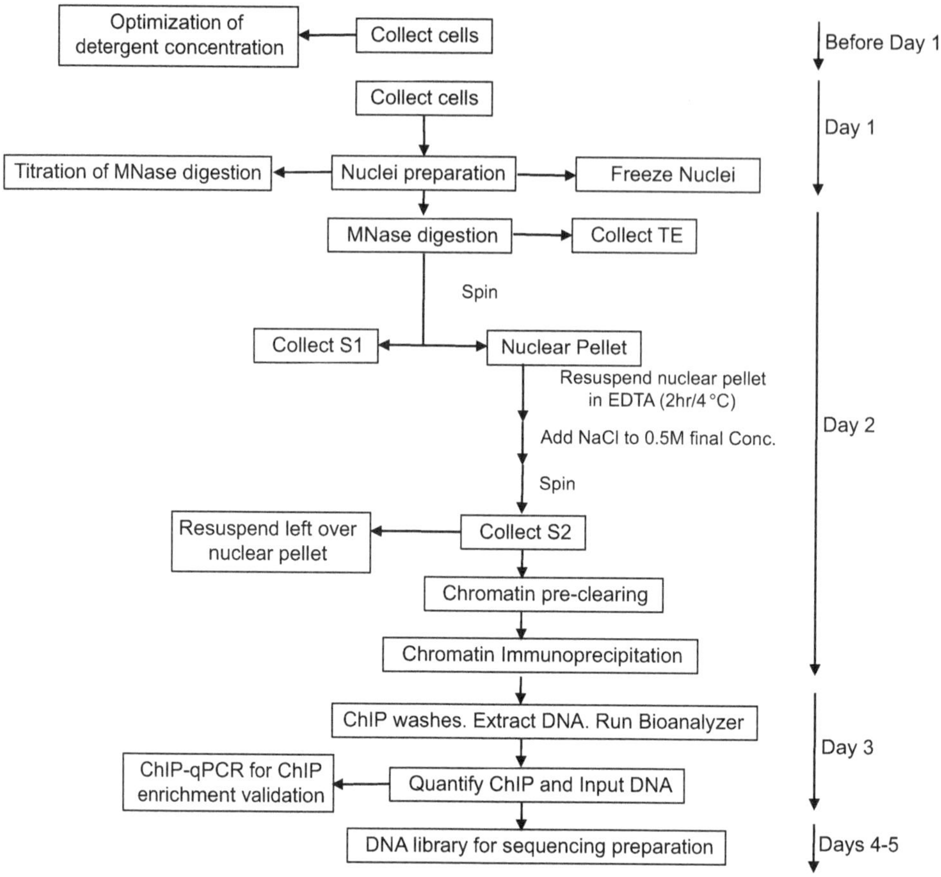

Fig. 1 Schematic flow-chart of ChIP-seq protocol. A schematic representation of the histone native ChIP protocol's conceptual organization. The main steps are shown and estimated for time required to complete. TE (total extract), S1 (soluble chromatin phase 1); S2 (soluble chromatin phase 2)

14. Take out 100–200 μL of cells with cut tip and transfer to 1.5 mL Eppendorf tube.
15. Spin cells at 300 × *g* for 5 min at 4 °C.
16. Flick pellet and then add 100 μL of ice-cold Buffer 1 to cells. Use cut tip and gently pipette up and down 4 times to disperse cells.
17. Add 100 μL of ice-cold Buffer 2 (containing variable concentrations of detergent, starting at 0.03%). Gently invert tubes 3–4 times to mix (*see* **Note 18**).
18. Evaluate nuclear preparation under a phase-contrast microscope (*see* **Note 19**). If none of the concentrations tested result in successful nuclei preparation, repeat **steps 14–18** with different detergent concentrations.

Day 1.

3.1.2 Cell Harvest and Nuclei Isolation

1. Aspirate media from plates (*see* **Notes 5–7**).
2. Rinse cells with pre-warmed 1× PBS, 1 mM $MgCl_2$ at RT. Tip plate to the side and remove residual PBS (*see* **Note 15**).
3. Add 1 mL pre-warmed Trypsin (*see* **Note 16**).
4. Stop the Trypsin by adding 10 mL of 4 °C media with FBS.
5. Gently collect cells into a 50 mL Falcon tube with 10 mL pipette. Collect up to 1×10^8 cells per tube.
6. Place cells on ice as soon as possible. From this point on, all reactions are done on ice.
7. Spin cells at 300 × *g* for 5 min at 4 °C.
8. Aspirate media. Flick pellet gently to disperse cells.
9. Resuspend cell pellet in 30–50 mL of ice-cold PBS and mix gently.
10. Count cells (*see* **Notes 17 and 20**).
11. Spin cells at 300 × *g* for 5 min at 4 °C.
12. Aspirate PBS. Flick pellet gently to disperse cells.
13. Resuspend cell pellet in 2 mL of ice-cold Buffer 1 (per $2–4 \times 10^7$ cells). Pipette up and down gently 4 times.
14. Transfer cells to 15 mL Falcon tube.
15. Add 2 mL of ice-cold Buffer 2 with appropriate concentration of IGEPAL (per $2–4 \times 10^7$ cells) and invert 3–4 times. Cells should be kept at this stage for no longer than 10 min.
16. Confirm nuclear preparation under microscope (*see* **Note 19**).
17. Collect 4 mL of cell suspension with 5 mL pipette.
18. Place the tip of the pipette on the upper part of the tube wall and slowly pipette 4 mL of cell suspension on top of 8 mL of Buffer 3 in Sarstedt tubes without disturbing the bottom layer.
19. Place Sarstedt tubes in adaptor and spin at 10,000 × *g* for 20 min, at 4 °C (*see* **Note 21**).
20. After centrifugation, tubes should look like as follows: Nuclei at the bottom, IGEPAL and cytoplasmic components at the top and sides of tube (Fig. 2).
21. Place tubes in ice bucket. From this point onward up to nuclei OD, all reactions are done in a cold room.
22. Use a plastic transfer pipette to discard top layer down to point 1 (Fig. 2). Move in a circular motion around the periphery of the sucrose layers as close as possible to the top. Make sure to remove all cell debris and IGEPAL without contacting the nuclei.

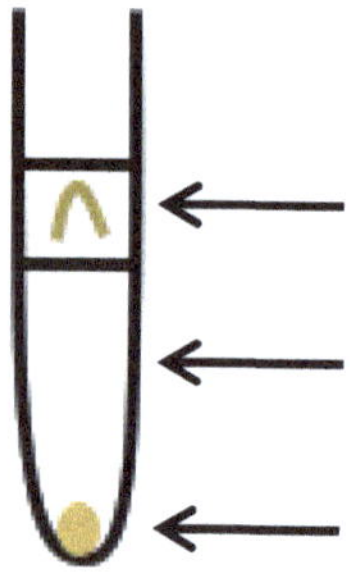

Fig. 2 Sucrose gradient separation. Depiction of nuclear preparation tubes post high-speed centrifugation for sucrose gradient separation. Nuclei are at the bottom (yellow) and IGEPAL and cytoplasmatic components are at the top (yellow). Reference for point 1 (Subheading 3.1.2, **step 22**) is shown

23. Take a kimwipe and wrap around tweezers. Stick the kimwipe/tweezers into the tubes and use it to wipe the residue from the sides of the tubes.
24. Remove the rest of the sucrose layers with a new transfer pipette. Make sure to remove as much liquid as possible without disturbing the pellet.
25. Use a new kimwipe/tweezers to remove the remaining sucrose at the bottom of the tube without disturbing the pellet.
26. Carefully add 150 μL of Buffer A to bottom of tube.
27. Resuspend cells by shaking. Make sure the pellet is completely dissolved (*see* **Note 22**).
28. Transfer resuspended nuclei to 1.5 mL Eppendorf tube with cut tip.
29. OD_{260} chromatin content in nuclear preparation using Nanodrop (*see* **Note 23**).
30. Calculate total chromatin collected and aliquot at 100 μg of chromatin in 1.5 mL Eppendorf tubes. Place aliquots at −80 °C. It is possible to continue to the next step with freshly prepared nuclei instead of freezing.

3.1.3 Titration of MNase Digestion to Generate a Nucleosome Ladder

1. Take a frozen aliquot of 100 μg of chromatin. Thaw chromatin on ice and dilute to 600 ng/μL with ice-cold Buffer A supplemented based on initial OD.
2. Mix nuclei well by gently flicking and inverting the tube.
3. Aliquot 20 μg of chromatin to 5× 1.5 mL Eppendorf tubes.
4. Add $CaCl_2$ to a final concentration of 3 mM based on final volume including MNase.
5. Titrate MNase by adding increasing amounts of MNase to each of the tubes (make a 5× or 10× dilution of MNase with Buffer A if needed) (*see* **Note 24**).

6. Flick the tube to mix, spin briefly, incubate at 37 °C for 10 min in Thermomixer at 500 rpm.
7. Place samples on ice.
8. Add EGTA to a final concentration of 10 mM. EGTA is added to chelate calcium ions to inhibit the MNase, while maintaining Mg^{2+}-dependent chromatin compaction and nuclear integrity [13].
9. Incubate samples on ice for 5 min.
10. Adjust reaction volume to 50 μL with 10 mM EDTA. Add 50 μL of stop mix buffer and 1 μL RNase A, incubate at 37 °C for 1 h in Thermomixer at 500 rpm.
11. Add 2 μL of Proteinase K and incubate at 37 °C for 2 h in Thermomixer at 500 rpm.
12. Purify DNA using PCR purification columns (elute in 30 μL of EB buffer).
13. OD_{260} DNA using Nanodrop. Dilute each sample to 4–6 ng/μL and evaluate the nucleosomal ladder on an Agilent High Sensitivity DNA chip using Agilent Technologies 2100 Bioanalyzer (Fig. 3) (*see* **Note 25**).

Day 2.

3.1.4 MNase Digestion and Soluble Chromatin Preparation

The steps to prepare chromatin suitable for ChIP, the S2, are depicted in Fig. 1. The success of the S2 preparation should be evaluated the first time this preparation is undertaken. Therefore, save aliquots (*see* **Note 26**) from **step** 7 below (TE, total extract), **step 9** (S1, soluble chromatin phase 1), step 17 (S2, soluble chromatin phase 2), and **step 18** (P: Pellet – insoluble chromatin) and process to quantify and visualize DNA and proteins.

1. Thaw frozen aliquots of 100 μg of chromatin on ice. Multiple aliquots of chromatin can be consolidated to 1× 1.5/2 mL Eppendorf tube for MNase digestion.

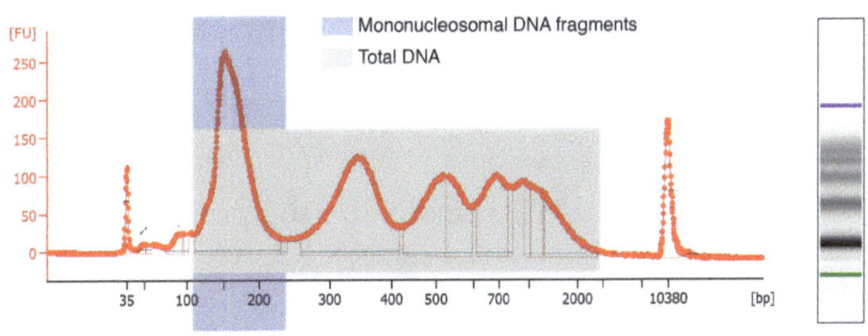

Fig. 3 Size distribution of DNA fragments from MNase digested chromatin. A representative Agilent Technologies High Sensitivity DNA chip trace of an optimal pattern of MNase digested chromatin for ChIP. Mononucleosomal DNA (blue) and total DNA (gray) are shown

2. Dilute chromatin to 600 ng/μL with ice-cold Buffer A supplemented. Mix chromatin well by gently flicking and inverting the tube.
3. Add $CaCl_2$ to a final concentration of 3 mM based on final volume including MNase.
4. Add MNase at the determined concentration (*see* **Note 27**).
5. Flick the tube to mix, spin briefly, incubate at 37 °C for 10 min in Thermomixer at 500 rpm.
6. Place samples on ice. Add EGTA to a final concentration of 10 mM.
7. Incubate samples on ice for 5 min. This will inhibit the MNase (This is the total extract tube, TE) (*see* **Note 26**).
8. Spin down TE at 500 × *g* for 7 min at 4 °C.
9. Remove the supernatant to a separate 1.5 mL tube without disturbing the pellet (label it **S1** soluble chromatin phase 1) (*see* **Note 26**).
10. Flick the TE tube to disperse nuclear pellet. To liberate the nuclear chromatin, resuspend this nuclear pellet in the original volume (**step 2**; including pellet volume) with ice-cold 10 mM EDTA pH 8 supplemented. Flick and invert the tube to resuspend nuclei.
11. Incubate for 2 h, rotating at 4 °C (*see* **Note 28**).
12. While the nuclear pellets are incubating, conjugate magnetic beads and IgG for preclearing of chromatin based on chromatin amounts and antibodies to be used. For each ChIP (25–100 μg chromatin) prepare IgG for preclearing: 3 mL PBS, 15 μL of magnetic beads and 15 μg of IgG per 25 μg chromatin in a 5 mL Eppendorf tube (*see* **Note 29**).
13. Conjugate magnetic beads and IgG rotating at 4 °C for 2 h.
14. After **step 11** is done, extract the chromatin by adding NaCl to a final concentration of 0.5 M (10% vol of 5 M solution). NaCl should be added stepwise, add 2–3 drops with a 200 μL pipette, close the tube and invert to mix, then repeat (*see* **Note 30**).
15. Incubate for additional 45 min, rotating at 4 °C.
16. Spin at max speed for 5 min at 4 °C.
17. Transfer the supernatant to a new low retention 1.5/2 mL tube (label it **S2**, soluble chromatin phase 2). Keep S2 fraction on ice and at 4 °C (*see* **Note 26**).
18. If you are testing the efficiency of soluble chromatin obtained, resuspend the pellet left in the tube with TE buffer to the original volume (**step 2**; including pellet volume, label the tube **P** (*see* **Note 26**). OD_{260} the 10 μL aliquots saved for TE, S1, S2, and P and calculate fraction of soluble chromatin (*see*

Note 31). Perform OD_{260} three times for S2 and then average to set up the desired number of ChIPs. Since S2 contains the histone fraction to be used for ChIP, it is important that the concentration is accurate.

19. If the digestion pattern is enhanced for mononucleosomes (Fig. 3), and the yield is above 40%, then continue with ChIP. Otherwise, it is recommended to repeat the soluble chromatin preparation.
20. While S2 is preclearing, thaw antibodies to be used for ChIP based on chromatin amounts and the specific antibodies. For each ChIP (25–100 μg chromatin) prepare antibodies for conjugation with beads: 4 mL PBS, 10 μL of magnetic beads and 1 μg of antibody per 10 μg chromatin in 5 mL Eppendorf tube (*see* **Notes 9–10**).

3.1.5 Immunoprecipitation

1. Dilute S2 to 100 ng/μL with Buffer B + 0.2% Tween.
2. Spin conjugated beads from Subheading 3.1.4 (12) at 800 × *g* for 5 min at 4 °C.
3. Aspirate PBS from beads (Do not place tubes with conjugated beads on magnetic rack).
4. Resuspend and collect beads with 500 μL diluted S2 with cut tip.
5. Preclear S2 for a minimum of 2 h rotating at 4 °C.
6. While S2 is preclearing, thaw antibodies to be used for ChIP based on chromatin amounts and the specific antibodies. For each ChIP (25–100 μg chromatin) prepare antibodies for conjugation with beads: 4 mL PBS, 10 μL of magnetic beads and 1 μg of antibody per 10 μg chromatin in 5 mL Eppendorf tube (*see* **Notes 9–10**).
7. Conjugate magnetic beads and antibodies rotating at 4 °C for a minimum of 2 h.
8. Spin bound antibody beads and precleared S2 at 800 × *g* for 5 min at 4 °C.
9. Aspirate PBS from antibody-bound beads (Do not put tubes with conjugated beads on magnetic rack).
10. Place precleared S2 containing tubes in magnetic rack.
11. Transfer 30 μL precleared S2 to a separate 1.5 mL Eppendorf tube for input control. Add 70 μL of buffer B no Tween and store at −80 °C.
12. Transfer precleared S2 to separate tubes according to the ChIP reactions.
13. Using cut tip, use precleared S2 to resuspend and collect conjugated antibodies.

14. Incubate O/N, rotating at 4 °C. Make sure the solution is moving from the bottom to the top of the tube while rotating (*see* **Note 28**).

Day 3

15. Short spin to pellet beads.
16. Wash beads 3× with 1 mL Buffer B + Tween. Rotate tubes for 3 min at 4 °C, followed by a short spin for each wash (*see* **Note 32**).
17. Wash beads with 1 mL Buffer B no Tween. Rotate tubes for 3 min at 4 °C, followed by a short spin.
18. Wash beads with 1 mL TE followed by a short spin.
19. Place tubes in magnetic rack. Remove the supernatant carefully without disturbing beads.
20. Remove tubes from magnetic rack.

3.1.6 Extraction of Immunoprecipitated DNA

1. Prepare fresh elution buffer: 800 μL water, 100 μL 10% SDS, and 100 μL 1 M NaHCO3.
2. Add 100 μL elution buffer to each ChIP sample, vortex, short spin and incubate in Thermomixer at 500 rpm for 30 min at 65 °C.
3. Place the tube on magnetic rack.
4. Transfer 100 μL into a new 1.5 mL Eppendorf tube.
5. Repeat elution with additional 100 μL elution buffer for each sample.
6. Place the tube on magnetic rack.
7. Collect 100 μL into the same tube to a total of 200 μL.
8. DNA is extracted from Input and ChIP samples in parallel as follows:
   (a) ChIP samples: add 12 μL 5 M NaCl and 1 μL RNase A, vortex, short spin and incubate in Thermomixer at 500 rpm for 1 h at 56 °C. Add 5 μL Proteinase K, vortex and incubate in Thermomixer at 500 rpm for 3 h at 56 °C.
   (b) Input sample: add 100 μL stop mix buffer, 1 μL RNase A, vortex and incubate in Thermomixer at 500 rpm for 1 h at 56 °C. Add 2 μL Proteinase K, vortex and incubate in Thermomixer at 500 rpm for 3 h at 56 °C.
9. Purify DNA using QIAquick PCR Purification Kit for input samples (elute in 30 μL EB buffer), and MinElute PCR Purification Kit for ChIP sample (elute in 15 μL EB buffer).
10. Quantify 1 μL of each ChIP DNA and input sample using Nanodrop or Qbit.
11. Dilute each sample to 4–6 ng/μL and run on an Agilent High Sensitivity DNA chip using Agilent Technologies 2100 Bioanalyzer. In most cases, the ChIP DNA shows a pattern

shifted toward high molecular DNA when compared to input (Fig. 4). We usually consider this shift as the first indication that the ChIP reaction worked. However, there should be enough material at the mononucleosomal range for sequencing library preparation (*see* **Note 11**).

12. Seal tubes and store at −20 °C.

3.2 Validation of ChIP Enrichment by Quantitive PCR (ChIP-qPCR)

1. For general experimental considerations please refer to **Notes 33–35**.

3.2.1 qPCR Amplification

1. Dilute 2–3 μL of ChIP and Input DNA to 4.5 μL × number of primer pairs × 2 technical replicates for each reaction.
2. Prepare PCR mix for each primer pair (5 μL x number of ChIP and input samples for each primer pair, add 0.5 μL of primer mix x number of reactions).
3. Add to 384-well PCR plate (we recommend using an automatic pipette):
   (a) 4.5 μL of diluted DNA to the bottom of the well.
   (b) 5.5 μL of PCR mix and primers to the top of the well making sure not to carry over DNA.
4. Run PCR program.
   Thermal profile (adjust based on specific primers used):
   (a) 50 °C—2 min.
   (b) 95 °C—10 min (according to Polymerase specifications)
   (c) 95 °C—15 s.
   (d) 60 °C—30 s.
   (e) Go To **step 3–39** more times.
   (f) 95 °C—15 s.
   (g) 65 °C—5 s.
   (h) 95 °C—5 s.
5. Evaluate dissociation curves and enrichment levels of ChIP over input DNA (Fig. 5).

3.3 Preparing DNA Libraries for Sequencing

1. For general experimental considerations please refer to **Notes 36–45**.
2. Experimental flow:
   Day 1—End repair (to generate blunt-end fragments), dA-tailing (to add an over-hang A) and adapter ligation.

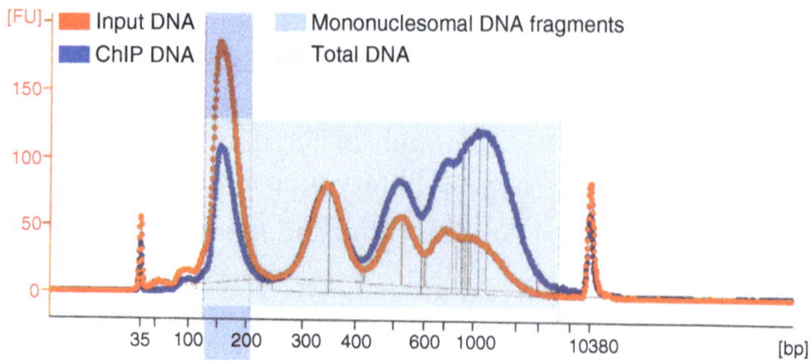

Fig. 4 Size distribution of DNA fragments from Input and ChIP samples. A representative Agilent Technologies High Sensitivity DNA chip trace of an input DNA (red), and ChIP DNA (blue) are shown. Note the inversed ratio of mononucleosomal DNA to higher molecular DNA between the two samples. Mononucleosomal DNA (blue) and total DNA (gray) are shown

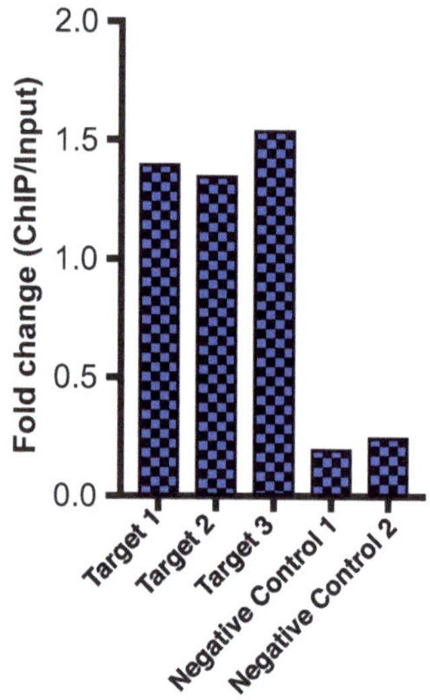

Fig. 5 ChIP-qPCR enrichment analysis. ChIP-qPCR for ChIP DNA at three enriched regions (Target 1–3), and two intergenic non-enriched regions (Negative Control 1–2). Fold enrichment of ChIP/Input is shown

Day 2—Size selection, amplification and DNA purification followed by library quantification. It is possible to stop the protocol after each step and store samples at −20 °C.

3.3.1 End Repair

1. Add to 0.5 mL PCR tube:
   (a) 8 ng of ChIP DNA in 40 μL H2O. (*see* **Note 36**).
   (b) 5 μL of 10× T4 DNA ligase buffer with 10 mM dATP.
   (c) 2 μL of 10 mM dNTPs.
   (d) 1 μL of DNA Polymerase Klenow. **Dilute to 1 U/μL.**

(e) 1 μL of T4 DNA Polymerase.

(f) 1 μL of T4 PNK.

2. Incubate in PCR machine for 30 min at 20 °C.
3. Purify DNA with PCR purification column (elute in 35 μL of EB buffer).

3.3.2 dA-Tailing

1. Add to 0.5 mL PCR tube:

(a) 34 μL of DNA from Subheading 3.3.1, **step 3**.

(b) 10 μL of 1 mM dATP.

(c) 5 μL of 10× NEB2 buffer (50 mM NaCl, 10 mM Tris-HCl, 10 mM $MgCl_2$, 1 mM DTT pH 7.9).

(d) 1 μL of Klenow (3′-5′ exo-).

2. Incubate in PCR machine for 30 min at 37 °C.
3. Purify DNA with MiniElute column (elute in 14 μL of EB buffer).

3.3.3 Adapter Ligation

1. Add to 0.5 mL PCR tube:

(a) 13 μL of DNA from Subheading 3.3.2, **step 3**.

(b) 15 μL of 2× Quick T4 DNA ligase buffer.

(c) 1 μL of adapters mix.

(d) 1 μL of Quick T4 DNA ligase.

2. Incubate in PCR machine for 15 min at 25 °C.
3. Purify DNA with MiniElute columns (elute in 15 μL of EB buffer).
4. Store samples at −20 °C or continue to size selection.

3.3.4 Size Selection

1. Place the gel apparatus in the cold room and ensure it is leveled.
2. Prepare proper volume of 2% low melt agarose gel in TAE (*see* **Note 45**). Pour agarose and place the comb. Make sure there are no bubbles. Let it cool.
3. In a clean electrophoresis chamber, place the gel and fill chamber with fresh 1× TAE buffer to cover the gel.
4. Take out samples from −20 °C, thaw and add 5 μL of 50% glycerol to each sample.
5. Load samples (from Subheading 3.3.3, **step 3**) onto gel using a 20 μL pipette with 10 μL tips (no filter). Samples are loaded in lanes 2 and 7 (8 wells comb) (Fig. 6).
6. Load 7 μL of 100 bp ladder in lane 4 and 7 μL of 25 bp ladder in lane 5.
7. Run at 100 V for 1 h.

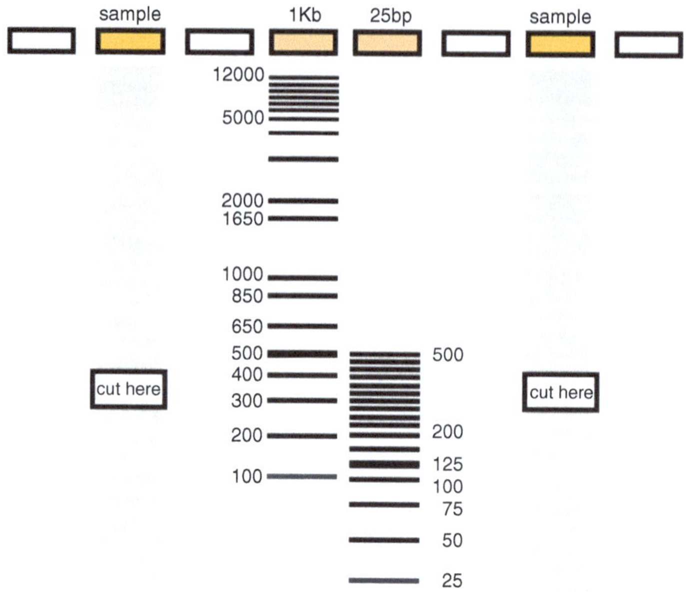

Fig. 6 Size selection gel loading scheme. Schematic of the agarose gel used for size selection in Subheading 3.3.4 of DNA sequencing library preparation. 1 Kb (lane 4) and 25 bp (lane 5) ladders are shown. DNA from two different libraries is loaded in (lanes 2 and 7). Fragments around the 300–400 bp range are excised for PCR amplification

8. Prepare staining solution; 5 μL of 10,000× gel star solution in 50 mL 1× TAE.
9. Stain for 40 min in a staining dish. This is light sensitive incubation that should be kept covered.
10. Place gel on blue light transilluminator. It is recommended to cover the light box with saran wrap to reduce DNA carryovers.
11. Weigh 2 mL tubes, one per sample.
12. Slice gel at the 300–400 bp marks (*see* **Note 41**) and transfer to 2 mL tubes. Use a clean razor blade for each sample. Gel slices should not exceed 400 mg.
13. Purify DNA using gel extraction kit at RT. Mix the sample every few minutes. Elute in 25 μL EB buffer.

3.3.5 PCR Amplification (Step 1)

1. Add to 0.5 mL PCR tube:
   (a) 24 μL of eluted DNA from Subheading 3.3.4, **step 13**.
   (b) 1 μL of primer mix.
   (c) 25 μL of 2× KAPA HiFi DNA polymerase mix.

2. Run PCR program.

   Thermal profile:

   (a) 98 °C—45 s.

   (b) 98 °C—15 s.

   (c) 65 °C—30 s.

   (d) 72 °C—30 s.

   (e) Go To **step 2–4** more times.

   (f) Hold at 10 °C.

3.3.6 Estimation of Number of Cycles for Library Amplification

1. Quantitative PCR reaction preparation; While the first PCR reaction is running, prepare mix for quantitative PCR reaction and aliquot to 0.2 mL PCR reaction tubes per the number of ChIP libraries:

   (a) 1 μL primer mix (same as in the previous amplification step).

   (b) 0.09 μL 100× SYBR green.

   (c) 7.5 μL 2× KAPA HiFi DNA polymerase mix.

   (d) 1.41 μL nuclease-free water.

2. Once the first PCR reaction is holding at 10 °C, open PCR machine lid, collect samples, vortex, short spin and return back to PCR machine.
3. Transfer 5 μL from each PCR reaction to the qPCR reaction mix.
4. Run qPCR program.

   Thermal profile for qPCR:

   (a) 98 °C—15 s.

   (b) 98 °C—15 s.

   (c) 65 °C—30 s.

   (d) 72 °C—30 s.

   (e) Go To **step 2–19** more times.

   (f) Hold at 10 °C.

5. Calculate number of additional cycles required for libraries amplification (Fig. 7). This step is required in order to generate libraries that are minimally amplified. Most PCR bias comes from later PCR cycles that occur during limited reagents concentrations [7]. In the amplification plot in Fig. 7, the maximum fluorescent intensity for sample A (green line) is 1750 and sample B (yellow line) is 1475. One third of 1750 is 583 and one third of 1475 is 491. According to the plot both 583 and 491 correspond to 8 cycles (round down). We find that amplifying one less cycle than the required number should suffice (7 cycles for both samples in this example [14]).

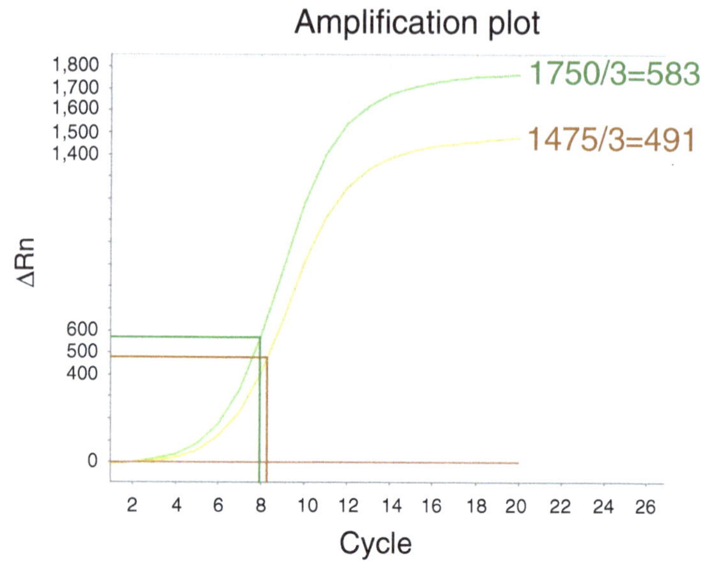

Fig. 7 qPCR for the determination of number of amplification cycles. Amplification plots demonstrating the estimated number of additional cycles to perform for two ChIP-seq libraries [14]

3.3.7 PCR Amplification (Step 2)

1. Add required number of cycles to each library (**steps 2–5** in the thermal profile (Subheading 3.3.6, **step 4**). Make sure that all libraries get the final 72 °C extension step for 1 min.
2. Purify-amplified libraries with Ampure XP beads:
   (a) Allow beads and samples to get to RT.
   (b) Mix beads by shaking and light vortexing. Make sure beads are well resuspended.
   (c) Add 45 μL slurry beads to samples with cut tip (1:1 ratio). Once the beads have been added to all samples pulse vortex 3–4 times.
   (d) Incubate at room temperature for 15 min.
   (e) Place samples on rack for at least 2 min or until the samples are clear of beads.
   (f) Remove suspension from beads (leave on magnetic rack).
   (g) Wash beads (leave on the magnetic rack) twice with 200 μL of <u>freshly</u> prepared 70% ethanol.
   (h) Take out ethanol, pulse spin, return tubes to magnetic rack, and remove the last bit of ethanol using a 10 μL tip.
   (i) Remove tubes from magnetic rack and allow beads to air dry (for 2–5 min).
   (j) Elute DNA with 15 μL EB buffer. Allow the beads to rehydrate at room temp for a few min; re-suspend by pipetting up and down 20–30 times.
   (k) Place tubes back on the magnetic rack and transfer the elute to a new tube.

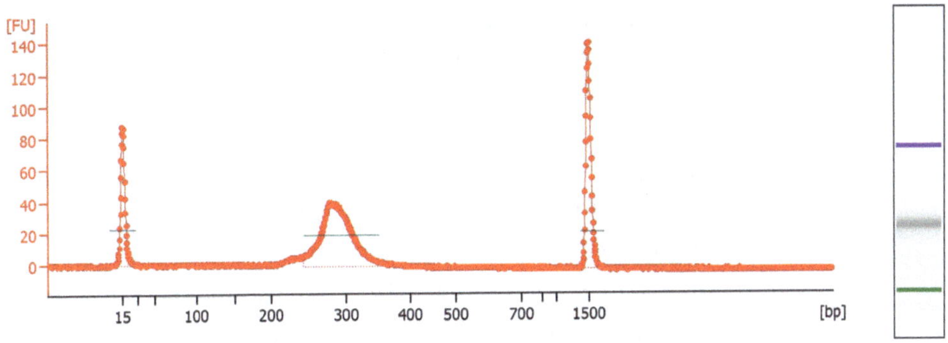

Fig. 8 Size distribution of DNA library for sequencing. A representative Agilent Technologies DNA 1000 chip trace of amplified DNA library for Illumina sequencing

3.3.8 DNA Library Size Estimation and Quantification

1. Take 1 μL for running an Agilent DNA 1000 chip using Agilent Technologies 2100 Bioanalyzer (Fig. 8).
2. Quantify 1 μL of each sequencing library using fluorometric-based methods (e.g., Qbit). Libraries are pooled at equal molarity and submitted for sequencing at approximately 10–15 nM or as specified by the sequencing core. Pooling ratio is based on required number of reads for each sample.
3. Seal tubes and store at −20 °C.

3.4 Native ChIP Protocol for 50 K Cells

Day 1

1. For each ChIP prepare IgG for preclearing: 250 μL PBS (PI+PMSF), 30 μL beads, and 3 μg of IgG in 1.5 mL low retention Eppendorf tube. Rotate at 4 °C for a minimum of 2 h (Experimental considerations).
2. Trypsinize 1 × 10 cm dish, collect into media, spin 250 × *g* for 3 min, resuspend in 4 mL ice-cold PBS.
3. Count cells.
4. Collect 50 K cells into 1.5 mL Eppendorf tube.
5. Spin 300 × *g* at 4 °C for 3 min.
6. Remove PBS and resuspend in 200 μL ice-cold Buffer 1 supplemented (1:1000 100 mM PMSF, 1:1000 0.5 M DTT, and 1:300 PI.
7. Incubate one ice for 5 min.
8. Add 200 μL ice-cold Buffer 2 with IGEPAL.
9. Incubate one ice for 5 min.
10. Spin 5000 × *g* at 4 °C for 5 min.
11. Carefully remove the supernatant (usually you can see the pellet).

12. Resuspend pellet in 100 μL of 10 mM Tris pH 8.8, 1 mM $CaCl_2$, 15 mM NaCl, 60 mM KCl, and 4 mM $MgCl_2$ supplemented (resuspension buffer).
13. Count nuclei.
14. OD_{260} chromatin content in nuclear preparation using Nanodrop (use resuspension buffer as blank).
15. Add MNase based on titration.
16. Incubate for 10 min at 37 °C in Thermomixer at 300 rpm.
17. Bring volume to 250 μL (very slow) to adjust to final: 20 mM Tris pH 8.8, 5 mM EDTA, 500 mM NaCl, and 0.2% Tween (ChIP buffer: 26.6 mM Tris pH 8.8, 5 mM EDTA, 823 mM NaCl, and 0.33% Tween).
18. Rotate at 4 °C for 2 h (cold room).
19. Spin at max speed for 10 min.
20. Collect the supernatant to a new tube. Resuspend the pellet in 250 μL of TE buffer.
21. OD_{260} chromatin (blank with 1:1.5; resuspension:ChIP buffers).
22. Skip this step unless you wish to evaluate MNase digestion or chromatin extraction: Bring SDS to 0.5%, add 2 μL RNase A, incubate at 37 °C for 1 h in Thermomixer at 500 rpm. Add 2 μL of Proteinase K, incubate at 37 °C for 2 h in Thermomixer at 500 rpm. Purify DNA using PCR purification column. Elute in 30 μL EB buffer. OD_{260} DNA using Nanodrop. Dilute each sample to 4–6 ng/μL and run on an Agilent High Sensitivity DNA chip using Agilent Technologies 2100 Bioanalyzer (*see* **Note 25**).
23. Spin down conjugated IgG (**step 1**) at 6000 rpm at 4 °C for 30 s, remove the supernatant, collect beads with 250 μL of chromatin with cut tip.
24. Preclear rotating at 4 °C for at least 2 h.
25. For each ChIP prepare bead-antibody binding: 250 μL PBS (PI+PMSF+DTT), 40 μL protein beads and 3 μg antibody in 1.5 mL low retention Eppendorf tube.
26. Rotate at 4 °C for a minimum of 2 h.
27. Spin precleared chromatin at 3500 × *g* in a 1.5/2 mL Eppendorf centrifuge.
28. Place the tube on magnetic rack and collect precleared chromatin to a new 1.5 mL low retention Eppendorf tube (save 15 μL for input and bring volume to 100 μL with TE).
29. Spin down the conjugated antibody (**step 25**) at 3500 × *g* at 4 °C for 30 s, remove the supernatant, collect beads with 250 μL of precleared chromatin (from **step 28**).

30. Rotate overnight at 4 °C.

Day 2

31. Spin down at 3500 × *g* at 4 °C for 30 s.
32. Discard the supernatant.
33. Wash with 250 μL Buffer B + 0.2% Tween.
34. Rotate for 3 min at 4 °C.
35. Repeat **steps 31–34** two more times.
36. Wash 1× with 250 μL Buffer B NO Tween supplemented (during this last wash transfer the beads to a new tube).
37. After last wash, short spin and place tube on magnetic rack to remove supernatant.
38. Continue with Subheading 3.1.6 (DNA extraction) in main protocol.

3.5 Native ChIP Protocol for Tissue

1. Cut ~100 mg tissue with razor blade into 1 mm pieces (~200 mg per sample).
2. Collect tissue into a 1 mL Dounce homogenizer.
3. Add 1 mL Buffer 1 supplemented (PI+PMSF+DTT).
4. Dounce 2–3 times. Evaluate nuclei. If needed, continue with 2–3 more dounces.
5. Transfer homogenate into 15 mL Falcon tube.
6. Repeat **steps 1–4**. Consolidate homogenates.
7. Continue with detergent titration (Subheading 3.1.1, **step 14**) or nuclear isolation (Subheading 3.1.2, **step 15**).

4 Notes

1. Once titrated and if producing similar digestion patterns as to what is shown in Fig. 3, MNase from any brand can be used.
2. If you need to titrate the concentration of detergent (Subheading 3.1.1) when adapting the protocol to a new cell line, prepare aliquots of Buffer 2 with a range of 0.03–0.2% of IGEPAL, otherwise prepare only the concentration needed.
3. In this protocol we use a combination of buffers, enzymes, and other reagents from multiple vendors instead of commercially library preparation kits, to reduce cost and introduce more flexibility into the preparation protocol. Alternatively, libraries amenable for sequencing can be generated using commercially available kits.
4. We recommend using Blue light transilluminator which is a safe alternative to UV transilluminators.

5. This protocol has been successfully applied to suspension cells (e.g., lymphoblastoid cell lines), and adherent cells (e.g., fibroblast cell lines). However, the protocol can be used with any cell type after adjusting the optimal detergent concentration and units of MNase required for chromatin digestion.
6. The required number of cells is dependent on the total number of ChIP reactions and the antibodies used. For ChIP with antibodies against histone modifications (e.g., H3K4me3) a relatively low amount of chromatin (~20 μg) should suffice. For histone variants H2A.Z or H3.3, a higher amount of chromatin (~80 μg) is required. In general, about 2–4 × 10^7 cells should yield 200–300 μg of **nuclear** chromatin (on average only about 50% of this chromatin is extracted and can be further used for ChIP). This protocol is adjusted for nuclear preparations from 2–4 × 10^7. If more chromatin is needed or a higher number of cells is available, nuclear preparations can be done in parallel, which can then be consolidated at the end of the nuclear preparation step. It is important to freeze nuclei in small aliquots (~100–150 μg) to eliminate repeated cycles of thawing and freezing.
7. Start nuclear preparation with a viable culture especially when working with cells in suspension (less obvious to distinguish between live and dead cells). It is recommended to estimate the percentage of live cells and proceed with the protocol if >85% of the cells are viable.
8. This protocol uses a mild detergent (i.e., IGEPAL) in Buffer 2 to disrupt the plasma membrane and permeabilize the nuclear membrane. The optimal detergent concentration is cell type-dependent and should be titrated as the first step. It is important to find the optimal detergent concentration that yields a high percentage of nuclei without a significant loss (i.e., cell/nuclei) by titrating the amount of detergent on small number of cells. Lower than optimal concentration will result in high percentage of intact cells leading to low chromatin yield in the extraction step. Higher than optimal levels will lead to significant loss of cells or significantly damaged nuclear membranes. In this case, the nuclei preparation will look as if all nuclei were lysed. In general, when evaluating the nuclei preparation, it is recommended to use Trypan Blue.
9. The amount of antibody to be used for ChIP is dependent on multiple variables (e.g., antibody's efficiency, target histone abundance, etc.). It is recommended to start with 1 μg of antibody per 10 μg of chromatin and further adjust based on chromatin amounts, antibody efficiency, and signal to noise ratio of enriched peaks post-sequencing.
10. We recommend using protein A beads and rabbit IgG (commercially available purified IgG from rabbit serum) for rabbit

polyclonal antibodies and protein A + G beads and mouse IgG for mouse monoclonal antibodies.

11. Selection of the appropriate reference control for immunoprecipitation is an important consideration. We use a fraction of the starting chromatin (input control) for each type of cell, as reference control post sequencing. In addition, comparing the patterns of MNase-digested chromatin of input DNA to immunoprecipitated DNA on an Agilent High Sensitivity DNA chip using Agilent Technologies 2100 Bioanalyzer is the first indication of the success of the immunoprecipitation. Finally, input DNA is also useful as a control for further ChIP-qPCRs to determine enrichment levels at positive and negative control chromosomal regions.
12. In order to minimize mechanical cell and nuclei rupture, we use wide-bore tips when pipetting cells, nuclei, undigested chromatin, and beads. These can be purchased from vendors or generated by cutting regular bore tips with a razor blade. Throughout, we recommend using low retention tips with barrier. For handling reactions with beads we recommend using low retention tubes and wide-bore tips.
13. We have now adapted this protocol to perform native ChIP from 50 K cells and frozen tissues. The steps that differ from this protocol are detailed in Subheadings 3.4 and 3.5 respectively.
14. In general, this protocol is highly sensitive to pipetting errors. Make sure that pipettes are calibrated and are not used at the volume extremes of their performance specifications.
15. Divalent cations are important to keep cells adherent during rinsing.
16. Cells should be trypsinized at RT for more sensitivity in order to prevent any excess damage to the plasma and nuclear membranes. Collect cells immediately after they start to detach.
17. For viewing cells under a phase-contrast microscope: Cut tip with blade. Mix 25 μL of cells with 25 μL Trypan blue. Put 10 μL of Trypan blue/cell dilution in a hemocytometer. Cells should be translucent.
18. The IGEPAL in Buffer 2 disrupt the cellular membrane while only permeabilizing the nuclear membrane. The optimal concentration is cell type dependent and therefore should be titrated first. For detergent titration, prepare Buffer 2 with increased amount of IGEPAL (e.g., 0.03%, 0.1% 0.15%, and 0.2%).
19. To evaluate nuclear preparation: Cut tip with blade. Mix 25 μL of cells with 25 μL Trypan blue. Put 10 μL of Trypan blue/ cells in a hemocytometer. After adding Buffer 2, nuclei should

be swollen, light blue, flat, with smooth circumference, and mostly in single nuclei suspension. If there are a lot of intact cells, IGEPAL concentration in Buffer 2 should be increased before repeating steps 1–18. Cells should not be kept in Buffer 1 + 2 for more than 10 min. If needed, cells could be further homogenized using a Dounce homogenizer to release nuclei. IGEPAL-CA-630 concentrations that are too high or repeated use of the Dounce homogenizer could result in significant loss of intact nuclei. It is recommended to count cells and then nuclei to confirm that there is no significance loss of cells during the nuclear preparation process.

20. For cell counting: Cut tip with blade. Mix 25 μL of cells with 25 μL Trypan blue and 50 μL 1× PBS. This is a 1:4 dilution of cells. Put 10 μL of Trypan blue/cell dilution in a hemocytometer. Count the number of cells in 3 squares of a 4 × 4 grid and take the average. Calculate number of cells [Average # of cells in 4 × 4 grid × 4 (dilution) × 10,000 (for # of cells in 1 mL) × 30 (volume) = total # of cells]. Cells can be counted with an automated counter as well however, it is highly recommended to look at the cells/nuclei under a phase contrast microscope to evaluate the culture and the nuclei quality.
21. The spin is done in high capacity centrifuge (Acceleration = max; Deceleration = Very slow. Should take about 10 min to stop). Total time should be around 30 min.
22. Be gentle when resuspending the nuclei in order to prevent disruption of the membranes. Do not pipette or vortex! Make sure the pellet is completely dissolved before transferring nuclei. This step could take some time, don't expect the nuclei to get into solution immediately.
23. Mix nuclei well by gently flicking and inverting the tube before each reading. Take three OD_{260} reading of the nuclei (readings should be within ~300–400 ng/μL range), and average for a final concentration. Concentration should be 600–1000 ng/μL. In case concentrations exceed 1000 ng/μL, dilute sample with Buffer A, and repeat reading. Use Buffer A to blank.
24. The MNase concentration needed for chromatin digestions should be determined based on the MNase used, the specific MNase batch, and cell type.
25. MNase titration should yield chromatin digested to different degrees with varying ratios of mononucleosomes (peak around 150 bp) (Fig. 3; blue rectangle), to a less digested chromatin (peaks at 300 bp, 450 bp etc.). The ideal digestion should; (a) have the majority of chromatin digested to mononucleosomes (145–160 bp), (b) smaller peaks at around 300 bp, 450 bp etc. (c) no shoulders on either side of the mononucleosomal peak, and (d) little to none over-digested chromatin at around

50–100 bp (Fig. 3). MNase digestion could also be evaluated on a 1.5% agarose gel (run 1 μg at 100 V for 15 min).

26. In order to evaluate the MNase digestion and estimate efficiency of chromatin extraction, aliquots from different extraction steps are saved. Transfer 10 μL (DNA extraction) and 25 μL (protein quantification/Western Blot) into two separate 1.5 mL Eppendorf tubes. Keep on ice.

27. MNase concentration as what was determined in the MNase titration step. MNase should be diluted the same as what was done for the titration step.

28. For successful extraction, reactions done in 1.5 mL tubes must have greater than 250 μL of diluted nuclei, otherwise the nuclei will not mix during rotation. For smaller volumes use 0.5/2 mL tubes. This may also be enhanced by a slower rotation.

29. For each ChIP (25–100 μg chromatin) prepare IgG for pre-clearing: 3 mL PBS, 15 μL of magnetic beads and 15 μg of IgG per 25 μg chromatin in a 5 mL Eppendorf tube. When the protocol is set up for the first time, we recommend going through the soluble chromatin preparation without pre-conjugating antibodies just in case the chromatin preparation fails.

30. Chromatin can become unstable in higher salt concentrations and therefore, it is important not to add a large volume of 5 M NaCl to the chromatin. The increased salt concentration enhances the soluble chromatin extraction from the nuclei.

31. Use this equation to plug in OD_{260} chromatin concentrations $\{[S2]/([S1] + [S2] + [P]) = \%\}$ to calculate the fraction of soluble chromatin. Ideally it should be ~55% or higher. It is recommended to assess the efficiency of the soluble chromatin preparation on a Western blot using H3 antibodies and further antibodies used for ChIP, with the collected fractions, at least when setting up the protocol. The TE and S2 fraction should consist of the majority of the proteins with lower amounts in the S1 and pellet fractions. At this step, DNA can be extracted from TE, S1, S2, and P, repeating **steps 10–13** in MNase digestion titration (Subheading 3.1.3), to confirm MNase ladder (*see* **Note 25**).

32. It is recommended to transfer chromatin to a new tube before the last wash in Buffer B (no Tween). Cut a low-retention tip and add 500 μL of Buffer B (no Tween, supplemented with DTT, PI and PMSF). Pipette to resuspend beads. Transfer beads to a new 1.5 mL low-retention tube. Use another tip to add 500 μL of Buffer B to the original tube, and use the first tip to wash the tube walls and tip. Transfer the 500 μL of Buffer B to the second tube (this effectively washes the first tube and first tip to get all beads out).

33. ChIP-qPCR using genomic DNA can serve as an accurate quantitative measurement for enrichment levels obtained in ChIP experiments and for evaluating the signal to noise ratio [6]. There are few aspects to consider for a successful ChIP-qPCR reaction including: (1) primer design (we recommend using dedicated software), (2) amplicon size (PCR primers should be designed to amplify amplicons that are smaller than mononucleosomal DNA fragments (100–120 bp)); the same primers can be further used to evaluate enrichment levels in the DNA libraries for sequencing if needed, (3) primer efficiency and specificity. We recommend testing all primer sets for efficiency and specificity with varying amounts of input DNA before amplifying ChIP DNA. After PCR amplification is complete, check the dissociation curve for each primer pair. Curves should produce a single sharp peak. In addition, DNA amplification should start mid-range through the PCR reaction (i.e., 15–25 cycles) for control DNA estimated to be at the same concentration as the ChIP sample, for all primers. We recommend choosing regions for positive and negative controls based on available data (e.g., ENCODE or GEO data sets) or functional predictions (e.g., promoter regions for H3K4me3 ChIP). Negative controls should be designed in intergenic regions with no functional predictions (e.g., gene annotations, ENCODE). We also recommend including technical and biological replicates based on the availability of ChIP and Input DNA.

34. DNA amounts can be very limited following ChIP. Dilute 2–3 μL of ChIP DNA to accommodate a limited qPCR reaction. Input DNA should be diluted to a concentration as close as possible to ChIP DNA concentration. When there is not enough DNA for both ChIP-qPCR and libraries for sequencing, ChIP-qPCR should be carried out on the sequencing library.

35. Enrichment values are measured for each primer over input control. We calculate the ΔCt values for each sample and then multiply by 1.93 which is the expected amplification constant (i.e., 1.93^[Ct (input) – Ct (ChIP)]. Positive control samples should present significant fold enrichment over negative controls (Fig. 5).

36. It is important to quantitate the ChIP and Input DNA and assess DNA quality prior to starting library preparation. Estimate the total DNA concentration (Fig. 3; gray rectangle) and the concentration of mononucleosomal DNA (Fig. 3; blue rectangle) using the Bioanalyzer "select region" option. For a standard library you should have about 8 ng of mononucleosomal DNA. It is possible however to prepare libraries from as low as 1–2 ng of mononucleosomal DNA (*see* **Note 44**).

37. Adapter dilution: Adapters should be used based on the starting total ChIP and input DNA amounts (Figs. 3 and 4; gray rectangles), and not based on starting mononucleosomal DNA amounts. We recommend adjusting the optimal adapter concentration when setting up the protocol for the first time. 1:40 dilution of a 15 μM adapter mix is a good starting point for titration. Lower than optimal adapter concentration will generate fragments with only one or no adapters ligated; higher than optimal concentration will result in high concentration of adapter dimers in the DNA sequencing libraries, which can be partially removed during DNA purification with Ampure XP beads (ligated adapters will appear as a peak at around 130 bp in the Bioanalyzer trace).
38. It is possible to add uniquely barcoded adapters to each ChIP in order to multiplex various ChIP-seq libraries onto one sequencing lane to reduce sequencing cost. Based on the desired coverage for each sample (depending on the target), the multiplexing strategy needs to be designed before the adapter ligation step and according to Illumina guidelines. It is recommended to sequence about 30 M reads for histone variants and histone modifications that occupy narrow genomic regions (e.g., H2AZ), and about 50–60 M reads for histone variants and histone modifications that occupy broad genomic regions (e.g., macroH2A). Input control (human) is usually sequenced to about 80–100 M reads.
39. Library preparation **steps 1–3** and the size selection step should be done in a designated area. If possible, separate from the area used for the PCR amplification, to minimize contamination.
40. When loading the gel for size selection, samples should be separated as much as possible. We usually load 2 samples (tagged with different barcodes—one on each side of the gel (Fig. 6) separated by two DNA size ladders (i.e., 100 bp and 25 bp) to minimize the possibility of cross contamination. Additional samples are loaded on separate gels.
41. Ligated DNA fragments do not run according to their true size due to the single stranded DNA tails that are part of the Illumina adapter strategy. DNA fragments after ligation, whose size is approximately 276 bp (150 bp (mononucleosomal insert) + 126 bp (adapters)), will run between the 300 and 400 bp marks. This should be taken into consideration when excising the targeted fragments for amplification (Fig. 6). We recommend further optimization when running the protocol for the first time. Multiple gel slices consisting DNA of different sizes can be excised and PCR amplified (see below) to determine the region on the gel containing mononucleosomal fragments.

42. When working with low amounts of ChIP DNA, size selection is performed based on the DNA ladders without actually visualizing the DNA. It is very important to have the gel apparatus leveled when pouring the gels, such that DNA in all lanes will run as similar as possible to the size markers.
43. When amplifying DNA libraries for sequencing, it is important to estimate the required number of cycles in order to generate DNA libraries that are minimally amplified. Most PCR bias comes from limited concentration of reagents in the later PCR cycles [7].
44. When preparing libraries from very low amount of DNA (estimated by Bioanalyzer), it is recommended to perform PCR amplification after adapter ligation, and then perform DNA size selection.
45. Volume of agarose should be determined based on the gel apparatus. Gels should be as thin as possible to accommodate loading of 20 μL samples.

References

1. Vardabasso C, Hasson D, Ratnakumar K et al (2014) Histone variants: emerging players in cancer biology. Cell Mol Life Sci 71(3): 379–404
2. Talbert PB, Henikoff S (2010) Histone variants--ancient wrap artists of the epigenome. Nat Rev Mol Cell Biol 11(4):264–275
3. Fontanals-Cirera B, Hasson D, Vardabasso C et al (2017) Harnessing BET inhibitor sensitivity reveals AMIGO2 as a melanoma survival gene. Mol Cell 68(4):731–744 e9
4. Hasson D, Panchenko T, Salimian KJ et al (2013) The octamer is the major form of CENP-A nucleosomes at human centromeres. Nat Struct Mol Biol 20(6):687–695
5. Roe JS, Marcan F, Rivera K et al (2015) BET Bromodomain inhibition suppresses the function of hematopoietic transcription factors in acute myeloid leukemia. Mol Cell 58(6):1028–1039
6. Tian B, Yang J, Brasier AR (2012) Two-step cross-linking for analysis of protein-chromatin interactions. Methods Mol Biol 809:105–120
7. Wiehle L, Breiling A (2016) Chromatin Immunoprecipitation. Methods Mol Biol 1480:7–21
8. Brind'Amour J, Liu S, Hudson M et al (2015) An ultra-low-input native ChIP-seq protocol for genome-wide profiling of rare cell populations. Nat Commun 6:6033
9. Ando S, Yang H, Nozaki N et al (2002) CENP-A, -B, and -C chromatin complex that contains the I-type alpha-satellite array constitutes the prekinetochore in HeLa cells. Mol Cell Biol 22(7):2229–2241
10. Alonso A, Mahmood R, Li S et al (2003) Genomic microarray analysis reveals distinct locations for the CENP-A binding domains in three human chromosome 13q32 neocentromeres. Hum Mol Genet 12(20):2711–2721
11. Vardabasso C, Gaspar-Maia A, Hasson D et al (2015) Histone variant H2A.Z.2 mediates proliferation and drug sensitivity of malignant melanoma. Mol Cell 59(1):75–88
12. Gaspar-Maia A, Qadeer Z, Hasson D et al (2013) MacroH2A histone variants act as a barrier upon reprogramming towards pluripotency. Nat Commun 4:1565
13. Sanders MM (1978) Fractionation of nucleosomes by salt elution from micrococcal nuclease-digested nuclei. J Cell Biol 79(1):97–109
14. Buenrostro JD, Wu B, Chang HY et al (2015) ATAC-seq: a method for assaying chromatin accessibility genome-wide. Curr Protoc Mol Biol 109:21.29.1–21.29.9

Chapter 6

How to Tackle Challenging ChIP-Seq, with Long-Range Cross-Linking, Using ATRX as an Example

Julia Truch, Jelena Telenius, Douglas R. Higgs, and Richard J. Gibbons

Abstract

Chromatin immunoprecipitation coupled with high-throughput, next-generation DNA sequencing (ChIP-seq) has enabled researchers to establish the genome-wide patterns of chromatin modifications and binding of chromatin-associated proteins. Well-established protocols produce robust ChIP-seq data for many proteins by sequencing the DNA obtained following immunoprecipitation of fragmented chromatin using a wide range of specific antibodies. In general, the quality of these data mainly depends on the specificity and avidity of the antibody used. However, even using optimal antibodies, ChIP-seq can become more challenging when the protein associates with chromatin via protein-protein interactions rather than directly binding DNA. An example of such a protein is the alpha-thalassaemia mental retardation X-linked (ATRX) protein; a chromatin remodeler that associates with the histone chaperone DAXX, in the deposition of the replication-independent histone variant H3.3 and plays an important role in maintaining chromatin integrity. Inherited mutations of ATRX cause syndromal mental retardation (ATR-X Syndrome) whereas acquired mutations are associated with myelodysplasia, acute myeloid leukemia (ATMDS syndrome), and a range of solid tumors. Therefore, high quality ChIP-seq data have been needed to analyze the genome-wide distribution of ATRX, to advance our understanding of its normal role and to comprehend how mutations contribute to human disease. Here, we describe an optimized ChIP-seq protocol for ATRX which can also be used to produce high quality data sets for other challenging proteins which are indirectly associated with DNA and complement the ChIP-seq toolkit for genome-wide analyses of histone chaperon complexes and associated chromatin remodelers. Although not a focus of this chapter, we will also provide some insight for the analysis of the large dataset generated by ChIP-seq. Even though this protocol has been fully optimized for ATRX, it should also provide guidance for efficient ChIP-seq analysis, using the appropriate antibodies, for other proteins interacting indirectly with DNA.

Key words Chromatin immunoprecipitation, High-throughput DNA sequencing, ChIP-seq, Alpha-thalassaemia mental retardation X-linked, ATRX, Histone variant H3.3, Library preparation, NGS, Chromatin remodeler, Double cross-linking, EGS

1 Introduction

Chromatin immunoprecipitation (ChIP) is now a key technique for analyzing protein-DNA interactions in vivo [1, 2]. The power of this technique comes from the ability of crosslinking agents, such as formaldehyde, to covalently join the protein-DNA complexes in

Guillermo A. Orsi and Geneviève Almouzni (eds.), *Histone Variants: Methods and Protocols*, Methods in Molecular Biology, vol. 1832, https://doi.org/10.1007/978-1-4939-8663-7_6, © Springer Science+Business Media, LLC, part of Springer Nature 2018

their native chromatin environment in living cells [3, 4]. The cross-linked chromatin is then sheared (by sonication or nuclease digestion) and immunoprecipitated using a specific antibody against a protein of interest. After reverse cross-linking of the ChIP sample, the target DNA fragments, pulled down with the protein of interest and enriched in the immunoprecipitate, are purified. This DNA enrichment can then simply be assessed, in a locus-specific manner, by quantitative Real Time PCR (qPCR): so-called ChIP-qPCR [5]. To appraise the ChIP data in defined segments of the genome and genome-wide, DNA fragments in the immunoprecipitate were, in the past, analyzed by labeling and hybridizing to a microarray, a technique known as ChIP-Chip [6, 7]. With the improvement of the next-generation sequencing (NGS) technologies, this method has now been supplanted and, today, DNA enrichment in ChIP experiments is assessed by high-throughput DNA sequencing (ChIP-seq) [8]. Using this technique it is now possible to study the genome-wide distribution of DNA-protein interactions at an economical cost. ChIP-seq has paved the way for many studies focusing on DNA-binding proteins, such as transcription factors, and histone modifications, as illustrated by the ENCODE project [9].

However, standard single cross-link ChIP experiments using a short spacer arm cross-linker agent such as formaldehyde (with an interaction range of 2 Å) have limitations and are most appropriate for proteins that directly bind DNA (nucleosomes and transcription factors) [3, 10, 11]. By contrast, assessing the genome-wide profiling of proteins indirectly bound to DNA and/or displaying hyperdynamic interactions can be more challenging and often requires additional cross-linking agents such as ethylene glycol-bis(succinimidylsuccinate) (EGS) with a spacer arm of approximately 16 Å [12–14].

This is the case of chromatin remodeler ATRX. This member of the SNF2 family of helicase/ATPases [15] is involved, in association with the histone chaperone DAXX, in the deposition of the replication-independent histone variant H3.3 in banks of repetitive sequences including telomeric DNA, pericentric DNA, and ribosomal DNA [16, 17]. Furthermore, ATRX colocalizes with variable tandem repeats which include GC-rich sequences predicted to form non-B DNA structures known as G-quadruplexes [18, 19]. ATRX has also been shown to associate with chromatin through its interactions with several proteins such as H3K9me3, HP1 [20], and MeCP2 [21].

Mutations in the *ATRX* gene were first identified as the cause of the ATR-X syndrome, characterized by severe intellectual disability, morphological abnormalities, and alpha-thalassaemia (a blood disorder due to defective production of alpha-globin) [22]. Since then, ATRX mutations have also been associated with a subset of cancers using the telomerase-independent Alternative Lengthening of Telomeres pathway to maintain their telomere length [23, 24]. Multiple lines of evidence have demonstrated the role of ATRX in expression of a subset of genes (e.g., the alpha-globin genes) [18, 25]. In addition, the loss of ATRX function affects diverse nuclear processes including DNA methylation, replication, genome stability, mitosis, and meiosis [26–31].

These observations emphasize the need to produce robust ATRX ChIP-seq data to identify the targets of this chromatin remodeler and help to define its role in health and disease. However, producing high quality ATRX-ChIP has been challenging for various reasons. (1) A large number of cells (10^8) are necessary. (2) A two-step cross-linking using EGS and formaldehyde is required, making the fixed chromatin more resistant to sonication. The use of stringent sonication parameters may have significant effects on the quality of the ChIP-seq data and exacerbate the artifacts such as the preferential fragmentation of DNA within regions of open chromatin, producing the artifactual appearance of a nuclease accessibility like-profile [32–34]. As a result, such data may produce a "noisy" background impeding their interpretation. Therefore, it is crucial to establish the optimal sonication. (3) The specificity and avidity of the antibody is also crucial and may be batch dependent. (4) The degree of DNA enrichment produced per ChIP is limited. (5) The small quantity of starting material available for the library preparation as well as the proportion of GC-rich sequences lead potentially to increases in PCR amplification bias. (6) The repetitive nature of many ATRX targets increases the complexity of the data analysis (e.g., issues with unique mapping of reads).

In this chapter, we describe the detailed methodology to carry out ChIP experiments requiring double cross-linking and to generate the subsequent libraries for high-throughput sequencing using the Illumina NextSeq500. Furthermore, we also discuss the quality control steps and provide some insights into the analysis of such data.

It should be noted that while this protocol has been optimized to generate high quality ATRX ChIP-seq data, it is potentially versatile enough to be adapted for other protein targets indirectly associated with DNA and/or with hyperdynamic interactions and requiring double-crosslinking such as DAXX [35], LKB1, the cofactor FOG-1 [14], HIRA, UBN1 and ASF1a [36].

2 Material

2.1 Crosslinking Reagent

1. Phosphate-buffered saline (PBS).
2. 38% formaldehyde solution.
3. EGS stock solution: make 500 mM EGS in Dimethyl sulfoxide (DMSO) stock solution immediately prior to use. Dissolve

0.046 g EGS in 200 μL of DMSO. Mix well until the powder is fully dissolved and the solution is clear. Use immediately (*see* **Note 1**).

4. Glycine stock solution: 1 M in ultrapure water. Dissolve 3.75 g of glycine powder in ultrapure water to a final volume of 50 mL. Mix until the powder is fully dissolved and the solution is clear. Store at 4 °C.

2.2 Bead Preparation

1. Dynabeads protein A and/or G.
2. Magnetic stand.
3. Primary antibody (*see* **Note 2**).
4. BSA solution: Make fresh on the day bovine serum albumin (BSA) solution 5 mg/mL in PBS. Dissolve 250 mg of BSA in PBS to a final volume of 50 mL. Mix until complete dissolution of the powder. Chill on ice.
5. PBS.

2.3 Lysis Buffers

1. Protease inhibitor cocktail tablets.
2. Lysis buffer 1 (100 mL) (100 mM Hepes-KOH, 140 mM NaCl, 1 mM EDTA, 10% glycerol, 0.5% NP-40 and 0.25% Triton X-100). In 76.25 mL of ultrapure water, add 10 mL of 1 M Hepes-KOH, pH 7.5, 2.8 mL of 5 M NaCl, 0.2 mL of 0.5 M Ethylenediaminetetraacetic acid (EDTA), 10 mL glycerol, 0.5 mL of 100% NP-40 and 0.25 mL of Triton X-100. Store at 4 °C. Protease inhibitor cocktail should be added to 1× final concentration on the day of use. Do not store the left over solution once the Protease inhibitor cocktail has been added.
3. Lysis buffer 2 (100 mL) (200 mM NaCl, 1 mM EDTA, 0.5 mM EGTA and 10 mM Tris-HCl). In 93.4 mL of ultrapure water add 4 mL of 5 M NaCl, 0.2 mL of 0.5 M EDTA, 0.4 mL of 125 mM Ethylene glycol-bis(2-aminoethylether)-N,N,N′,N′-tetraacetic acid (EGTA) and 2 mL of 0.5 M Tris pH 8. Store at 4 °C (*see* **Note 3**). Protease inhibitor cocktail should be added to 1× final concentration on the day of use. Do not store the left over solution once the Protease inhibitor cocktail has been added.
4. Lysis buffer 3 (100 mL) (1 mM EDTA, 0.5 mM EGTA, 10 mM Tris-HCl at pH 8, 100 mM NaCl, 0.1% Na-Deoxycholate and 0.5% N-lauroyl sarcosine). In 50 mL of ultrapure water add 0.2 mL of 0.5 M EDTA, 0.4 mL of 125 mM EGTA, 2 mL of 500 mM Tris-HCl at pH 8, 2 mL of 5 M NaCl, 0.1 g of Na-Deoxycholate and 500 mg of N-lauroyl sarcosine. Then adjust the volume to 100 mL with ultrapure water. Store at 4 °C. Protease inhibitor cocktail should be added to 1× final concentration on the day of use. Do not store the left over solution once the Protease inhibitor cocktail has been added.

2.4 Sonication

1. Covaris S-series.
2. Covaris milliTUBE 1 mL AFA Fiber.
3. 10% Triton X-100.

2.5 Washes, Elution, and Reverse Cross-Linking

1. Wash buffer (100 mL) (50 mM Hepes at pH 7.6, 1 mM EDTA, 0.7% Na-Deoxycholate, 1% NP-40 and 0.5 M LiCl). In 50 mL of ultrapure water add 10 mL of 500 mM Hepes at pH 7.6, 200 μL of 0.5 M EDTA, 0.7 g of Na-Deoxycholate, 1 mL of NP-40 and 6.25 mL 8 M LiCl (or 2.12 g powder). Then adjust the volume to 100 mL with ultrapure water. Store at 4 °C. Protease inhibitor cocktail should be added to 1× final concentration on the day of use. Do not store the left over solution once the Protease inhibitor cocktail has been added.
2. Elution buffer (5 mL) (0.5% SDS and 100 mM NaHCO3). Make fresh on the day elution buffer by adding 125 μL of 20% SDS and 500 μL of 1 M NaHCO3 in 4.375 mL of ultrapure water. Keep at room temperature. Scale up if needed depending on the number of ChIP samples.
3. PBS. Protease inhibitor cocktail should be added to 1× final concentration on the day of use. Do not store the left over solution once the Protease inhibitor cocktail has been added.
4. 5 M NaCl.

2.6 DNA Purification

1. Proteinase K 20 mg/mL.
2. 1 M Tris, pH 6.5.
3. 0.5 M EDTA.
4. Phenol-chloroform-isoamylalcohol (25:24:1, v/v/v).
5. Ethanol 100% and 70%.
6. 3 M Sodium Acetate.
7. Glycogen.

2.7 DNA Quantification

1. Qubit dsDNA HS Assay Kit.

2.8 Sonication Validation

1. 2% agarose gel.
2. Loading dye.
3. SYBR safe DNA gel stain.
4. DNA Ladder.

2.9 ChIP Validation by qPCR

1. Fast SYBR Green Master Mix.
2. Primers for positive and negative control regions (*see* **Note 4**).

2.10 Library Preparation and Quantification and Sequencing

1. NEBNext Ultra DNA Library Prep Kit for Illumina (E7370S/L).
2. NEBNext Multiplex Oligos for Illumina (Index Primers Set 1) (E7335S/L).

3. AMPure XP Beads.
4. 10 mM Tris-HCl, pH 7.5–8.0.
5. 0.1× TE, pH 8.0.
6. Ethanol 80% (made fresh on the day).
7. High Sensitivity D1000 ScreenTape.
8. High Sensitivity D1000 Reagents.
9. Agilent 2200 TapeStation.
10. KAPA Library Quantification Kit.
11. NextSeq 500/550 High Output Kit v2 (150 cycles) or Mid Output Kit v2 (150 cycles) (if using NextSeq500 as Illumina sequencing platform).
12. NextSeq500 (or other Illumina Sequencing platform).

3 Method

3.1 ATRX ChIP

The ATRX ChIP protocol has been optimized based on Law et al. [18]. The experimental workflow is shown in Fig. 1.

3.1.1 Cell Cross-Linking

Both room temperature and cold PBS are required for the cross-linking step.

1. Use 10^8 live cells per ChIP (upon harvest, the cell viability should be ≥90%).
2. Wash twice with room temperature PBS.
3. Resuspend the cells in a 50 mL falcon tube in 20 mL room temperature PBS. Make sure that the cells are in single cell suspension. Add 80 μL of freshly made 500 mM EGS stock solution per 20 mL of PBS (final concentration of 2 mM) (*see* **Note 5**).
4. Incubate at room temperature for 45 min on a roller with constant mixing (*see* **Note 6**).
5. Add 526 μL of 38% formaldehyde solution per 20 mL of cell suspension (1% final) (*see* **Note 7**).
6. Incubate at room temperature for 20 min on a roller with constant mixing.
7. Add 2.56 mL of 1 M glycine stock solution per 20.5 mL cross-linked cell suspension (125 mM final) to quench the cross-linking reagents (formaldehyde and EGS). Mix well and incubate for 2 min on the roller at room temperature.
8. Spin the sample at 240 × *g* for 5 min at 4 °C.
9. Discard the supernatant and gently wash the pellet with 10 mL of 4 °C PBS on ice.

10. Spin the sample at 240 × *g* for 5 min at 4 °C.
11. Discard the supernatant and repeat the wash by gently resuspending the pellet with 10 mL of 4 °C PBS on ice.
12. Spin the sample at 240 × *g* for 5 min at 4 °C.
13. Discard the supernatant (*see* **Note 8**).
14. Place the washed pellet on ice.

3.1.2 Dynabead Preparation

The preparation of the Dynabeads will require 5 mg/mL BSA solution and 4 °C PBS. Work on ice.

Each sample requires three 1.5 mL DNA LoBind microcentrifuge tubes: one for the preclearing during Subheading 3.1.4 Chromatin Immunoprecipitation **step 1**, one for the ChIP of the targeted protein (in this case, ATRX), and the last one for the IgG control (both for chromatin immunoprecipitation during Subheading 3.1.4 Chromatin Immunoprecipitation **step 4**).

1. To one 1.5 mL DNA LoBind microcentrifuge tube for the preclearing, add 100 μL of magnetic beads (Dynabeads Protein A or G depending on the antibodies used for the ChIP) (*see* **Note 9**).
2. To two 1.5 mL DNA LoBind microcentrifuge tubes, add 50 μL of magnetic beads (*see* **Note 9**). One of these will be for the ATRX immunoprecipitation (IP) and the other for the IgG control.
3. For each tube, wash the beads with 1 mL of 5 mg/mL BSA solution, quick spin. Place the tube on the magnetic stand, wait for 3 min until the solution is clear, and remove the supernatant.
4. Repeat the previous step three times (4 washes in total) (*see* **Note 10**).
5. Resuspend the 100 μL of washed beads for preclearing in 300 μL of 5 mg/mL BSA solution. Keep them at 4 °C under constant agitation until needed.
6. For the ATRX IP: resuspend one of the 50 μL of washed beads in 300 μL of cold PBS containing 10 μg of ATRX antibody (*see* **Note 2**). Incubate for 3–4 h on a roller at 4 °C.
7. For IgG control: resuspend the other 50 μL of washed beads in 300 μL of cold PBS containing 10 μg of IgG antibody (*see* **Note 2**). Incubate for 3–4 h on a roller at 4 °C.
8. Before adding to the precleared cell lysate (*see* Subheading 3.1.4 Chromatin Immunoprecipitation **step 4**), wash the beads incubated with the ATRX and IgG antibodies, 4 times in 5 mg/mL BSA solution and resuspend the beads in 100 μL of PBS at 4 °C.

3.1.3 Cell Sonication

The cell sonication step is done at 4 °C unless specified otherwise. Work on ice.

This is an essential step for a successful ATRX ChIP. It will require optimization as the sonication settings are cell-type dependent (*see* **Note 11**) [37].

1. Add protease inhibitor cocktail (1× final) to all lysis buffers before use.
2. Chill centrifuge to 4 °C.
3. If the cross-linked cell pellets were frozen at the **step 13** of the Subheading 3.1.1 Cell cross-linking, thaw the pellets on ice (*see* **Note 8**).
4. Resuspend each cell pellet in 10 mL of 4 °C lysis buffer 1 containing 1× protease inhibitor cocktail.
5. Rock samples at 4 °C for 10 min.
6. Centrifuge the samples at 950 × *g* for 2 min at 4 °C.
7. Discard the supernatant and resuspend cells in 10 mL of 4 °C lysis buffer 2 containing 1× protease inhibitor cocktail.
8. Centrifuge the samples at 1500 × *g* for 2 min at 4 °C.
9. Discard the supernatant and resuspend the pellet in 3 mL of 4 °C lysis buffer 3 containing 1× protease inhibitor cocktail.
10. Syringe at least 10 times each sample with a G27 needle (*see* **Note 12**).
11. Divide each sample in three milliTUBE 1 mL tubes with AFA fiber for the Covaris (*see* **Note 13** and Fig. 1).
12. Sonicate the samples to get fragments between 100 and 250 bp (Fig. 2) (*see* **Note 11** and Subheading 3.1.6 DNA purification **step 12**).
13. After sonication, transfer the sample from each 1 mL tube with AFA fiber into a 1.5 mL DNA LoBind microcentrifuge tube.
14. Add 1/10 volume (100 μL to each 1 mL of sonicated sample) of 10% Triton X-100.
15. Spin at 18400 × *g* for 10 min at 4 °C.
16. Collect the supernatant and for each sample (divided during **step 11**) (Fig. 1), combine the three 1 mL aliquots together into a graduated 14 mL falcon tube. As volume loss may occur during the sonication step, top up to 3 mL with 4 °C lysis buffer 3 containing protease inhibitors if necessary.
17. Take 50 μL of sonicated chromatin from each sample as input and store this aliquot at −20 °C (*see* **Note 14**).

3.1.4 Chromatin Immunoprecipitation

The chromatin immunoprecipitation step is done at 4 °C unless specified otherwise. Work on ice.

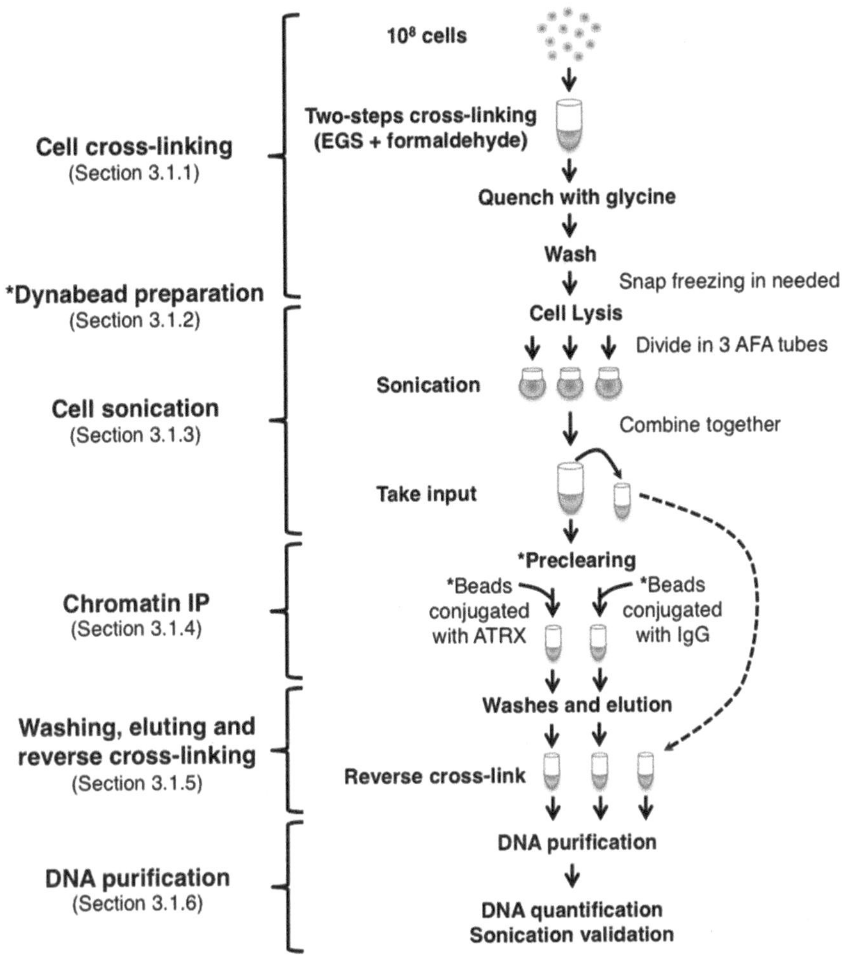

Fig. 1 Flowchart highlighting the main steps of the ATRX ChIP protocol. The beads prepared during the Subheading 3.1.2 Dynabead preparation step are highlighted by (*)

1. Preclear the lysate by adding 300 μL of the washed preclearing beads (Subheading 3.1.2 Dynabead Preparation, **step 5**) to the 3 mL of cell lysate and incubating the samples on roller at 4 °C for 1 h.
2. Split the 3 mL precleared samples into two 1.5 mL DNA LoBind microcentrifuge tubes (Fig. 1).
3. Precipitate the beads from the chromatin using a magnetic stand and collect one 1.5 mL fraction of supernatant for the ATRX IP and the second 1.5 mL fraction of supernatant for the IgG control, each into a new 5 mL falcon tube (Fig. 1).
4. Add the washed beads conjugated with the ATRX antibody (from the Subheading 3.1.2 Dynabead Preparation, **step 8**) to the tube for the ATRX IP containing the first 1.5 mL pre-cleared lysate and add the washed beads conjugated with the

IgG (from the Subheading 3.1.2 Dynabead Preparation, **step 8**) to the tube for the IgG control containing the second 1.5 mL precleared lysate (Fig. 1).

5. Incubate overnight at 4 °C on a roller.

3.1.5 Washing, Eluting, and Reverse Cross-Linking

The washing steps are done at 4 °C unless specified otherwise. Work on ice. It is recommended to perform the washing steps in a 4 °C cold room.

1. Add protease inhibitor cocktail to the wash buffer to a final concentration of 1×.
2. Transfer each sample (lysates and beads) into a new 1.5 mL DNA LoBind microcentrifuge tube (the targeted chromatin is now bound to the beads).
3. Precipitate the beads using the magnetic stand and discard the supernatant.
4. Wash the beads by adding 1 mL of wash buffer containing 1× protease inhibitor cocktail at 4 °C. Incubate for 5 min on the roller at 4 °C. Spin down shortly. Precipitate the beads on the magnetic stand and discard the supernatant (*see* **Note 15**).
5. Repeat the above wash step four times (five washes in total) (*see* **Note 15**).
6. Wash the beads with 1 mL of PBS containing 1× protease inhibitor cocktail at 4 °C. Precipitate the beads on the magnetic stand and discard the supernatant (*see* **Note 15**).
7. Elute the beads at room temperature by adding 250 μL of freshly made elution buffer. Incubate at room temperature for 15 min on the roller.
8. Precipitate the beads on the magnetic stand and collect the eluate in a new 1.5 mL DNA LoBind microcentrifuge tube.
9. Repeat the elution by resuspending the beads with another 250 μL of freshly made elution buffer. Using a ThermoMixer, incubate at 65 °C for 20 min with mixing cycles of 40 s ON, 20 s OFF at 1400 rpm.
10. Precipitate the beads on the magnetic stand, collect the eluate, and combine it with the previous round of elution from **step 8** (500 μL in total).
11. Reverse cross-link by adding 20 μL of 5 M NaCl to each 500 μL elute and incubate at 65 °C for 4 h (*see* **Note 16**).
12. Defrost the 50 μL input for each sample (from Subheading 3.1.3 Cell sonication, **step 17**). Add 450 μL of freshly made elution buffer at room temperature. Reverse cross-link by adding 20 μL of 5 M NaCl to each 500 μL eluate and incubate at 65 °C for 4 h (*see* **Note 16**).

3.1.6 DNA Purification

1. To each tube (ATRX IP, IgG control, and input) add 20 μL of 1 M Tris pH 6.5, 10 μL of 0.5 M EDTA, and 2 μL of 20 mg/mL Proteinase K and incubate for 1 h at 45 °C.
2. Add 500 μL of phenol-chloroform-isoamylalcohol (25:24:1, v/v/v), mix well by vortexing and spin at 17900 × *g* for 5 min at room temperature (*see* **Note 17**).
3. Collect the top phase containing the DNA in a new tube (*see* **Note 18**).
4. Add 1 mL of 100% EtOH, 50 μL of 3 M Sodium Acetate and 1 μL of Glycogen (20 μg). Incubate at −20 °C overnight (*see* **Note 19**).
5. Spin down at 17900 × *g* for 1 h at 4 °C.
6. A tiny pellet should be visible. Discard the supernatant carefully without disturbing the pellet.
7. Wash the pellet by adding 1 mL of 70% EtOH, mix and spin at 17900 × *g* for 10 min at room temperature.
8. Carefully discard as much supernatant as possible and allow the pellet to dry for about 15 min (*see* **Note 20**).
9. For the ATRX ChIP and IgG control samples: resuspend the pellets in 20 μL of ultrapure water.
10. For the input sample: resuspend the pellet in 50 μL of ultrapure water.
11. Incubate at room temperature for 30 min to allow the DNA to go into solution. Mix and spin down.
12. On a 2% agarose gel, run 12 μL of input to check the sonication (Fig. 2) (*see* **Note 21**).
13. Quantify the samples using the Qubit dsDNA HS Assay Kit with 1 μL of each sample (*see* **Note 22**).
14. Store the samples at −20 °C.

3.2 ChIP Validation by qPCR

In order to assess the ChIP quality, the ATRX ChIP enrichment over the IgG negative control has to be checked both at a locus enriched for ATRX (positive control) and a locus depleted for ATRX (negative control) (*see* **Note 4**). At all loci, the IgG ChIP enrichment should remain low, whereas the ATRX enrichment should display a significant increase at the positive locus and drop close to the IgG level at the negative locus (*see* **Note 23** and Fig. 3). It is recommended to analyze the data as a percentage of input.

1. For each sample (ATRX, IgG and Input) take a 4 μL aliquot and dilute it in 36 μL of Sigma water in a new tube (1 in 10 dilution) (*see* **Note 24**). For each qPCR reaction, add 5 μL of diluted sample with 6.5 μL of PCR grade water, 0.5 μL of 10 μM Forward primer, 0.5 μL of 10 μM Reverse primer, and 12.5 μL of Fast SYBR Green Master Mix (25 μL total).

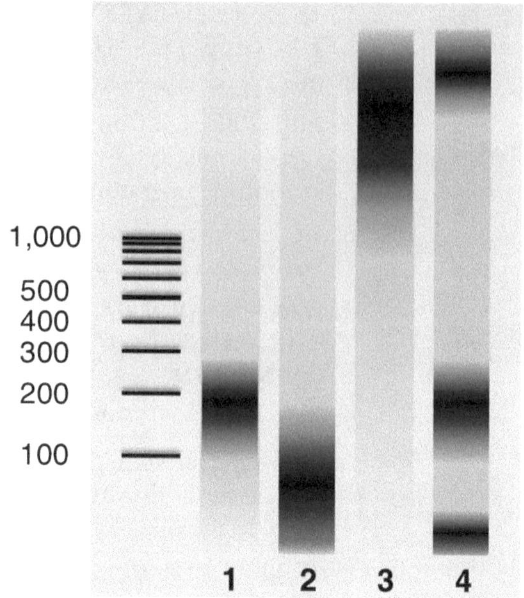

Fig. 2 Schematic example of a sonication control on a 2% agarose gel. Lane 1: Efficient sonication done on a cross-linked sample that has been frozen for less than 48 h. The majority of the DNA fragments are comprised between 100 and 300 bp. Lane 2: Example of an over-sonicated sample requiring less stringent sonication settings (e.g., shorter sonication time). Lane 3: Example of an under-sonicated sample requiring more stringent sonication conditions (e.g., increase sonication time and/or intensity). Lane 4: Artifact example often observed with sonicated cross-linked sample that has been frozen for weeks before sonication. Despite the formation of a fraction of properly sonicated fragments (between 100 and 300 bp), note the presence of both over-sonicated and under-sonicated chromatin

2. Using a Real-Time PCR Detection System, first incubate the sample at 90 °C for 20 s (holding stage), then for 40 cycles incubate at 95 °C for 3 s followed by 30 s at 60 °C for annealing/extension. Performing a melting curve step is also recommended for quality control.
3. If the samples pass all the validation steps (Fig. 3), they can be used for ChIP-qPCR or sequencing.

3.3 Library Preparation

Do not size select your sample for the library preparation. Only the ATRX ChIP and the input samples will be used for the following step. The IgG ChIP is mainly useful for the quality check by qPCR or for any further qPCR analysis if desired.

The library preparation is done using the NEBNext Ultra DNA Library Prep Kit for Illumina (E7370S/L) and the NEBNext Multiplex Oligos for Illumina (E7335S/L) following New England Biolabs's instructions, adapted to the ATRX ChIP as described below. This protocol will produce library to be sequenced on Illumina platform (here, we use the NextSeq500 as an example).

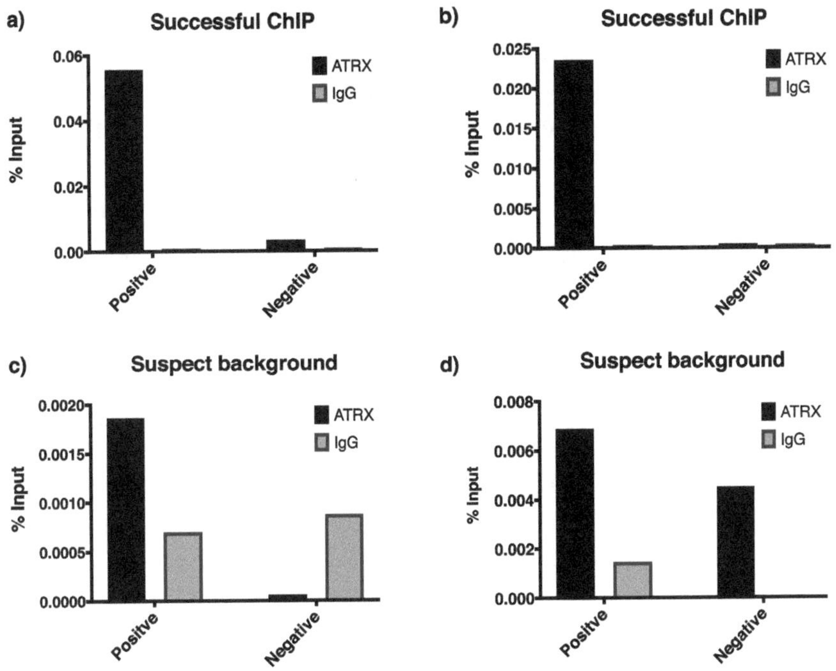

Fig. 3 Examples of ChIP validation by qPCR. (**a**) and (**b**) Successful ChIP experiments showing a strong ATRX enrichment at the positive control and a basal level at the negative control. Note that the ChIP efficiency (%Input) can vary between experiments (usually between 0.02 and 0.06% input for a successful ChIP). The IgG control, however, remains very low at both loci. (**c**) Example of a high IgG background, which may result in a very high background in the ChIP-seq data. The ChIP efficiency is very low. The batch of antibody used may not display a high avidity. The ChIP yield was too low (< 0.5 ng/mL) to perform a library preparation (***see* Note 22**). (**d**) Example of a batch of antibody displaying a low specificity for the protein of interest. The ChIP-seq data may have a high background and may not be interpretable (as shown in Fig. 5d). In both (**c**) and (**d**) examples, it is strongly advised to re-start the ChIP instead of pursuing the library preparation. All ChIP experiments were done using lymphoblastoid cell lines and different batches of antibody. Positive: rDNA, Negative: RHBDF1 (***see* Note 4**)

The manufacturer's instructions require at least 5 ng of ChIP purified DNA per sample. A single ATRX ChIP sample, however, is not expected to produce a sufficient yield to reach this threshold. Nevertheless, the 15 μL sample (after qPCR validation) should result in at least 2 ng of fragmented DNA. If it is the case, the sample can be processed as if a 5 ng sample and produce an acceptable library. Alternatively, up to three ATRX ChIP replicates (45 μL total) can be pooled and processed as a single 5 ng fragmented DNA sample.

3.3.1 DNA End Preparation: DNA End Repair, 5' Phosphorylation, and dA-Tailling

1. Dilute the 15 μL of ATRX ChIP sample (or more if combining replicates) with ultrapure water to give a final volume of 55.5 μL (*see* **Note 25**).
2. Dilute 5 ng of input sample (based on the measurement obtained at the Subheading 3.1.6 DNA purification **step 13**)

in a sufficient volume of ultrapure water to a final volume of 55.5 μL of diluted fragmented DNA.

3. To each 55.5 μL diluted fragmented DNA sample, add 3 μL of End Prep Enzyme Mix and 6.5 μL of End Repair Reaction Buffer (10×) for a final volume of 65 μL.
4. Mix by pipetting and quickly spin the samples.
5. Using a thermocycler, heat the samples for 30 min at 20 °C and then for 30 min at 65 °C. Hold the samples at 4 °C until the next step.

3.3.2 Adaptor Ligation

1. Near to the end of the 30 min incubation at 65 °C, make a fresh 10-fold dilution of the NEBNext Adaptor for Illumina (15 μM stock solution) in 10 mM Tris-HCl to a final concentration of 1.5 μM. Use immediately and discard any surplus.
2. To each 65 μL end prep sample, add 15 μL of Blunt/TA Ligase Master Mix, 2.5 μL of 1.5 μM NEBNext Adaptor for Illumina, and 1 μL of ligation enhancer for a final volume of 83.5 μL.
3. Mix by pipetting and quickly spin the samples.
4. Using a thermocycler, incubate the samples for 15 min at 20 °C.
5. Immediately add 3 μL of USER Enzyme to each 83.5 μL adapter ligated reaction.
6. Mix by pipetting and quickly spin the samples.
7. Incubate the samples for 15 min at 37 °C (*see* **Note 26**).

3.3.3 Cleanup of Adaptor-Ligated DNA without Size Selection

1. Make a fresh solution of 80% ethanol by mixing 40 mL of 100% EtOH with 10 mL of ultrapure water.
2. Transfer the 86.5 μL of ligation reaction in a new 1.5 mL DNA LoBind microcentrifuge tube.
3. Add 86.5 μL of resuspended AMPure XP Beads (*see* **Note 27**).
4. Mix by pipetting 15 times and incubate at room temperature for 5 min.
5. Briefly spin the samples and precipitate the beads using a magnetic stand (*see* **Note 28**).
6. Remove the supernatant (*see* **Note 29**).
7. Wash the beads by adding 200 μL of fresh 80% ethanol. Incubate 30 s at room temperature and carefully discard the supernatant without disturbing the beads (during this step, do not remove the samples from the magnetic stand).
8. Repeat the previous step (two washes in total).
9. Quickly spin the tube and precipitate the beads on the magnetic stand and remove any residual ethanol (without removing the tubes from the magnet).

10. Allow the beads to air dry for 5 min (cap opened) (*see* **Note 30**).
11. Remove the tube from the magnet and add 17 μL of 10 mM Tris-HCl to elute the DNA target. Mix by pipetting 15 times and incubate at room temperature for 2 min.
12. Briefly spin the tube and precipitate the beads on the magnetic stand (*see* **Note 28**).
13. Carefully transfer 15 μL of the supernatant to a new PCR tube (without disturbing the beads) for amplification.

3.3.4 PCR Enrichment of Adaptor Ligated DNA

1. For each 15 μL of cleaned adapted ligated DNA reaction, add 25 μL of NEBNext Q5 Hot Start HiFi PCR Master Mix (*see* **Note 26**), and 5 μL of 10 μM Universal PCR Primer.
2. Add 5 μL of the appropriate 10 μM Index Primer for a final volume of 50 μL (*see* **Note 31**).
3. Using a thermocycler, incubate the samples for 30 s at 98 °C (initial denaturation) followed by 12 cycles of denaturation at 98 °C for 10 s and annealing/extension at 65 °C for 75 s. After 12 cycles of denaturation and annealing/extension, incubate at 65 °C for 5 min. Hold the reaction at 4 °C.

3.3.5 Cleanup of PCR Amplification

1. Transfer the 50 μL of PCR amplified reaction to a new 1.5 mL DNA LoBind microcentrifuge tube.
2. Add 45 μL of resuspend AMPure XP Beads (*see* **Note 27**).
3. Mix by pipetting 15 times and incubate at room temperature for 5 min.
4. Briefly spin the samples and precipitate the beads using a magnetic stand (*see* **Note 28**).
5. Remove the supernatant (*see* **Note 29**).
6. Wash the beads by adding 200 μL of fresh 80% ethanol. Incubate for 30 s at room temperature and carefully discard the supernatant without disturbing the beads (by keeping the samples on the magnetic stand).
7. Repeat the previous step (two washes in total).
8. Briefly spin the tube and precipitate the beads on the magnetic stand and remove any left-over of ethanol (without removing the tubes from the magnet).
9. Allow the beads to air dry for 5 min (cap opened) (*see* **Note 30**).
10. Remove the tube from the magnet and add 33 μL of 0.1× TE to elute the DNA target.
11. Mix by pipetting 15 times and incubate at room temperature for 2 min.

12. Briefly spin the tube and precipitate the beads on the magnetic stand (*see* **Note 28**).
13. Carefully transfer 28 μL of supernatant (library) to a new 1.5 mL DNA LoBind microcentrifuge tube without disturbing the beads.
14. Dilute 2 μL of the library in 2 μL of 0.1× TE and check the size distribution of the library on the TapeStation using a High Sensitivity D1000 ScreenTape (Fig. 4).
15. Store the library at −20 °C.

3.4 Library Quantification and Sequencing

An accurate quantification of the libraries is essential for a successful sequencing.

1. Quantify the library using the Qubit dsDNA HS Assay Kit.
2. Based on the Qubit quantification and the average fragment size of the library (obtained in Subheading 3.3.5 Cleanup of PCR amplification, **step 14**), make 15 μL of 12 nM library from the stock library and store the rest at −20 °C (*see* **Note 32**).
3. Quantify the 12 nM library using the KAPA library Quantification kit by making a 1:1000 and a 1:2000 dilution in 10 mM Tris-HCl (pH 8.0), 0.05% Tween 20.
4. Based on the KAPA library Quantification, make 5 μL of 4 nM library and store the rest at −20 °C.

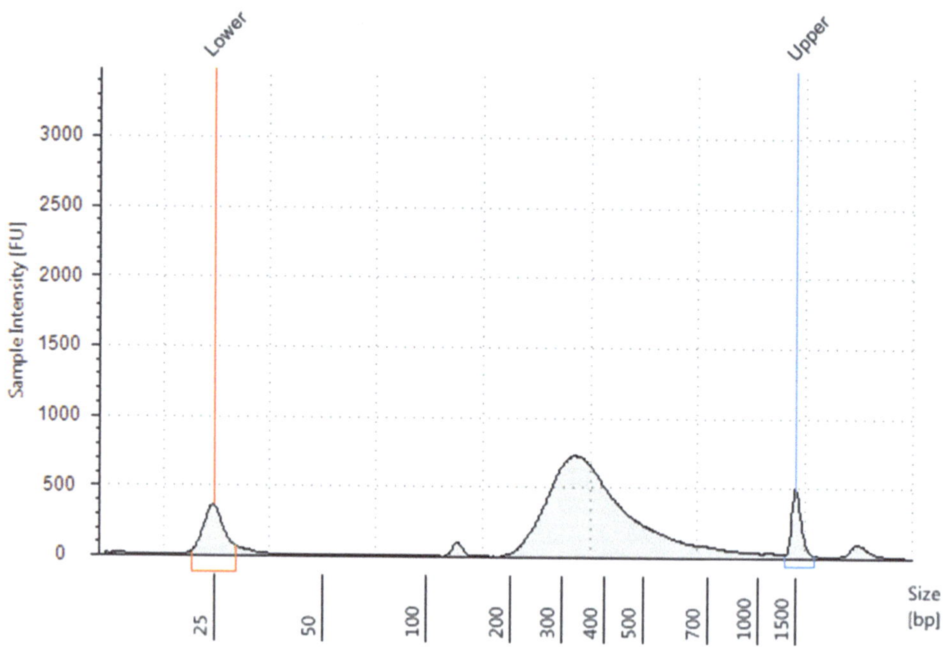

Fig. 4 Example of ATRX library run on the TapeStation using a High Sensitivity D1000 ScreenTape. The library was made from 2.5 ng of ATRX ChIP material from a lymphoblastoid cell line

5. If several libraries are sequenced together, pool the 4 nM libraries with different indexes together (*see* **Note 31**).
6. Use a NextSeq 500/550 High Output Kit v2, 150 cycles (or Mid Output Kit v2, 150 cycles depending on the number of samples) and sequence using the NextSeq500 from Illumina. It is recommended to have a sequencing depth of at least 20 millions uniquely mapped reads per sample with an input coverage similar to or higher than its related ATRX ChIP-seq sample [34] (*see* **Note 33**). A successful ATRX ChIP-seq should display a significant enrichment above background and a very low input background (Fig. 5).

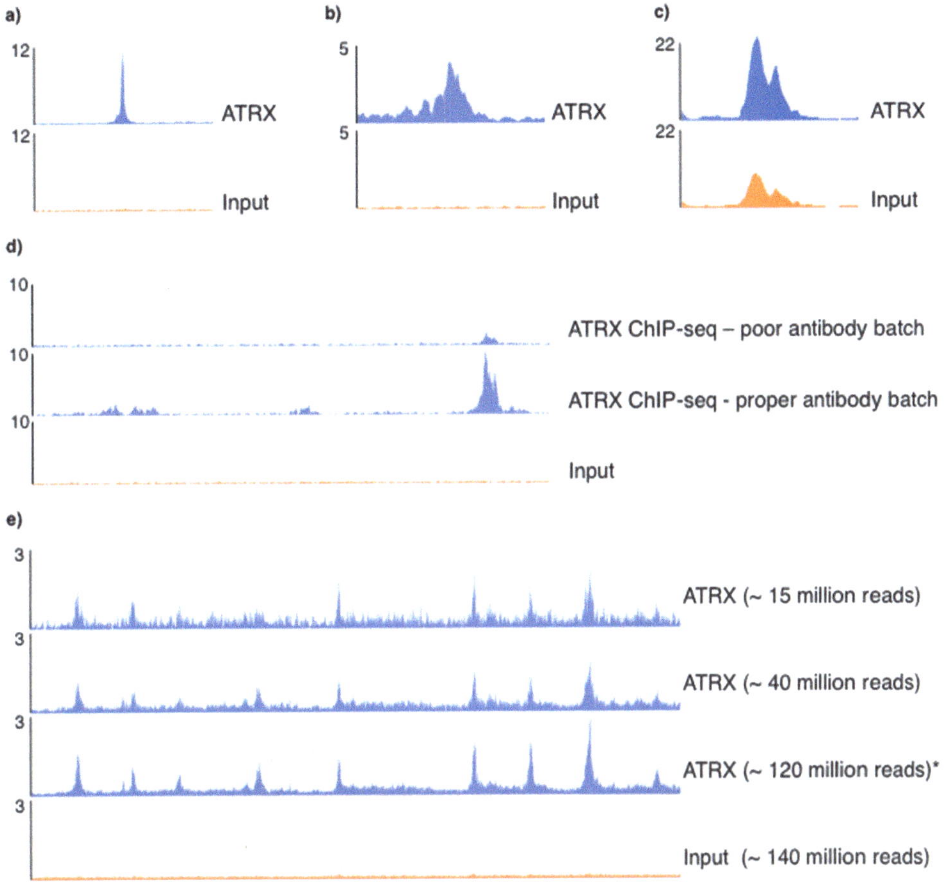

Fig. 5 Examples of ATRX ChIP-seq data visualized on the UCSC genome browser. ATRX ChIP-seqs were carried out using lymphoblastoid cell lines and mapped on hg19 using Bowtie2. (**a**) and (**b**) Examples of ATRX enrichment over input highlighting successful ChIP-seqs at different loci. Notice the variability in peak shapes. (**c**) Example of a ChIP-seq artifact. The enrichment in the ATRX track is probably not real based on the background observed in the input track. (**d**) Example of antibody batch effect on the quality of the ChIP-seq experiment. The two ATRX ChIP-seq data done on the same sample, following the same protocol using two different batches of ATRX antibody. The top track was produced with a poor batch of ATRX antibody as observed in Fig. 3d. The enrichment is poor and the background is high. (**e**) Example of an ATRX-binding rich region illustrating the importance of the sequencing depth. Deeper sequencing can improve the resolution, providing more information for further analysis. (*) Pool of three independent biological replicates highlighting the reproducibility of the data

3.5 Analysis of the ChIP-seq Data

Each ChIP-seq experiment generates a large amount of data that is not trivial to analyze and bioinformatic expertise is essential for a comprehensive analysis. Here, we attempt to provide some basic guidance for this process. It should be mentioned, however, that there is no single way to perform ChIP-seq data analysis and the tools mentioned below are not exhaustive.

When processing ChIP-seq data, a few essential steps have to be considered:

Analysis of the raw sequencing data

1. Mapping the sequences to the reference genome: A widely used tool for mapping is Bowtie (1 and 2) [38, 39] but other mappers such as BWA [40] could be considered. Several versions of a given reference genome may be available (for instance, the human genome—hg18 and hg19) and it is essential to map all the replicates to the same version of the genome. Additional parameters, such sequence variations (e.g., SNP) with the reference genome, should be considered [33]. In the context of the ATRX ChIP-seq data analysis, in which a large number of sequences are known to be repetitive, subsequent analysis of enrichment at repetitive elements for individual repeat types [41] or using the REPEATOME [42] (if applicable) should be considered.
2. Trimming of the adapter sequences: As mentioned during library preparation, Illumina adapters were added to the sample prior to sequencing (*see* Subheading 3.3.2 Adaptor ligation) and, subsequently, during the data analysis, residual adapter sequences could impair the mapping analysis. Therefore, depending on the mapper used, the trimming of the sequences should be considered using tools such as trim-galore [43].
3. Assess the data quality at each stage of the process: the generation of quality control reports such as fastQC reports [44] is a good practice for each step of the data processing and mapping analysis. Further quality controls such as cross-contamination check using tools such as FastQ Screen (http://www.bioinformatics.babraham.ac.uk/projects/fastq_screen/. Accessed 18 July 2017) could be relevant.
4. Remove the PCR duplicates: PCR duplicates are artifacts generated during the library amplification step, and may create artificial enrichments. They should not be considered in subsequent analysis such as enrichment analysis and should be removed using tools such as Samtools [45] and Picard (http://broadinstitute.github.io/picard/. Accessed 18 July 2017).
5. Remove the signal artifact blacklisted regions [46]: these regions produce artifact signals with high signal/read counts due to sequence mappability shortcoming and should not be considered biologically relevant.

6. Visualize the data on a genome browser such as UCSC [47] (Fig. 5).

Analysis of the ChIP-seq data per se.

1. Assess the quality of the ChIP-seq using the ATRX ChIP and Input samples (e.g., assess the coverage, the enrichment of the ChIP over input) using packages such as Deeptools [48].
2. Data normalization: The data normalization such as correction for the sequencing depth or GC bias may be critical for downstream analysis. Versatile packages such as Deeptools [48] and Homer [49] could be useful. Depending on the types of samples (e.g., different genders), it could be relevant to take mitochondrial and/or sex chromosomes mapping reads out of the normalization of the ChIP signals.
3. Peak calling: Central goal of a ChIP-seq experiment, the enrichment analysis characterizing the genomic regions targeted by the protein of interest can be supported by a vast numbers of packages including MACS [50] and HOMER [49]. In the context of ATRX ChIP-seq, the peak calling is relatively challenging due to the complexity of the ATRX enrichment (diversity of peak shapes and density across the genome) (Fig. 5). Fine-tuning of the peak caller parameters and the use of an input are required to achieve a high quality peak calling. It is essential to visually assess the peak calling using tools like MIG [51].
4. Assess for reproducibility: for a robust analysis, it is essential to produce biological ChIP-seq replicates and assess their reproducibility (Fig. 5).

4 Notes

1. As EGS is water-insoluble, it has to be first dissolved in DMSO before being added to aqueous solution. It is essential to prepare the 500 mM EGS stock solution just before using it and discard any leftover as DMSO promotes the hydrolyzation of the EGS NHS ester moiety which becomes non-reactive (see the manufacturer's instructions for Thermo Scientific EGS, cat. 21,565).
2. Successful ATRX ChIP-seqs have been performed using the ab97508 anti-ATRX antibody (Rabbit polyclonal) from Abcam or the H300 anti-ATRX antibody (Rabbit polyclonal) from Santa Cruz Biotechnology. These antibodies could be substituted by antibodies targeting proteins requiring a long-range cross-linking. It cannot be guaranteed, however, that the substitute antibodies could produce a successful ChIP-seq following this protocol and further optimization may be required.

Furthermore, it is important to pay attention to the protein isoform and select the most appropriate antibody based on the experimental design. In the case of ATRX, the selective targeting of the full-length protein requires an antibody raised against the C-terminal end. Always take note of the antibody batch, as the efficiency can be batch dependent (especially with polyclonal antibodies). Use the IgG antibody from the same species as the one used to raise the antibody for the protein of interest. Use rabbit IgG in the case of the ATRX ChIP if the antibody used is ab97508 anti-ATRX antibody (Rabbit polyclonal) from Abcam or the H300 anti-ATRX antibody (Rabbit polyclonal) from Santa Cruz Biotechnology.

3. To dissolve EGTA powder, the solution needs to be adjusted at pH 8 using NaOH.
4. These loci may be cell type dependent. However, rDNA is known as an ATRX target in several human cell types and may be a good positive control to test [18].

   Potential primer pairs for the qPCR validation of ChIPs performed on human-derived samples:

   ATRX ChIP positive control for human-derived sample (rDNA)

   Positive-hg-F: TTCAAAGCCCCATTCGTATGC

   Positive-hg-R: AGTTTTCAGCCCCAACACACC

   ATRX ChIP negative control for human-derived sample (RHBDF1)

   Negative-hg-F: GAGATGCTGGAGTCAGGACCAT

   Negative-hg-R: AGGAGTCAGGAGCAGCAGTCA
5. It is critical that the PBS temperature is above 20 °C (up to 30 °C) as cold PBS could affect the cross-linking efficiency. For an efficient cross-linking, it is essential that the samples are in single cell suspension prior to the addition of the crosslinking agent. Upon the addition of the 500 mM EGS stock solution, the solution will become cloudy and clear again very quickly. After the EGS dispersion, no EGS precipitate should be visible. If some EGS precipitates once the solution is clear, check that the PBS was at room temperature.
6. Precool the centrifuge at 4 °C for **step 8** of the Subheading 3.1.1 Cell cross-linking.
7. Use the laminar flow cabinet to add the formaldehyde. Do not store an open bottle of 38% formaldehyde solution for a long period of time (usually not more than 6 months to a year).
8. This is a safe stopping point: samples can be snap frozen on dry ice and ethanol and stored at −80 °C for up to 3 days. However,

do not store the cross-linked pellets for a longer period of time as it may affect the sonication efficiency (*see* Fig. 2) [52].

9. The choice between the Dynabeads protein A and G is dependent on the affinity of the protein A/G for the IgG types from different species. Choose the Dynabeads protein A if using the ab97508 anti-ATRX antibody (Rabbit polyclonal) from Abcam or the H300 anti-ATRX antibody (Rabbit polyclonal) from Santa Cruz Biotechnology. Alternatively, a mix of Dynabeads protein A and G can be used. If this protocol is adapted to another target protein requiring long-range crosslinking, check the species with the antibody has been raised as well as the IgG type to select the most appropriated Dynabeads protein type (for further information, refer to: NEB, Tools & Resources, Affinity of Protein A/G for IgG Types from Different Species: https://www.neb.com/tools-and-resources/selection-charts/affinity-of-protein-ag-for-igg-types-from-different-species. Accessed 9 Jun 2017).
10. Do not let the beads dry in between washes.
11. As the chromatin undergoes a double cross-linking (first long-range cross-linking using EGS followed by a second cross-linking using formaldehyde), essential for an efficient and reliable ATRX ChIP, it became more resistant to sonication. More stringent sonication parameters are therefore required to achieve similar results than with a single cross-linked sample using only 1% formaldehyde for 10 min (as used in standard single cross-link protocol [10, 11]). It is strongly recommended to optimize the sonication for each cell type. Be careful to not over-sonicate or under-sonicate the DNA (*see* Fig. 2) as this may affect the quality of the ChIP-seq. Possible settings to try as a starting point: Duty = 5% and Intensity = 5 at 4 °C for 18–32 min, Duty = 10% and Intensity = 5 at 4 °C for 18–32 min, Duty = 12.5% and Intensity = 5.2 at 4 °C for 18–32 min. For each setting, assessing the sonication after 18, 22, 25, 28 and 32 min should be informative.
12. Syringing the sample helps to improve the sonication efficiency.
13. The Covaris has been chosen for the sonication step as it offers a significant flexibility of settings to design an optimal programme for each cell type as well as a good reproducibility. Alternatively, the sonication could also be carried out using a Bioruptor.
14. The input is a very important control. It is used to check the sonication of the sample, correct for the percentage input during the ChIP-qPCR experiments, and show the background in the ChIP-seq data.

15. Alternatively, the washes can be done using the wash buffers from the Chromatin Immunoprecipitation Assay Kit from Merck (cat. 17–295) after adding the protease inhibitor cocktails to 1× final concentration. In that case, the first wash is done with 1 mL of Low Salt Immune Complex wash buffer; the second wash is done with 1 mL of High Salt Immune Complex wash buffer; the third wash is done with 1 mL of LiCl Immune Complex wash buffer, and the two last washes are done with 1 mL of TE buffer. After the second wash with the TE buffer, skip the **steps 4**, **5**, and **6** of the Subheading 3.1.5 Washing, eluting and reverse cross-linking and resume at the **step 7**.
16. Alternatively, the sample can be incubated overnight.
17. Take extra care to avoid contact with phenol and use the laminar flow cabinet. Alternatively, the Qiagen MinElute PCR Purification kit (cat. 28,004) may be used. However, note that DNA purification following this method will not retain the fragments smaller than 70 bp and may alter some of the qPCR analysis. This alternative is only recommended if the sequencing input displays a strong background due to a failure to generate an optimal sonication (presence of a significant fraction of over-sonicated fragments) (Fig. 2). In these circumstances, using the Qiagen minElute purification kit will allow removing the over-sonicated fragments and decrease the background generated during the library preparation and sequencing.
18. Be careful not to contaminate the top phase by touching the middle interphase.
19. At this stage, the samples can stay at −20 °C for up to 48 h. Alternatively, quick precipitation can also be carried out at −80 °C for at least 4 h.
20. Once dry the pellet may become transparent and almost invisible. Care must be taken to not overdry the pellet, as it will be difficult to dissolve.
21. A successful sonication should appear as a smear from 100 to 250 bp without significant over or under sonication (Fig. 2). Further quality checks can be done by running 1 μL of input on the TapeStation using a D1000 tape.
22. ATRX ChIP sample should be quantifiable and give at least 140 ng/mL, IgG ChIP sample may not be quantifiable (reading: "too low < 0.5 ng/mL"), the input should be >50,000 ng/mL).
23. If not the case, there is a high probability that the ChIP failed (Fig. 3). One common issue may come from the batch of antibody. In that case, repeat the ChIP experiment with a new batch of antibody. In addition, ensure that the sonication is

optimal and that the cell viability was >90%. If the issue persists, use a fresh solution of formaldehyde and make fresh buffers. Alternative positive/negative loci can be tested.

24. This is enough to run each qPCR experiment in triplicate.
25. If several ATRX ChIP samples are combined, the volume of ultrapure water to add should be adjusted to reach a final volume of 55.5 μL of fragmented DNA.
26. A potential presence of a precipitate may be observed in the NEBNext Q5 Hot Start HiFi PCR Master Mix. Equilibrate the NEBNext Q5 Hot Start HiFi PCR Master Mix at room temperature and mix gently by inverting the tube to ensure an optimal performance.
27. Be sure to thoroughly resuspend the AMPure XP Beads by vortexing before use.
28. Incubate the sample for 5 min on the magnetic stand until the solution is clear and the beads are fully precipitated.
29. Be cautious and do not disturb the beads. If some beads were aspirated along with the supernatant, pipet down the sample and wait another 3 min to allow the beads to precipitate, then remove the supernatant carefully (with the tubes on the magnetic stand).
30. Be careful and do not over dry the beads as it may decrease yield of recover DNA target.
31. Each index number will give a different barcode. Using a different index number for each individual sample allows sequencing these samples together as a pool and extract the data for each individual sample after sequencing using the unique barcode associated with each DNA fragment. It is important to think in advance which samples will be pooled together to judiciously select the appropriate indexes. For best results, it is recommended to use indexes 6 and 12 when only two samples are pooled. Use indexes 4, 6 and 12 for a pool of three samples and the indexes 2,4,5,6,7 and 12 for a pool of six samples.
32. Making a pre-dilution to 12 nM on a fraction of the stock library before the KAPA quantification allows:
    (a) A better comparison between different libraries.
    (b) A more accurate dilution to a 4 nM library as the dilution factor will be decreased compared with the stock library.
    (c) Ensuring to be in the range of the KAPA standard by doing 1:1000 and 1:2000 dilutions for the KAPA quantification.

    The Qubit quantification is expressed as ng/μL. To obtain the molarity of the library, the following calculation is needed:

Library molarity (M) = Library concentration (g/L) / (Average fragment size of the library (bp) * Average molecular weight of 1 bp).

Then, dilute an aliquot of the stock library to obtain 15 μL of 12 nM dilute library:

Volume of stock library aliquot (μL) = 15 × 12/stock library molarity (nM).

Add 0.1× TE to the stock library aliquot to obtain a final volume of 15 μL.

33. As ATRX binds to repeats, generating longer reads (compared to the 75-cycle kit) increases the probability to map single reads overlapping with repeats. The libraries prepared with NEBNext Ultra DNA Library Prep Kit for Illumina (E7370S/L) and the NEBNext Multiplex Oligos for Illumina are compatible with the Illumina sequencing platform. In this protocol we use the example of the NextSeq500 but other Illumina platforms can be used. The sequencing coverage is an essential factor that can influence peak calling. Sufficient sequencing depths for ChIP sample and its related input are required for good quality peak calling. The sequencing depth of the input should be similar to or higher than its related ChIP-seq sample [34].

Acknowledgments

We would like to thank David Clynes and Thomas Kent for a critical reading of this manuscript. This work was supported by the Medical Research Council [grant number MC_UU_12025/unit programme MC_UU_12009/3].

References

1. Kuo M-H, Allis CD (1999) In vivo cross-linking and immunoprecipitation for studying dynamic protein: DNA associations in a chromatin environment. Methods 19(3): 425–433
2. Landt SG, Marinov GK, Kundaje A et al (2012) ChIP-seq guidelines and practices of the ENCODE and modENCODE consortia. Genome Res 22(9):1813–1831
3. Orlando V, Strutt H, Paro R (1997) Analysis of chromatin structure byin VivoFormaldehyde cross-linking. Methods 11(2):205–214
4. Park PJ (2009) ChIP–seq: advantages and challenges of a maturing technology. Nat Rev Genet 10(10):669–680
5. Collas P, Dahl JA (2008) Chop it, ChIP it, check it: the current status of chromatin immunoprecipitation. Front Biosci 13(17):929–943
6. Ren B, Robert F, Wyrick JJ et al (2000) Genome-wide location and function of DNA binding proteins. Science 290(5500):2306–2309
7. Lee TI, Johnstone SE, Young RA (2006) Chromatin immunoprecipitation and microarray-based analysis of protein location. Nat Protoc 1(2):729
8. Furey TS (2012) ChIP–seq and beyond: new and improved methodologies to detect and characterize protein–DNA interactions. Nat Rev Genet 13(12):840–852

9. Consortium EP (2012) An integrated encyclopedia of DNA elements in the human genome. Nature 489(7414):57–74
10. O'Geen H, Echipare L, Farnham PJ (2011) Using ChIP-seq technology to generate high-resolution profiles of histone modifications. Epigenetics Protocols 791:265–286
11. O'Geen H, Frietze S, Farnham PJ (2010) Using ChIP-seq technology to identify targets of zinc finger transcription factors. In: Engineered zinc finger proteins, Methods and Protocols. Humana Press, New York, pp 437–455
12. Gavrilov A, Razin SV, Cavalli G (2014) In vivo formaldehyde cross-linking: it is time for black box analysis. Briefings in functional genomics 14:163–165
13. Tian B, Yang J, Brasier AR (2012) Two-step cross-linking for analysis of protein–chromatin interactions. In: Transcriptional regulation, Methods and protocols. Humana Press, New York, pp 105–120
14. Zeng P, Vakoc CR, Chen Z et al (2006) In vivo dual cross-linking for identification of indirect DNA-associated proteins by chromatin immunoprecipitation. BioTechniques 41(6):694
15. Picketts DJ, Higgs DR, Bachoo S et al (1996) ATRX encodes a novel member of the SNF2 family of proteins: mutations point to a common mechanism underlying the ATR-X syndrome. Hum Mol Genet 5(12):1899–1907
16. Goldberg AD, Banaszynski LA, Noh K-M et al (2010) Distinct factors control histone variant H3 3 localization at specific genomic regions. Cell 140(5):678–691
17. Lewis PW, Elsaesser SJ, Noh K-M et al (2010) Daxx is an H3. 3-specific histone chaperone and cooperates with ATRX in replication-independent chromatin assembly at telomeres. Proc Natl Acad Sci 107(32):14075–14080
18. Law MJ, Lower KM, Voon HPJ et al (2010) ATR-X syndrome protein targets tandem repeats and influences allele-specific expression in a size-dependent manner. Cell 143(3):367–378. https://doi.org/10.1016/j.cell.2010.09.023
19. Nguyen DT, Voon HPJ, Xella B et al (2017) The chromatin remodelling factor ATRX suppresses R-loops in transcribed telomeric repeats. EMBO Rep 18(6):914–928
20. Eustermann S, Yang J-C, Law MJ et al (2011) Combinatorial readout of histone H3 modifications specifies localization of ATRX to heterochromatin. Nat Struct Mol Biol 18(7):777–782
21. Nan X, Hou J, Maclean A et al (2007) Interaction between chromatin proteins MECP2 and ATRX is disrupted by mutations that cause inherited mental retardation. Proc Natl Acad Sci U S A 104(8):2709–2714. https://doi.org/10.1073/pnas.0608056104
22. Gibbons RJ, Picketts DJ, Villard L et al (1995) Mutations in a putative global transcriptional regulator cause X-linked mental retardation with α-thalassemia (ATR-X syndrome). Cell 80(6):837–845
23. Lovejoy CA, Li W, Reisenweber S et al (2012) Loss of ATRX, genome instability, and an altered DNA damage response are hallmarks of the alternative lengthening of telomeres pathway. PLoS Genet 8(7):e1002772
24. Schwartzentruber J, Korshunov A, Liu X-Y et al (2012) Driver mutations in histone H3 3 and chromatin remodelling genes in paediatric glioblastoma. Nature 482(7384):226–231
25. Voon HP, Hughes JR, Rode C et al (2015) ATRX plays a key role in maintaining silencing at interstitial heterochromatic loci and imprinted genes. Cell Rep 11(3):405–418
26. Gibbons RJ, McDowell TL, Raman S et al (2000) Mutations in ATRX, encoding a SWI/SNF-like protein, cause diverse changes in the pattern of DNA methylation. Nat Genet 24(4):368–371
27. Clynes D, Jelinska C, Xella B et al (2014) ATRX dysfunction induces replication defects in primary mouse cells. PLoS One 9(3):e92915
28. De La Fuente R, Baumann C, Viveiros MM (2011) Role of ATRX in chromatin structure and function: implications for chromosome instability and human disease. Reproduction (Cambridge, England) 142(2):221–234. https://doi.org/10.1530/rep-10-0380
29. Huh MS, O'Dea TP, Ouazia D et al (2012) Compromised genomic integrity impedes muscle growth after Atrx inactivation. J Clin Invest 122(12):4412
30. Ritchie K, Seah C, Moulin J et al (2008) Loss of ATRX leads to chromosome cohesion and congression defects. J Cell Biol 180(2):315–324
31. De La Fuente R, Viveiros MM, Wigglesworth K et al (2004) ATRX, a member of the SNF2 family of helicase/ATPases, is required for chromosome alignment and meiotic spindle organization in metaphase II stage mouse oocytes. Dev Biol 272(1):1–14
32. Teytelman L, Özaydın B, Zill O et al (2009) Impact of chromatin structures on DNA processing for genomic analyses. PLoS One 4(8):e6700
33. Meyer CA, Liu XS (2014) Identifying and mitigating bias in next-generation sequencing

methods for chromatin biology. Nat Rev Genet 15(11):709–721

34. Chen Y, Negre N, Li Q et al (2012) Systematic evaluation of factors influencing ChIP-seq fidelity. Nat Methods 9(6):609–614
35. Drané P, Ouararhni K, Depaux A et al (2010) The death-associated protein DAXX is a novel histone chaperone involved in the replication-independent deposition of H3. 3. Genes Dev 24(12):1253–1265
36. Pchelintsev NA, McBryan T, Rai TS et al (2013) Placing the HIRA histone chaperone complex in the chromatin landscape. Cell Rep 3(4):1012–1019
37. Kidder BL, Hu G, Zhao K (2011) ChIP-Seq: technical considerations for obtaining high-quality data. Nat Immunol 12(10):918–922
38. Langmead B, Trapnell C, Pop M et al (2009) Ultrafast and memory-efficient alignment of short DNA sequences to the human genome. Genome Biol 10(3):R25
39. Langmead B, Salzberg SL (2012) Fast gapped-read alignment with bowtie 2. Nat Methods 9(4):357–359
40. Li H, Durbin R (2010) Fast and accurate long-read alignment with burrows–wheeler transform. Bioinformatics 26(5):589–595
41. Day DS, Luquette LJ, Park PJ et al (2010) Estimating enrichment of repetitive elements from high-throughput sequence data. Genome Biol 11(6):R69
42. Woo T-H, Hong T-H, Kim S-S et al (2007) REPEATOME: a database for repeat element comparative analysis in human and chimpanzee. Genomics & Informatics 5(4):179–187
43. Krueger F (2015) Trim Galore. A wrapper tool around Cutadapt and FastQC to consistently apply quality and adapter trimming to FastQ files
44. Andrews S (2010) FastQC: a quality control tool for high throughput sequence data
45. Li H, Handsaker B, Wysoker A et al (2009) The sequence alignment/map format and SAMtools. Bioinformatics 25(16):2078–2079
46. Kundaje A (2013) A comprehensive collection of signal artifact blacklist regions in the human genome. ENCODE[hg19-blacklist-README doc-EBI]. Available online at: https://sites.google.com/site/anshulkundaje/projects/blacklists
47. Kent WJ, Sugnet CW, Furey TS et al (2002) The human genome browser at UCSC. Genome Res 12(6):996–1006
48. Ramírez F, Dündar F, Diehl S et al (2014) deepTools: a flexible platform for exploring deep-sequencing data. Nucleic Acids Res 42(W1):W187–W191
49. Heinz S, Benner C, Spann N et al (2010) Simple combinations of lineage-determining transcription factors prime cis-regulatory elements required for macrophage and B cell identities. Mol Cell 38(4):576–589
50. Zhang Y, Liu T, Meyer CA et al (2008) Model-based analysis of ChIP-Seq (MACS). Genome Biol 9(9):R137
51. McGowan SJ, Hughes JR, Han Z-P et al (2013) MIG: multi-image genome viewer. Bioinformatics 29(19):2477–2478
52. Bortz PDS, Wamhoff BR (2011) Chromatin immunoprecipitation (ChIP): revisiting the efficacy of sample preparation, sonication, quantification of sheared DNA, and analysis via PCR. PLoS One 6(10):e26015

Chapter 7

time-ChIP: A Method to Determine Long-Term Locus-Specific Nucleosome Inheritance

Wojciech Siwek, Mariluz Gómez-Rodríguez, Daniel Sobral, Ivan R. Corrêa Jr, and Lars E. T. Jansen

Abstract

Understanding chromatin dynamics is essential to define the contribution of chromatin to heritable gene silencing and the long-term maintenance of gene expression. Here we present a detailed protocol for time-ChIP, a novel method to measure histone turnover at high resolution across long timescales. This method is based on the SNAP-tag, a self-labeling enzyme that can be pulse labeled with small molecules in cells. Upon pulse biotinylation of a cohort of SNAP-tagged histones we can determine their abundance and fate across a chase period using a biotin-specific chromatin pulldown followed by DNA sequencing or quantitative PCR. This method is unique in its ability to trace the long-term fate of a chromatin bound histone pool, genome wide. In addition to a step by step protocol, we outline advantages and limitations of the method in relation to other existing techniques. time-ChIP can define regions of high and low histone turnover and identify the location of pools of long lived histones.

Key words Epigenetic memory, Histone variants, Chromatin dynamics, SNAP-tag, time-ChIP, Turnover

1 Introduction

1.1 State of the Art

The nucleosome is the basic unit of chromatin which contributes to the maintenance of gene expression states in the form of histone variants and modifications [1]. There is evidence for such specific chromatin structures to be heritable [2–6]. To achieve this, histones can either be stably retained in chromatin in cis across multiple cell division cycles or key modifications can be copied to new molecules with high rates ensuring stability of the mark. Mathematical modeling has shown that, in fact, a bistable chromatin state can be achieved even if individual chromatin components are highly dynamic [7]. On the other hand experimental work in

Wojciech Siwek and Mariluz Gómez-Rodríguez contributed equally to this work.

Guillermo A. Orsi and Geneviève Almouzni (eds.), *Histone Variants: Methods and Protocols*, Methods in Molecular Biology, vol. 1832, https://doi.org/10.1007/978-1-4939-8663-7_7, © Springer Science+Business Media, LLC, part of Springer Nature 2018

human cells has shown that replenishment of silent chromatin marks following genome duplication is a slow process that can take until the next cell cycle [8]. Much attention has been given to histone modifications and their regulation but in order to understand how histones can act as carriers of epigenetic information it is necessary to understand their association and dissociation rates in relation to the duration of the cell cycle. Histone H3 variants have a central role in epigenetic maintenance [9, 10]. Therefore, the dynamic property of these histones at genes is key to understanding the maintenance of gene expression states. We will review the currently available methods that have been used to assess histone dynamics in cellular chromatin. Two distinct types of approaches exist, one probing the behavior of bulk chromatin, the other gives insight into nucleosome kinetics at specific loci.

1.2 Methods to Measure Dynamics of Bulk Chromatin

Classic experiments that assess dynamics of bulk chromatin initially exploited pulse-chase labeling of proteomes with radioactive and stable isotope-labeled amino acids followed by formaldehyde crosslinking and mononucleosome purification [11]. These experiments demonstrated that the core H3/H4 heterotetramer is more stable in chromatin than the H2A/H2B dimers whose high turnover is dependent on replication and transcription [11].

With the development of modern mass spectrometers, these pulse-chase experiments were taken a step further. Quantitative pulse SILAC (stable isotope labeling by amino acids in cell culture) experiments combined with force expression of tagged histones and purification of mononucleosomes confirmed that the H2A/H2B dimers are dynamic in chromatin but also distinguished between different H3 histone variants where H3.3 turns over faster than its canonical homolog H3.1. The authors additionally uncovered that a significant proportion of H3.3-H4 tetramers undergo splitting during the cell cycle in sharp contrast to the H3.1-H4 complex [12]. A more recent variant of a mass spectrometry-based measurement of chromatin dynamics was established by combining pulse SILAC with nascent chromatin capture. This elegant strategy uses biotin-dUTP labeling of nascent DNA coupled to strong crosslinking to purify recently replicated chromatin in synchronized cells [13]. This analysis confirmed that bulk chromatin is largely recycled onto daughter DNA during replication. By tracking turnover rates of specific, modified peptides a subsequent study determined the posttranslational modification status of histones which showed that some modifications such as H3K36me1 are acquired almost immediately after replication, while others (e.g., H3K9me3 and H3K27me3) are not fully replenished until the following cell cycle [8].

An orthogonal approach to determining bulk chromatin dynamics in living cells has been to use fluorescently tagged histones under the control of a heat-shock promoter in *Drosophila melanogaster*. This approach has been used to determine the chromatin incorporation dynamics of several histone variants in cell lines. In particular, the centromeric H3 variant, CID [14] and H3.3 [15] were both shown to assemble into chromatin outside of S phase using this strategy.

A more sophisticated approach was used more recently to determine H3 histone kinetics in living flies. The system uses a histone variant fused to a fluorescent protein in a way that after induction of a recombinase, one florescent tag is exchanged with another (i.e., green to red switch). The approach offered both spatial (the histone fusion is under a tissue-specific Gal4 driver), and temporal control (the recombinase is induced by heat-shock). By employing this system in the *D. melanogaster* male germline asymmetric cell division was shown to strongly correlate with asymmetric distribution of the H3.1 but not the H3.3 histone variant [16].

The above strategies depend on induction of protein synthesis resulting in a relatively long temporal delay before dynamics can be assessed. To avoid this drawback, photo bleaching (fluorescent recovery after photo bleaching, FRAP) of chromatin-bound histones tagged with fluorescent proteins can be used. Determining the rate of fluorescence recovery following a bleach pulse provides a measure of the rate of histone reincorporation within the bleached nuclear area. These experiments measured recovery rates up to a few hours and confirmed that H2B turns over much faster than the core H3 and H4 histones [17].

Alternatively, photo switchable or photo activatable protein variants have been used to allow fluorescence activation of a cohort of molecules. In this case, rather than determining association kinetics, the dissociation rate of pulsed molecules is followed over time. The technology was applied to the centromeric H3 histone variant (CENP-A) to show that it has extremely slow turnover at the centromere while it is quite dynamic at other loci [18]. A similar strategy was used for the H2A histone to show that the turnover of this protein in chromatin slows down after DNA damage [19].

While fluorescence bleaching or activation techniques follow local cohorts of molecules in cells, an alternative pulse labeling method has been developed to label whole-cell populations of proteins based on the SNAP-tag. SNAP is a self-labeling enzyme derived from the human O^6-alkylguanine DNA alkyltransferase, a DNA repair enzyme that de-alkylates guanines in DNA by transfer of the alkyl group to a reactive cysteine within the enzyme. This suicide-mode of SNAP is engineered to catalyze its own covalent binding to cell permeable O^6-benzylguanine (BG) derivatives [20]. During this reaction a benzyl-ring is transferred to SNAP in a covalent and irreversible manner conjugating a label such as a fluorescent dye to the substrate [20–22]. We and others have extensively employed the SNAP-tag for fluorescent labeling of the centromere-bound histone H3 variant CENP-A [23–30]. By pulse labeling total cellular CENP-A followed by a chase we can determine the turnover rate of this histone. Because cells are labeled and chased in culture before being processed for imaging, one can follow populations of cells over long time periods. While photo bleaching and photo activating methods are applicable for short timescale (fast turnover measurements) SNAP-tag pulse labeling complements these at the long timescale range. Using this strategy, we showed that CENP-A exhibits extreme stability in chromatin with no detectable turnover during multiple mitotic divisions [18, 30].

In summary, bulk biochemical methods or those based on fluorescence indicate that histones can be long lived in chromatin and that different nucleosome components have different dynamics in living cells. However, these methods do not provide any positional information or genomic spatial resolution which is restricted by the optical resolution limit. Below we describe complementary approaches that determine dynamics at specific loci, a parameter that is crucial to be able to link nucleosome stability to any role in the control of gene expression.

1.3 Methods to Measure Locus-Specific Chromatin Dynamics

Defining locus-specific histone turnover is typically reliant on protocols based on chromatin immunoprecipitation (ChIP) or related chromatin pull-down methods. Initial efforts used inducible promoters in budding yeast coupled to ChIP to measure incorporation rates across the genome [31]. In this early insightful study, a Myc-tagged version of H3 histone was constitutively expressed while a second FLAG-tagged version was pulse inducible. The ratio of the ChIP signal of the two differentially tagged histones was determined genome-wide and showed that histone exchange on genes is dependent on transcription and that promoters are the regions with the highest histone incorporation rates [31]. A similar strategy was used in human cells. The H3.3 histone variant was force expressed either as an HA or FLAG-tagged version in a sequential manner to distinguish between the old and new pools of H3.3 proteins. Next, mononucleosomes were isolated and subjected to two subsequent chromatin immunoprecipitation steps for HA and Flag tags. In this way nucleosomes carrying both old and new H3.3 were purified and bound DNA was sequenced. This approach detected hybrid nucleosomes on active genes as well as on cell-specific enhancers which showed the highest degree of H3-H4 tetramer splitting events [32] as was previously shown to occur in yeast [33].

An elegant complementary approach named CATCH-IT (Covalent attachment of tags to capture histones and identify turnover) [34] is based on pulse labeling of endogenous proteins without the need for overexpression or tagging. The nascent proteome is pulse labeled using a substitute of methionine (azidohomoalanine, Aha). After a chase period a cycloaddition reaction of biotin is performed, followed by high salt chromatin isolation to strip any non-histone proteins and pull down of biotin labeled nucleosomes using immobilized streptavidin. This, combined with genome-wide analysis, provides association rates of histones per locus. The results showed that histones incorporate preferentially at active genes, enhancer elements, and origins of replication, and that the rate of their turnover correlates with gene expression [34] as previously suggested and shown [31, 35, 36]. The caveat of this approach is that no specific histone variants can be analyzed. Further, because this method is based on measuring dynamics of nascent proteins at short time scales it is not suited for detection of stable pools of nucleosomes.

A methodology developed in the yeast *Saccharomyces cerevisiae* circumvents the limitation of analyzing incorporation rates and allows detection of turnover rates of ancestral pools of specific histones. This system called recombination-induced tag exchange (RITE) uses a constitutively epitope-tagged version of H3 at its endogenous genomic locus. The locus is engineered such that upon activation of Cre recombinase (controlled by β-estradiol) one epitope tag is genetically and irreversibly exchanged with another one [37], an approach similar to the one, described above, to monitor histone dynamics during asymmetric cell division of the *D. melanogaster* male germ cells [16]. The RITE system applied to HA/T7 tag exchange on histone H3 revealed replication independent histone replacement with preferential retention of old H3 pools at 5′ ends of long poorly transcribed genes and proposed an RNA PolII dependent retrograde movement of nucleosomes on genes [38]. Interestingly, a proportion of H3 histones at these loci were shown to survive through multiple cell division cycles.

In summary, the methods described above, each with their specific advantages, have provided insight into several fundamental features of chromatin. Nevertheless, they suffer from specific drawbacks. For example, bulk (mass spectrometry or microscopy-based) measurements cannot directly link the dynamic behavior of nucleosomes to gene expression status. The locus-specific methods can be linked to gene expression but they suffer from transcription and translation delays (inducible genes). Further, when analyzing a short pulse of nascent proteins, the assay is biased to the detection of regions with fast turnover and is blind to the long-term retention of an ancestral histone pool (CATCH-IT). The RITE system offers a solution to the latter problem, however it also suffers from a delayed response to Cre induction and, as of yet, cannot be used in mammalian systems due to lack of efficiency and strict control of Cre recombinase.

1.4 time-ChIP

Here, we advance and complement the existing techniques by developing and employing a novel method that allows for the detection of nucleosome retention at high resolution in human cells at long time-scales. Rather than analyzing histone incorporation rates, this method aims to measure the stability of previously incorporated ancestral histones. The cornerstone of the method is the self-labeling SNAP-tag. As outlined above, SNAP can be pulse labeled in cells, typically by using fluorescent dyes coupled to imaging [20–22]. Here, we modify this approach by developing a pulse-chase affinity purification strategy, based on a biotin-conjugated SNAP substrate (Fig. 1a). Biotin-mediated pulse labeling of SNAP-tagged histones allows us to isolate, and directly measure, histone retention in chromatin at specific loci in human cells. We call this method time-ChIP as the pull down strategy is akin to a chromatin immunoprecipitation (ChIP) experiment with the added temporal component to determine dynamics of nucleosome occupancy. In brief, following pulse labeling of living cells, the excess label is washed out, cells are chased, nuclei are isolated, and chromatin is enzymatically fragmented. Soluble biotinylated chromatin is then isolated on immobilized streptavidin and processed for quantitative PCR or sequencing (Fig. 1b).

1.5 Biotinylated SNAP Substrates

SNAP biotinylation can be performed using commercially available BG-Biotin (New England Biolabs) (Fig. 2a). Recently, we have developed an improved substrate to achieve more efficient and specific SNAP biotinylation in chromatin in living cells. This optimized substrate (CP-Biotin; Fig. 2b) differs from the established BG-Biotin by its CP (chloropyrimidine) scaffold which has been shown to be more permeant to cell membranes than BG derivatives [39] and results in improved biotinylation of histones in chromatin (Fig. 2c). CP-Biotin was synthesized by reacting CP-NH_2 [39] with biotin amidohexanoic acid *N*-hydroxysuccinimide ester as described [21, 40, 41] (Fig. 2b) (*see* **Note 1** for detailed methodology).

1.6 Outline and Initial Characterization of the Method

To perform chromatin dynamics measurements with time-ChIP, HeLa cell lines stably expressing H3.1-SNAP or H3.3-SNAP [28, 30] are in vivo pulse labeled with BG-Biotin (or CP-Biotin), followed by a chase period to allow for histone turnover. At specific times following pulse labeling, cells are subjected to gentle lysis to isolate nuclei. These nuclei are then MNase treated and soluble chromatin is isolated. Histone-SNAP-biotinylated nucleosomes are then captured on immobilized streptavidin beads and DNA is

Fig. 1 (continued) somes decays over time. At specific time points, cells are lysed, nuclei are isolated and chromatin is liberated by MNase treatment resulting in soluble chromatin of 1–3 nucleosomes in length. Biotinylated nucleosomes are isolated and purified on streptavidin beads. (**c**) CP-Biotin treatment results in a biotinylation dependent pulldown of SNAP-tagged histone. HeLa H3.3-SNAP-3XHA cells were pulse biotinylated or left untreated and soluble chromatin was isolated on streptavidin beads. Immunoblot probed for the H3.3-SNAP-3XHA protein by anti-hemagglutinin (HA) antibody. In—Input; Un—Unbound; B—Bound. (**d**) Dose-response of biotinylation. Biotinylated cells were mixed with unbiotinylated cells at indicated percentages processed as outlined in panel B and total DNA yield was quantified by PicoGreen. Genomic DNA that is recovered correlates with degree of biotinylation (*n* = 3–6 exp.; bars are SEM)

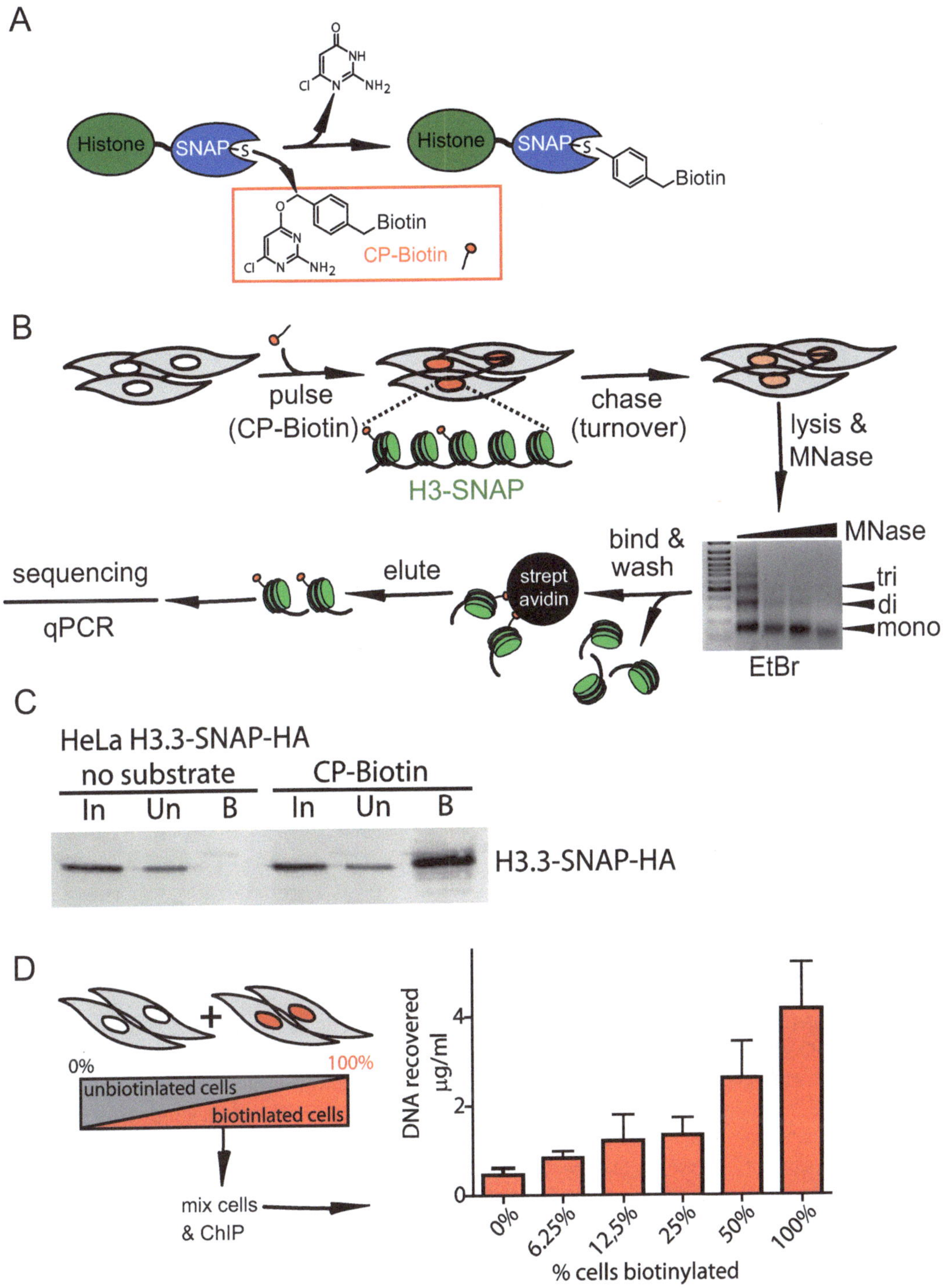

Fig. 1 time-ChIP, a novel method to determine chromatin retention at high resolution, genome-wide. (**a**) Reaction scheme of SNAP with CP-Biotin resulting in the covalent self-labeling of SNAP with biotin through a reactive cysteine (S). (**b**) Outline of pulse-chase time-ChIP assay. Cells expressing SNAP-tagged histone are pulse labeled with CP-Biotin. Following a chase period, the fraction of biotinylated histones retained at nucleo

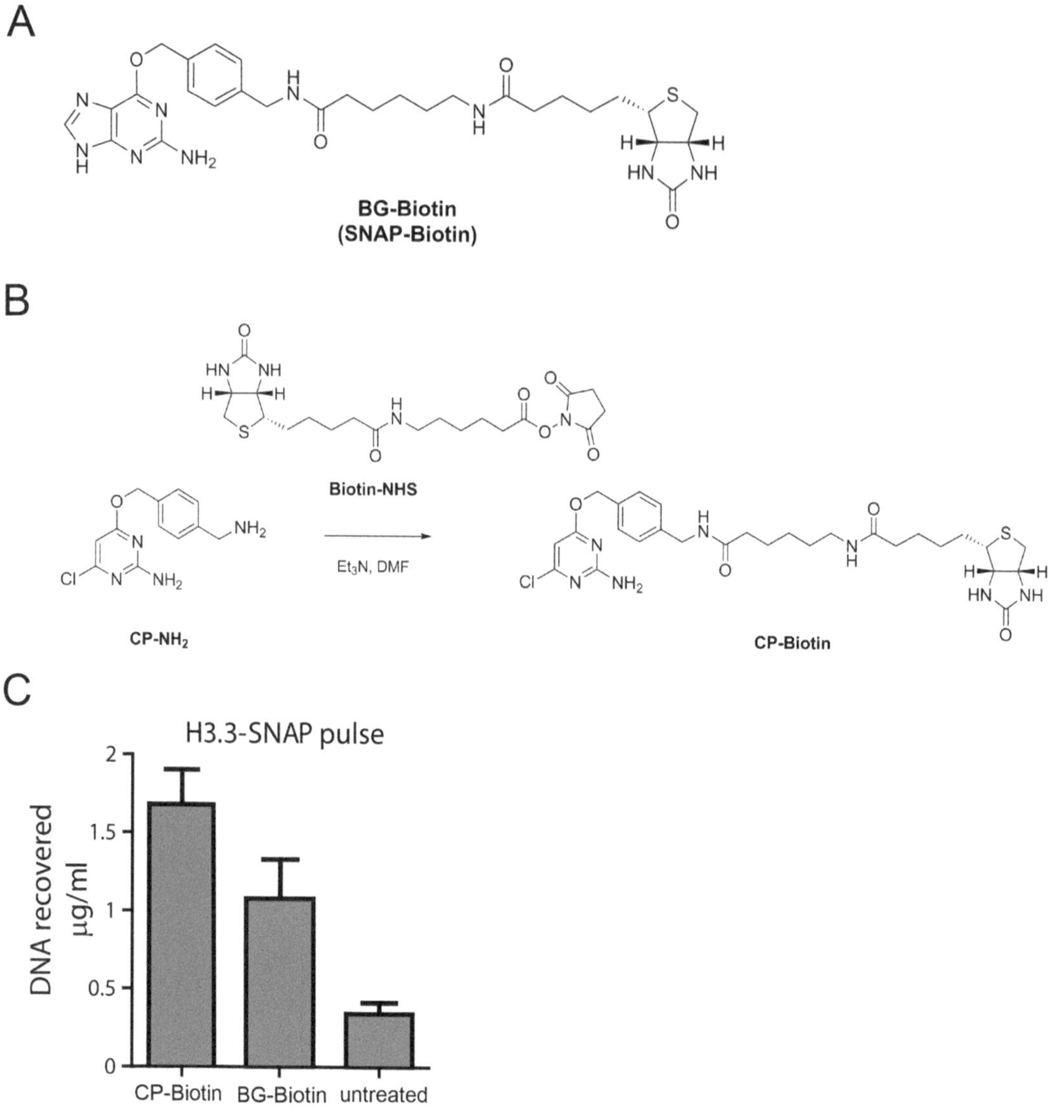

Fig. 2 Synthesis of CP-Biotin. Chemical structures of (**a**) BG-Biotin (commercially available as SNAP-Biotin, New England Biolabs (NEB), Ipswich, MA). (**b**) Benzylchloropyrimidine-NH2 (CP-NH2), biotin amidohexanoic acid N-hydroxysuccinimide ester (Biotin-NHS) and CP-Biotin. CP-Biotin was synthesized by reacting CP-NH2 with Biotin-NHS in the presence of anhydrous N,N-dimethylformamide (DMF) and trimethylamine (Et3N). (**c**) Comparison of labeling efficiency of BG-Biotin (NEB) and CP-Biotin (this study). time-ChIP pulse labeling of HeLa cells expressing H3.3-SNAP with either CP-Biotin, BG-Biotin or DMSO control. Chromatin was liberated by MNase treatment and isolated in a biotin dependent manner as described for Fig. 1. Recovered DNA was quantified by PicoGreen measurement. Averages and SEM are plotted from three experiments

purified. Quantification of biotin histone retention at specific loci is determined by quantitative PCR or sequencing (Fig. 1b). Western blot analysis shows SNAP-HA-tagged histones are specifically isolated in a biotin pulse label dependent manner (Fig. 1c). To determine the dynamic range of the assay we mix in vivo labeled HeLa H3.1–SNAP cells with unlabeled cells at different ratios. The genomic DNA that we recovered correlates with the degree of biotinylation and allowed us to detect as little as ~6% histone retention (Fig. 1d).

1.7 time-ChIP-qPCR

Analysis of biotin-dependent isolated DNA can be performed either by qPCR or sequencing methods. For qPCR we use standard SYBR green protocols with the following key adaptations in order to determine the linearity and dynamic range at which histone retention can be measured. First, a dilution series of input DNA is made to determine the linear range of the PCR reaction. Next the percentage input (% input) is calculated by subtracting the cut off cycles for the measured IP from the corrected input as presented in detail with the formula at Subheading 3.7, **step 8**. In addition to the qPCR standard curve, a time-ChIP standard curve is generated for each pulse-chase experiment to determine the dose-response of the qPCR to the fraction of biotinylated nucleosomes. This determines how much DNA is recovered at different chase time points relative to the initial pool present at the pulse. To this end, we mix known ratios of biotin-pulse labeled and non-labeled cells at the beginning of the protocol in a 2-fold ratio series. Cells are then lysed, MNase treated and subjected to biotin pulldown as described in detail in the Methods section. A linear relationship between IP values and the level of chromatin biotinylation is expected. In Fig. 3, a typical result is shown where dilution of biotin-pulse labeled cells expressing H3.1-SNAP results in a correspondingly reduced % input signal. In Fig. 3a, an example of an active gene locus (RPL13) is shown that is probed with a specific primer set hybridizing to a coding exon. Dilution of biotinylated chromatin results in a corresponding reduction in IP signal. This is used to produce a standard curve upon which chase signals are interpolated (Fig. 3b) resulting in a measure of histone retention expressed as a fraction of the signal present at steady state (Fig. 3c).

1.8 H3.1-SNAP Is Locally Retained in Chromatin

time-ChIP-qPCR measurements of ancestral H3.1-SNAP show that a proportion of H3.1 can be retained in cis on the DNA even during continued transcription and replication (Fig. 3c). Presumably DNA unwinding during these processes results in nucleosome disruption. Histone retention would therefore involve sliding or local recapture of ancestral histones. However, it is formally possible that nucleosomes are lost and recaptured distally in trans. This would result in “apparent” retention rather than actual in cis stability of nucleosomes (Fig. 4a). Such recapture of old histones in trans is unlikely because the soluble pool size is a small fraction of the total chromatin bound pool. Therefore, recapture of histones would represent a minor part of the biotin signal in chromatin. Indeed, histones with a chromatin signature are not detected in the soluble histone pool [42]. Nevertheless, we directly tested whether the apparent histone retention signal could be the

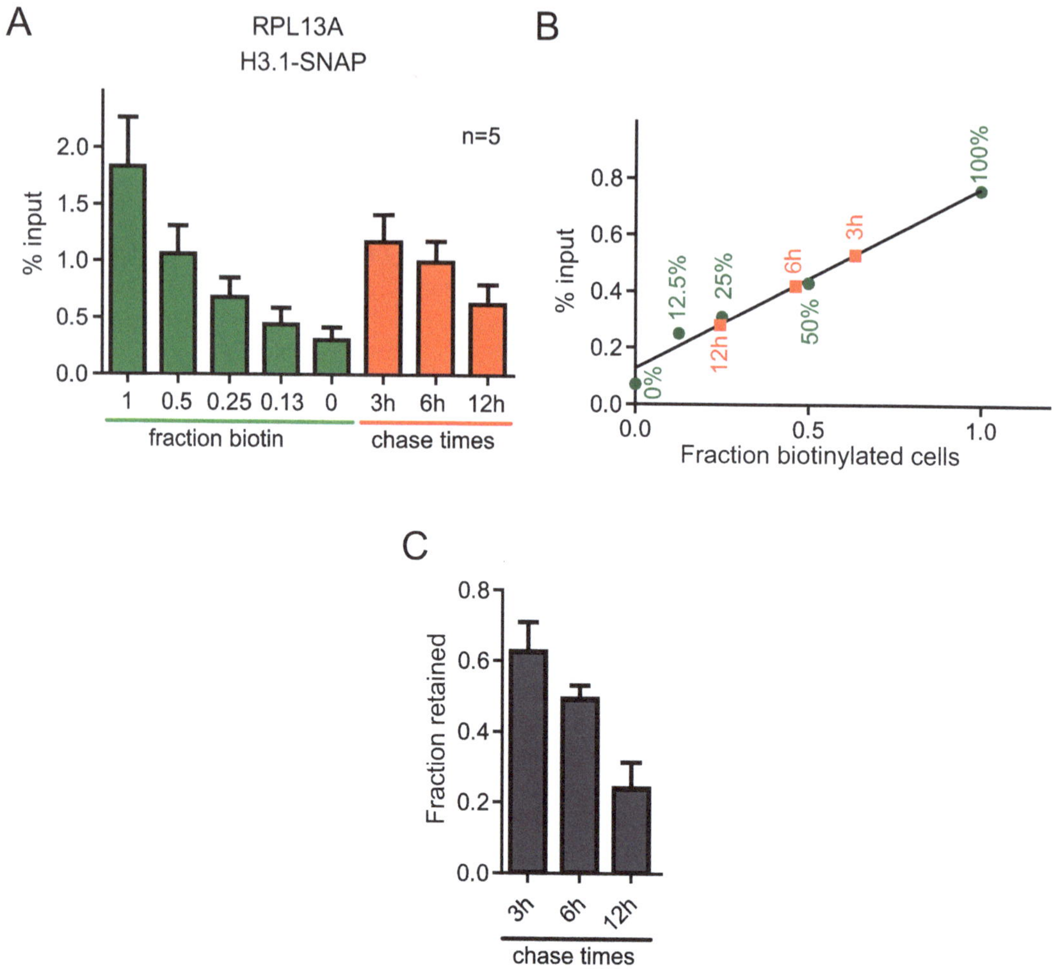

Fig. 3 time-ChIP-qPCR analysis quantifies histone retention rates. (**a**) HeLa cells expressing H3.1-SNAP were pulse labeled and mixed with mock labeled cells at indicated ratios (1, 0.5, 0.25, etc). A proportion of pulse labeled cells were chased for 3, 6 and 12 h before being processed for histone-biotin purification. Biotin:mock labeled cell mixtures (green bars) and chase samples (red bars) were analyzed by qPCR for the active house-keeping locus RPL13A and % input was determined. IP efficiency correlates with the degree of biotinylation. Mean and SEM of five replicate experiments are shown. (**b**) Example of linear regression of time-ChIP standard curve derived from biotin:mock labeled cell mixtures. Chase time points are interpolated on the regression line to obtain corresponding "fraction biotin" values. A plot of a single replicate is shown. (**c**) For each replicate the fraction of biotin retention is determined by interpolation to the time-ChIP standard (obtained independently for each replicate). Mean and SEM of retention values are plotted for each chase time point. In this example, H3.1-SNAP is present at 62%, 48%, and 24% after 3, 6, and 12 h respectively

result of dynamic reincorporation. H3.1 assembly is strictly dependent on DNA synthesis [15, 28]. Therefore, blockage of DNA replication will prevent incorporation of histone H3.1. As a consequence, if the observed H3.1 retention would be the result of reincorporation of distally recycled old histones, then retention would be no longer observed. In contrast, treatment of cells with thymidine to prevent DNA replication-dependent turnover during a

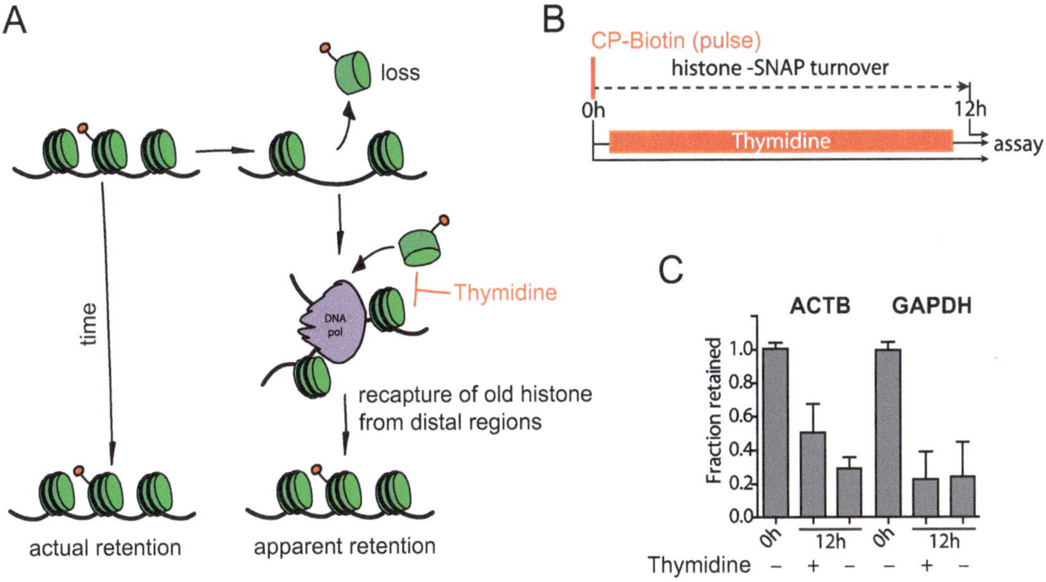

Fig. 4 H3.1-SNAP is locally retained in chromatin. (**a**) Possible scenarios of nucleosome retention or loss. Biotinylated histones may be detected either following stable retention of nucleosomes or potentially by reincorporation during DNA replication of old histones, excised from other regions of the genome. (**b**) Thymidine prevents DNA replication. H3.1-SNAP cells were treated with thymidine for the duration of the chase period (12 h) preventing replication coupled assembly of H3.1 chromatin. (**c**) time-ChIP-qPCR of H3.1-SNAP cells for ACTB and GAPDH genes for indicated times with or without thymidine-induced blockage of DNA replication. Averages and SEM are plotted from at least three experiments

12 h H3.1-SNAP chase did not result in an elevated histone turnover (Fig. 4b, c), indicating that H3.1 histones, once incorporated, are locally retained in chromatin.

1.9 time-ChIP-Seq

To uncover histone dynamics genome wide we coupled our time-ChIP method to high-throughput sequencing. As a proof of concept experiment, we decided to perform the measurement for the H3.3 histone variant as it presents a known characteristic distribution across the genome linked to gene activity [35]. At each time point (0, 12, and 24 h) and for an input sample we sequenced approximately 200 million reads, 50 bp length in single end mode. We mapped the data to GRCh38 reference genome, combined the reads into 1000 bp bins, normalized to read count, and subtracted the input signal. We next intersected the data with active (H3K9ac, H3K27ac) and inactive (H3K9me3, H3K27me3) parts of the genome and found faster exchange to correlate with active marks, indicating that our H3.3-SNAP fusion protein is behaving as expected (Fig. 5a). We also note that, in general, there is less H3.3 accumulation on silent chromatin domains. Direct visualization of H3.3 occupancy as a function of genomic loci, across time shows gradual loss of H3.3 (Fig. 5b). We have recently employed this approach in the context of mouse embryonic stem cells to

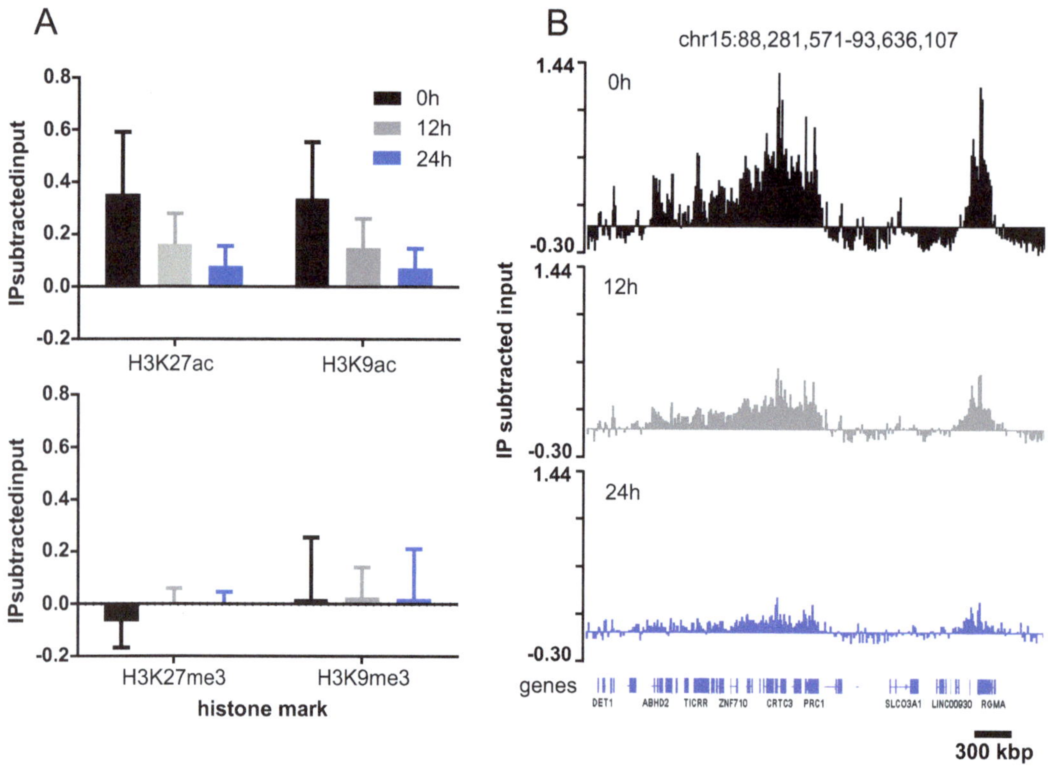

Fig. 5 time-ChIP coupled to high-throughput sequencing. Cells expressing H3.3-SNAP were biotinylated and chased for 0, 12 and 24 h in culture. Biotinylated H3.3-SNAP chromatin was fragmented by MNase, isolated and associated DNA was purified and subjected to Illumina sequencing (EMBL, Heidelberg, Germany). The reads were mapped to GRCh38 and normalized to read count. The input signal was subtracted from the pull down data. (**a**) Processed data for time points 0, 12 and 24 h for four different genomic domains: H3K27ac, H3K9ac representing active chromatin and H3K27me3, H3K9me3 representing silent chromatin. (**b**) Visual representation of time-ChIP-seq data for a genomic window showing histone H3.3-SNAP enrichment and turnover

map H3.3 dynamics genome wide [43]. In that study, we confirmed fast turnover of H3.3 at promoters and enhancers as previously reported [32, 34]. Interestingly, we found that regions of Polycomb activity reduce H3.3 dynamics, correlating with transcriptional silencing. We additionally uncovered novel regions of the genome with rapid histone turnover linked to cellular differentiation which may point to previously uncharacterized enhancers [43]. Overall these results show that the method can be combined with high-throughput sequencing to gain insight into locus-specific chromatin dynamics.

1.10 time-ChIP Summary and Comparison with Existing Methods

In this work we describe a novel method we name time-ChIP, capable of measuring local histone dynamics and inheritance with both quantitative PCR as well as high-throughput sequencing. The method has several key advantages:

1. The creation of a labeled pool is not dependent on de novo transcription and/or translation as the existing pools of protein are labeled. However, there is a delay of approximately 1 h due to the SNAP labeling procedure.
2. Both association and long-term dissociation kinetics can be measured when used in quench-chase-pulse or pulse-chase setup, respectively (*see* **Note 2**).
3. time-ChIP is based on the use of genetically encoded tags allowing specific histones or histone variants to be assessed.
4. It is applicable to mammalian systems and was already used to determine histone dynamics in differentiating mouse ES cells [43].

The main disadvantage of time-ChIP is low efficiency of biotin labeling (due to limited membrane permeability of the substrate) which requires compensation with relatively high cell numbers. We have addressed this problem, in part, by developing an enhanced biotin label (CP-Biotin, Fig. 2).

In Table 1 we list an extensive comparison of time-ChIP with existing methods and outline the specific features of each strategy.

2 Materials

2.1 Specialized Equipment

1. Dounce homogenizer, 15 mL volume, tight pestle.
2. Orbital shaker.
3. Magnetic stand designed for 1.5 mL tubes.

2.2 Culture Media

1. HeLa cell culture medium: DMEM (Dulbecco's Modified Eagle's medium High Glucose w/o L-Glutamine w/o Sodium Pyruvate), 10% newborn calf serum, 2 mM L-Glutamine, 100 μg/mL Penicillin, 100 μg/mL Streptomycin.

2.3 Buffers

1. Dulbecco's Phosphate-buffered saline (D-PBS) w/o Calcium, w/o Magnesium, cell culture grade.
2. Cell homogenization buffer: 3.75 mM Tris–HCl pH 7.5, 20 mM KCl, 0.5 mM EDTA, 0.5 mM DTT, 0.05 mM Spermidine, 0.125 mM Spermine, 0.1% Digitonin (recrystallized from 50% purity, *see* **Note 3** for recrystallization protocol), 1 mM PMSF, protease inhibitor cocktail.
3. Washing buffer A: 20 mM HEPES sodium salt pH 7.7, 20 mM KCl, 0.5 mM EDTA, 0.5 mM DTT, 0.5 mM PMSF, protease inhibitor cocktail.
4. Washing buffer B: 20 mM HEPES sodium salt pH 7.7, 20 mM KCl, 0.5 mM EDTA, 0.5 mM DTT, 0.5 mM PMSF, protease inhibitor cocktail, 300 mM NaCl.

Table 1
Comparison of methods to analyze histone dynamics

Method	Feature									
	Timescale	Association kinetics	Dissociation kinetics	Histone variants	Locus specific	Genetic perturbation	Applicable to animal cells	Specialized small molecules	Other comments	References
Pulse labeling with radioactive and stable isotopes	Minutes to days	Yes[b]	Yes	No	No[i]	No	Yes	Radio and density labeled amino acids	Early method to study chromatin dynamics	[11]
Pulse labeling with stable isotopes combined with mass spectrometry	Minutes to days	Yes[b]	Yes	Yes	No	No	Yes	Density labeled amino acids, inducers of gene expression, biotin-dUTP[k]	Variations of the method exist with diverse purification steps	[8, 12, 13]
Florescence measurements after photo bleaching or photo switching	Seconds to minutes	Yes	Yes[e]	Yes	No[i]	Yes	Yes	No	Photo toxicity and bleaching limits the time scale at which dynamics can be measured	[17–19]

SNAP-based fluorescent pulse labeling	Minutes to hours	Yes[d]	Yes[g]	Yes	No	Yes	Yes	BG- or CP-fluorophore	Ideal for assessing slow dynamics based on imaging. Resolution limited	[21–30]
Covalent attachment of tags to capture histones and identify turnover (CATCH-IT)	Minutes to hours	Yes[b]	Yes[f]	No	Yes	No	Yes	Aha[l], biotin-alkyne	Assay is designed to detect fast dynamics, stable chromatin is not detected	[34]
Pulsed expression of tagged histones	Hours to days	Yes[c]	Yes[h]	Yes	Yes	Yes	Yes	Inducers of gene expression, β-estradiol[m]	RITE[j] uses an engineered, small molecule controlled Cre recombinase	[14–16, 31–33, 35]

(continued)

Table 1
(continued)

Method	Feature									
	Timescale	Association kinetics	Dissociation kinetics	Histone variants	Locus specific	Genetic perturbation	Applicable to animal cells	Specialized small molecules	Other comments	References
time-ChIP	Hours to days[a]	Yes[d]	Yes[g]	Yes	Yes	Yes	Yes	BG- or CP-biotin	Requires relatively high cell numbers due to low efficiency of biotin labeling	[43], this study

[a]Faster than pulsed expression of tagged histones
[b]Association dynamics is dependent on the protein translation rate
[c]Association dynamics is dependent on transcription and translation rates
[d]When combined with quench-chase-pulse assay
[e]When photo switchable proteins are used
[f]Dissociation dynamics is limited to the recently incorporated histone pool potentially biasing to fast dynamics
[g]When combined with pulse-chase assay
[h]Steady state cannot be reached with recombination induced tag exchange (RITE)
[i]There is no certainty that the measured histones constitute chromatin
[j]RITE shown only in yeast but equivalent Cre-mediated tag exchange system has been used in *Drosophila melanogaster* [16]
[k]Biotin-dUTP is used for nascent chromatin capture [13]
[l]Aha, azidohomoalanine is a methionine substitute
[m]When using recombination induced tag exchange (RITE) to control Cre recombinase [37, 38]

5. Washing buffer C: 20 mM HEPES sodium salt pH 7.7, 20 mM KCl, 5 mM EDTA, 0.5 mM DTT, 0.5 mM PMSF, protease inhibitor cocktail, 500 mM NaCl, 0.5% NP-40 (nonyl phenoxypolyethoxylethanol).
6. Beads blocking buffer: 20 mM HEPES sodium salt pH 7.7, 20 mM KCl, 5 mM EDTA, 0.5 mM DTT, 0.5 mM PMSF, protease inhibitor cocktail, 300 mM NaCl, 0.05% NP-40, 50 mg/mL bovine serum albumin, 200 μg/mL yeast tRNA.
7. RNA removal buffer: 10 mM Tris–HCl pH 7.4, 100 mM NaCl, 10 mM EDTA, 0.5% SDS.
8. TE buffer: 10 mM Tris–HCl pH 8.0, 1 mM EDTA.

2.4 Solutions

1. BG-Biotin (SNAP-Biotin, New England Biolabs) dissolved in DMSO at a concentration of 4 mM stored at −80 °C long term, working stocks are kept at −20 °C (although kept frozen for long-term storage, we do not find the substrate to be temperature sensitive in the short term, hours-days).
2. CP-Biotin: dried compound dissolved in DMSO at a concentration of 4 mM stored at
   −80 °C (not commercially available).
3. 300 mM $CaCl_2$.
4. 500 mM EGTA.
5. 10% nonyl phenoxypolyethoxylethanol (NP-40).
6. Phenol:chloroform:isoamyl alcohol (25:24:1) solution pH 8.
7. Chloroform:isoamyl alcohol (24:1).

2.5 Enzymes

1. Trypsin, cell culture grade.
2. Micrococal nuclease (MNase) (Roche or equivalent).
3. RNaseA, DNase free.

2.6 Antibodies

1. Primary antibody: anti-HA, clone HA11 (Bio-Legend or equivalent).
2. Secondary antibody: anti-mouse IRDye 800CW (Li-COR or equivalent).

2.7 Other Materials

1. Streptavidin Magnetic Beads (Pierce or equivalent).
2. MinElute Reaction Cleanup Kit (QIAGEN).
3. Quant-iT PicoGreen dsDNA assay kit (Invitrogen).
4. PerfeCTa SYBR Green FastMix ROX (Quanta Biosciences).
5. NEBNext ChIP-Seq Library Prep Reagent Set for Illumina (NEB).

3 Methods

3.1 General Considerations

All solutions and equipment in contact with living cells should be sterile and aseptic techniques should be used throughout the procedure.

HeLa cell lines expressing histone-SNAP fusions were cultured in HeLa cell culture medium at 37 °C and 5% CO_2. Cells were grown to a maximum confluency of 80%. Proliferating cultures were maintained by washing with D-PBS, trypsinization and dilution into fresh culture medium.

DNA vectors containing the SNAP-tag and the protocols for retroviral mediated transduction, monoclonal selection and characterization of cell lines are described in detail in Bodor et al. [29].

Here, we describe a native time-ChIP protocol with chromatin solubilization by MNase treatment. Please note that an alternative protocol with formaldehyde crosslinking and sonication is also applicable (*see* **Note 4**).

3.2 SNAP Pulse-Chase

All operations at this stage of the protocol, including centrifugation steps, should be performed at room temperature to avoid cell stress.

The pulse labeling procedure can be performed with either commercially available BG-Biotin or an improved CP-Biotin label developed by Ivan Correa (New England Biolabs, Ipswich, MA), not commercially available (*see* **Note 1**).

1. Grow, harvest, and pellet by centrifugation (500 × *g*, 5 min) $5*10^8$ cells expressing a histone-SNAP fusion of choice. Cell counts can be determined by a Scepter 2.0 Handheld Automated Cell Counter or other methods of choice.
2. Resuspend the cell pellet in 4 mL culture medium supplemented with 10 μM BG-Biotin (or the same concentration of CP-Biotin).
3. Incubate the cell suspension for 1 h at 37 °C in a water bath. To ensure proper labeling, mix cells every 10 min.
4. Add 20 mL medium and centrifuge at 500 × *g* for 5 min.
5. Wash 2 times with 10 mL culture medium; centrifuge 500 × *g* for 5 min between each suspension step.
6. Resuspend the cell pellet in 30 mL culture medium.
7. Count and collect 5×10^7 cells. Incubate for 30 min at 37 °C in a water bath, mixing every 10 min (this step allows excess BG-Biotin to be released from cells).
8. Centrifuge the cells at 500 × *g* for 5 min.
9. Snap freeze the cell pellet in liquid nitrogen and store at −80 °C until further use (*see* Subheading 3.3).

10. For chase experiments, re-seed the remainder of the labeled cells (all minus 5×10^7 collected in Subheading 3.2, **step** 7) in fresh medium and allow to proliferate for the desired time. Typical time points include 0, 6, 12, 24, 48 h after pulse labeling.
11. At each time point collect 5×10^7 cells and harvest as described for Subheading 3.2, **steps 8** and **9**.

(protocol can be suspended at this point)

3.3 Soluble Nucleosome Preparation

All subsequent operations should be performed for all samples (pulse and chase time points) in parallel. All steps should be performed at 4 °C or on ice unless stated differently.

1. Before starting, chill centrifuges and following materials to 4 °C or on ice: Dounce homogenizer (15 mL with tight pestle), cell homogenization buffer, washing buffer A and washing buffer B.
2. Resuspend frozen cell pellets (5×10^7 cells) in 15 mL of cell homogenization buffer.
3. Lyse the cells with precooled Dounce homogenizer with 10 strokes. Be careful not to create air bubbles.
4. Centrifuge nuclei at 500 × *g* for 5 min and discard the supernatant.
5. Resuspend nuclei in 15 mL of cell homogenization buffer.
6. Repeat the homogenization step as in Subheading 3.3, **step 3**.
7. Centrifuge the lysate at 500 × *g* for 5 min and discard the supernatant.
8. Wash the nuclei twice with 15 mL of washing buffer A and centrifuge at 500 × *g* for 5 min.
9. Resuspend the pellet in 7.5 mL of washing buffer B.
10. Centrifuge nuclei at 500 × *g* for 10 min at 4 °C.
11. Resuspend in 500 μL of washing buffer B and transfer to a 1.5 mL tube.
12. Add $CaCl_2$ to a final concentration of 3 mM (5 μL from 300 mM stock).
13. Add MNase to final concentration of 800 U/mL (the correct MNase concentration may need to be experimentally defined, *see* **Note 5**).
14. Incubate in an orbital shaker at room temperature for 1 h.
15. Add EGTA to final concentration of 5 mM to stop the reaction (5 μL from 500 mM stock).
16. Add NP-40 to final concentration of 0.5% and mix to solubilize nuclear membranes (25 μL from 10% stock).
17. Centrifuge the lysate at 10,000 × *g* for 15 min at 4 °C.

18. Transfer the supernatant (containing solubilized chromatin) to a fresh 1.5 mL tube. Store the pellet fraction as a control, *see* **Note 6**.
19. Determine the concentration, i.e., optical density (OD) of soluble chromatin at 276 nm. A standard spectrophotometer such as NanoDrop can be used for this purpose.
20. Normalize sample to a final concentration of 1 OD_{A276}, with washing buffer B, ensuring a final volume of at least 400 μL.
21. Collect a 30 μL input sample for DNA isolation at Subheading 3.6 (this is an essential step required for downstream analysis).
22. Collect a sample of 10 μL for agarose gel analysis of MNase treated DNA.
23. Optionally a 10 μL sample can be taken for Western Blot analysis to determine the efficiency of the pull down in Subheading 3.5 (*see* **Note 7**).
24. Store the rest of the chromatin sample (about 350 μL) at −80 °C until Subheading 3.4.
25. Purify the DNA from the sample collected in Subheading 3.3, **step 22** as in Subheading 3.6 and separate DNA fragments by standard agarose gel electrophoresis to assess the efficiency of MNase digestion. The majority of DNA should be digested to mononucleosome fragments (*see* Fig. 1b). If the DNA is under- or over-digested, MNase concentration should be optimized (*see* **Note 5**).

(protocol can be suspended at this point)

3.4 Purification of Biotin-Labeled SNAP-Tagged Nucleosomes

All operations in this step of the protocol should be performed for all chase time points in parallel.

1. Transfer 20 μL (0.2 mg) of homogenized streptavidin magnetic beads into a 1.5 mL tube per sample (pulse or chase time point). Do not allow beads to dry as this greatly reduces their binding capacity.
2. Place the tubes into a magnetic stand designed for 1.5 mL tubes to concentrate beads at the tube wall and remove the supernatant.
3. Wash the beads with 1 mL of washing buffer A for 5 min with rotation on an orbital shaker and remove the supernatant by using the magnetic rack.
4. Repeat washing step with 1 mL of buffer A as in Subheading 3.4, **step 3**.
5. Repeat washing step with 1 mL of buffer C.
6. Repeat washing step with 1 mL of buffer B.
7. Add 300 μL of beads blocking buffer and incubate 1 h at 4 °C with rotation on an orbital shaker.

8. Carefully remove the supernatant by using the magnetic rack.
9. Add 310 μL of purified nucleosomes at a concentration of 1 OD_{A276} (from Subheading 3.3, 24) and incubate overnight at 4 °C with rotation.
10. The next day, place the tubes on a magnetic rack, allow the beads to concentrate and collect the unbound fraction. Store this fraction as a control, *see* **Note 7**.
11. Wash the beads 3 times with 1 mL washing buffer B for 5 min at room temperature with rotation on an orbital shaker.
12. Between every wash step remove the supernatant on a magnetic rack.
13. Add 310 μL of RNA removal buffer supplemented with 100 μg/mL of RNaseA to the samples and 270 μL of the same buffer to the previously collected input sample (see Subheading 3.3, **step 21**).
14. Incubate the samples for 10 min at room temperature.
15. Resuspend the samples with the beads and collect 10 μL for western blot analysis. This step is required to assess the efficiency of the biotin pulldown.
16. Freeze the rest of the beads as well as the input samples (300 μL each) in −80 °C for further processing.

(protocol can be suspended at this point)

3.5 Immunoblotting

In brief, to assess the efficiency of the biotin pulldown, samples (collected at Subheading 3.4, **step 15**) can be separated by SDS-PAGE and probed by immunoblotting using standard procedures. Our SNAP-tagged constructs carry a triple hemagglutinin (HA) epitope tag for this purpose. Samples can be separated on an SDS-PAGE gel (8–12%, depending on the expected size. For histones we typically use 12%) and transferred to Hybond PVDF membranes. Blots are probed with a monoclonal mouse anti-HA antibody at 1:2000 dilution and an anti-mouse secondary antibody at 1:10000 dilution. Fluorescence is detected and quantified with an Odyssey Image Analyzer. Alternatively, standard HRP-conjugated secondary antibodies and ECL detection on photosensitive film can be used.

(protocol can be suspended at this point)

3.6 DNA Purification

1. Add 300 μL of phenol:chloroform:isoamyl alcohol (25:24:1) solution (use buffered phenol at pH 8, *see* **Note 8**) to samples collected in Subheading 3.4, **step 16** and mix by inverting the tubes.
2. Centrifuge at 20,000 × *g* for 5 min at room temperature.
3. Remove the upper aqueous phase and transfer to a new tube.
4. Add 300 μL of chloroform:isoamyl alcohol (24:1) solution and mix.

5. Centrifuge at 20,000 × *g* for 5 min at room temperature.
6. Remove the upper aqueous phase and purify the DNA with Qiagen MinElute Reaction Cleanup Kit according to the manufacturer's instructions (the use of this kit is required to obtain sufficient DNA purity for subsequent high-throughput sequencing steps).
7. Elute DNA in 30 μL TE buffer.
8. Determine the concentration of the recovered DNA with Quant-iT PicoGreen dsDNA assay kit to assess the efficiency of DNA purification.

   (protocol can be suspended at this point)

3.7 Quantitative PCR

Perform all quantitative PCR reactions with technical triplicates to ensure high accuracy of the measurement.

1. Perform tenfold serial dilutions on purified input samples up to 10^{-6}. This step is required to calculate the percentage input as well as to measure the linear response of the quantitative PCR reaction.
2. Prepare a master mix of PerfeCTa SYBR Green FastMix ROX (Quanta) for each desired primer pair. Account for IP sample as well as input DNA dilutions (up to 10^{-6}) with technical triplicates for each reaction for all performed chase times. Primers are used at a concentration of 300 nM. Store the master mix in the dark at 4 °C.
3. Pipette 6 μL of master mix (containing SYBR Green FastMix and primers) for a desired number of reactions into a quantitative PCR plate.
4. To the distributed primer master mixtures pipette 4 μL of each input dilution and pulled down DNA.
5. Seal the plate with optical sealing film. Use gloves to ensure the seal remains clean, not to perturb the qPCR measurement.
6. Centrifuge the plate at 500 × *g* for 30 s at room temperature to ensure PCR volume is homogeneous and concentrated at the bottom of the plate.
7. Perform a real-time PCR measurement. For our experiments we used the 7900HT Fast Real-time PCR System from Applied Biosystems.
8. Perform the quantification of percentage input by the following formula:

100×2^[(Ct_{input} × Log2(dilution factor)-(Ct_{IP})] where dilution factor is 10 in our experiments.

3.8 Preparation of time-ChIP-qPCR Calibration Curve

To obtain a quantitative measure of the fraction of retained nucleosomes or the turnover rate a time-ChIP standard curve should be generated.

1. Perform SNAP-labeling as in point Subheading 3.2 but without any chase period.
2. At the same time mock label a portion of cells with DMSO.
3. Mix defined numbers of in vivo biotin labeled cells with the mock treated cells in a twofold ratio series (100%, 50%, 25%, 12,5%, etc. of biotinylated cells).
4. Perform the subsequent steps: soluble nucleosome preparation, biotin-labeled SNAP-nucleosomes purification, DNA purification, quantitative PCR measurements of percentage input as described above for all twofold ratio series.
5. Make a standard curve of percent input for the different ratio mixes (as in Fig. 3b) and define linear range.
6. Interpolate the time-ChIP-qPCR data obtained in Subheading 3.7 on the time-ChIP calibration curve (*see* Fig. 3b for example) to obtain the fraction of retained nucleosomes or the turnover rates (by interpolation on the linear regression line) (*see* Fig. 3c for example).

3.9 Deep Sequencing

Sequencing can be performed on any platform. In our case, library preparation and sequencing was performed at the EMBL core facility in Heidelberg. DNA material from Subheading 3.6, **step 7** was used to generate sequencing libraries with NEBNext ChIP-Seq Library Prep Reagent Set for Illumina (NEB). Please note that other methods to generate sequencing libraries can be used. Sequencing was performed on a HiSeq2000, every chase time point on an individual lane, in a single end mode with 50 bp read length (200 million reads per sample).

1. After obtaining the FASTQ files resulting from sequencing, test the quality of sequencing with the FASTQC program.
2. Filter out the reads with average quality below 20 with FASTX-Toolkit.
3. Remove the sequencing adaptors with cutadapt [44].
4. Align the data for each time point to the human reference genome (GRCh38) using bowtie2 [45].
5. Remove the duplicated reads with rmdup [46].
6. Combine the sequencing reads into 1000 bp bins, normalize to read count and subtract the input sequencing data from all the time points of the chase experiment with bamcompare [47].

7. Obtain chosen genomic coordinate data from ENCODE or similar database. In our analysis we downloaded ChIP-seq peak coordinates for H3K9me3 (ENCFF712ATO), H3K9ac (ENCFF510LKP), H3K27me3 (ENCFF512TQI), H3K27ac (ENCFF392EDT) from ENCODE. For intersection analysis we used the top 10% peaks based on the peak score [48].
8. Calculate the average number of normalized reads falling inside each genomic domain and each chase time point to obtain access to dynamic properties of the selected chromatin feature (*see* Fig. 5a).
9. Generate a bigwig output file from the bamcompare program for direct visualization with genome viewer. We used Integrative Genomics Viewer (IGV) for this purpose [49], (*see* Fig. 5b).

4 Notes

1. Details for synthesis of CP-Biotin. CP-NH_2 (2.9 mg, 11.0 μmol) is dissolved in anhydrous DMF (1 mL). Biotin amidohexanoic acid *N*-hydroxysuccinimide ester (5.0 mg, 11.0 μmol) and triethylamine (1.5 μL, 11.0 μmol) is added and the reaction mixture stirred overnight at room temperature. The solvent is removed under vacuum and the product purified by reversed-phase high-performance liquid chromatography (RP-HPLC). RP-HPLC is performed on Agilent 1100 semi-preparative scale purification system on a VYDAC 218TP series C18 polymeric reversed-phase column (22 × 250 mm, 10 μm particle size) at a flow rate of 20 mL/min with a binary water/acetonitrile gradient and monitored by UV-visible absorbance at 280 nm. Yield: 52%. Mass spectrum is recorded by electrospray ionization (ESI) on an Agilent 6210 Time-of-Flight (TOF) system. ESI-MS m/z 604.2454 $[M-H]^+$ (calc. For $C_{28}H_{38}ClN_7O_4S$, m/z 604.2468).
2. In the described time-ChIP protocol we are measuring the dissociation kinetics of histones in a pulse-chase setup. By preincubating cells with an unconjugated SNAP substrate (SNAP-Cell Block, NEB), preexisting SNAP can be quenched. Following a period of new histone synthesis, nascent histones can be selectively biotinylated. We previously used this so-called quench-chase-pulse assay in the context of fluorescent pulse labeling of histones [23, 28, 29]. This strategy can in principle be adopted to measure the association kinetics of a chosen histone variant by time-ChIP.
3. Digitonin recrystallization. Digitonin (0.1 g, 50% purity) was dissolved in 1.3 mL of ethanol (96%) at 75 °C. Next, the solution was incubated on ice for 15 minutes and centrifuged

10,000 × *g* for 5 min at 4 °C. The ethanol solution was removed and the procedure was repeated two times. The pellet was dried in a vacuum concentrator for 10 min. The dried pellet was dissolved in cell homogenization buffer and used in the cell lysis procedure.

4. Here we describe a "native chromatin" time-ChIP protocol with MNase treatment as a method to liberate mononucleosomes from chromatin. Alternatively, chromatin can be formaldehyde crosslinked and solubilized by sonication. We piloted this alternative protocol and show by western blot that it can be effectively used in combination with time-ChIP (*see* Fig. 6 for more details).

5. MNase treatment should lead to chromatin samples highly enriched for mononucleosomes. This is a crucial step of the protocol and subsequent procedures depend on the quality of this digestion. A titration experiment should be performed

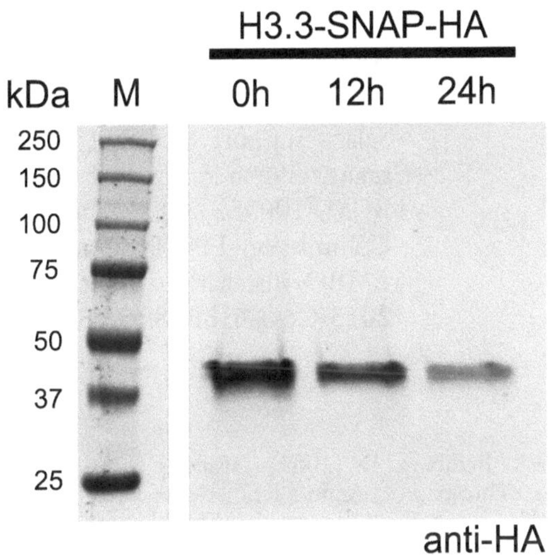

Fig. 6 time-ChIP can be used with a chromatin crosslinking protocol. Cells expressing H3.3-SNAP-HA were biotinylated and chased for 0, 12, and 24 h in culture. Equal numbers of cells ware collected for each time point. Biotinylated H3.3-SNAP cells were crosslinked with 1% formaldehyde, fragmented by sonication, and subjected to a pull down on streptavidin magnetic beads. The purified material was de-crosslinked by overnight incubation at 65 °C. Samples were separated by SDS-PAGE and analyzed by western blot by probing for the HA tag. We find a biotin-specific isolation of H3.3-SNAP chromatin whose signal diminished during the chase period, indicating turnover

with varying MNase concentration to determine the optimal MNase conditions to maximize the mononucleosome yield (*see* Fig. 1b for example of titration).

6. To assess the yield of soluble chromatin by MNase, collect the pellet fraction after enzyme treatment. Perform DNA isolation and compare the amounts of extracted DNA to the DNA from the soluble input fraction with Quant-iT PicoGreen dsDNA assay kit.
7. To assess the binding efficiency of magnetic beads perform a western blot to compare the amount of unbound, input and pull down fractions (*see* Fig. 1c).
8. The pH 8 of the phenol:chlorophorm:isoamyl alcohol (25:24:1) is crucial for efficient DNA isolation.

Acknowledgments

We thank João Mata (IGC) for technical support, Brenda Baker (New England Biolabs) for assistance with synthesis of CP-Biotin. Bianka Baying and Vladimir Benes (EMBL genecore) for NGS support. We are grateful to Alekos Athanasiadis (IGC) and EpiLab members for helpful comments on the manuscript.

This work is supported by Fundação para a Ciência e a Tecnologia (FCT) postdoctoral fellowship SFRH/BPD/117179/2016 (to WS) and doctoral fellowship SFRH/BD/33567/2008 (to MRG). Salary support to LETJ is provided by an "Investigador FCT" grant. Further support was provided by FCT grants BIA-BCM/100557/2008 and BIA-PRO/100537/2008, European Commission FP7 Program, Marie Curie Reintegration grant, an EMBO installation grant and an ERC consolidator grant ERC-2013-CoG-615638 to LETJ.

References

1. Campos EI, Reinberg D (2009) Histones: annotating chromatin. Annu Rev Genet 43:559–599. https://doi.org/10.1146/annurev.genet.032608.103928
2. Ragunathan K, Jih G, Moazed D (2015) Epigenetics. Epigenetic inheritance uncoupled from sequence-specific recruitment. Science 348:1258699. https://doi.org/10.1126/science.1258699
3. Audergon PNCB, Catania S, Kagansky A et al (2015) Restricted epigenetic inheritance of H3K9 methylation. Science 348:132–135. https://doi.org/10.1126/science.1260638
4. Hansen KH, Bracken AP, Pasini D et al (2008) A model for transmission of the H3K27me3 epigenetic mark. Nat Cell Biol 10:1291–1300. https://doi.org/10.1038/ncb1787
5. Margueron R, Justin N, Ohno K et al (2009) Role of the polycomb protein EED in the propagation of repressive histone marks. Nature 461:762–767. https://doi.org/10.1038/nature08398
6. Berry S, Hartley M, Olsson TSG et al (2015) Local chromatin environment of a Polycomb target gene instructs its own epigenetic inheritance. elife 4:e07205. https://doi.org/10.7554/eLife.07205

7. Dodd IB, Micheelsen MA, Sneppen K, Thon G (2007) Theoretical analysis of epigenetic cell memory by nucleosome modification. Cell 129:813–822. https://doi.org/10.1016/j.cell.2007.02.053
8. Alabert C, Barth TK, Reverón-Gómez N et al (2015) Two distinct modes for propagation of histone PTMs across the cell cycle. Genes Dev 29:585–590. https://doi.org/10.1101/gad.256354.114
9. Akiyama T, Suzuki O, Matsuda J, Aoki F (2011) Dynamic replacement of histone H3 variants reprograms epigenetic marks in early mouse embryos. PLoS Genet 7:e1002279. https://doi.org/10.1371/journal.pgen.1002279
10. Ng RK, Gurdon JB (2008) Epigenetic memory of an active gene state depends on histone H3.3 incorporation into chromatin in the absence of transcription. Nat Cell Biol 10:102–109. https://doi.org/10.1038/ncb1674
11. Jackson V (1990) In vivo studies on the dynamics of histone-DNA interaction: evidence for nucleosome dissolution during replication and transcription and a low level of dissolution independent of both. Biochemistry 29:719–731. https://doi.org/10.1021/bi00455a019
12. Xu M, Long C, Chen X et al (2010) Partitioning of histone H3-H4 tetramers during DNA replication–dependent chromatin assembly. Science 328:94–98. https://doi.org/10.1126/science.1178994
13. Alabert C, Bukowski-Wills J-C, Lee S-B et al (2014) Nascent chromatin capture proteomics determines chromatin dynamics during DNA replication and identifies unknown fork components. Nat Cell Biol 16:281–293. https://doi.org/10.1038/ncb2918
14. Ahmad K, Henikoff S (2001) Centromeres are specialized replication domains in heterochromatin. J Cell Biol 153:101–110
15. Ahmad K, Henikoff S (2002) The histone variant H3.3 marks active chromatin by replication-independent nucleosome assembly. Mol Cell 9:1191–1200. https://doi.org/10.1016/S1097-2765(02)00542-7
16. Tran V, Lim C, Xie J, Chen X (2012) Asymmetric division of Drosophila male germline stem cell shows asymmetric histone distribution. Science 338:679–682. https://doi.org/10.1126/science.1226028
17. Kimura H, Cook PR (2001) Kinetics of Core histones in living human cells little exchange of H3 and H4 and some rapid exchange of H2b. J Cell Biol 153:1341–1354. https://doi.org/10.1083/jcb.153.7.1341
18. Falk SJ, Guo LY, Sekulic N et al (2015) CENP-C reshapes and stabilizes CENP-A nucleosomes at the centromere. Science 348:699–703. https://doi.org/10.1126/science.1259308
19. Liu J, Vidi P-A, Lelievre SA, Irudayaraj JMK (2015) Nanoscale histone localization in live cells reveals reduced chromatin mobility in response to DNA damage. J Cell Sci 128:599–604. https://doi.org/10.1242/jcs.161885
20. Damoiseaux R, Keppler A, Johnsson K (2001) Synthesis and applications of chemical probes for human O6-alkylguanine-DNA alkyltransferase. Chembiochem A Eur J Chem Biol 2:285–287
21. Keppler A, Pick H, Arrivoli C et al (2004) Labeling of fusion proteins with synthetic fluorophores in live cells. Proc Natl Acad Sci U S A 101:9955–9959. https://doi.org/10.1073/pnas.0401923101
22. Keppler A, Gendreizig S, Gronemeyer T et al (2003) A general method for the covalent labeling of fusion proteins with small molecules in vivo. Nat Biotechnol 21:86–89. https://doi.org/10.1038/nbt765
23. Jansen LET, Black BE, Foltz DR, Cleveland DW (2007) Propagation of centromeric chromatin requires exit from mitosis. J Cell Biol 176:795–805. https://doi.org/10.1083/jcb.200701066
24. Silva MCC, Bodor DL, Stellfox ME et al (2012) Cdk activity couples epigenetic centromere inheritance to cell cycle progression. Dev Cell 22:52–63. https://doi.org/10.1016/j.devcel.2011.10.014
25. Carroll CW, Silva MCC, Godek KM et al (2009) Centromere assembly requires the direct recognition of CENP-A nucleosomes by CENP-N. Nat Cell Biol 11:896–902. https://doi.org/10.1038/ncb1899
26. Stankovic A, Guo LY, Mata JF et al (2017) A dual inhibitory mechanism sufficient to maintain cell-cycle-restricted CENP-A assembly. Mol Cell 65:231–246. https://doi.org/10.1016/j.molcel.2016.11.021
27. Bergmann JH, Rodríguez MG, Martins NMC et al (2011) Epigenetic engineering shows H3K4me2 is required for HJURP targeting and CENP-A assembly on a synthetic human kinetochore. EMBO J 30:328–340. https://doi.org/10.1038/emboj.2010.329
28. Ray-Gallet D, Woolfe A, Vassias I et al (2011) Dynamics of histone H3 deposition in vivo reveal a nucleosome gap-filling mechanism for H3.3 to maintain chromatin integrity. Mol Cell 44:928–941. https://doi.org/10.1016/j.molcel.2011.12.006

29. Bodor DL, Rodríguez MG, Moreno N, Jansen LET (2012) Analysis of protein turnover by quantitative SNAP-based pulse-chase imaging. Curr Protoc Cell Biol Chapter 8:Unit8.8. https://doi.org/10.1002/0471143030.cb0808s55
30. Bodor DL, Valente LP, Mata JF et al (2013) Assembly in G1 phase and long-term stability are unique intrinsic features of CENP-A nucleosomes. Mol Biol Cell 24:923–932. https://doi.org/10.1091/mbc.E13-01-0034
31. Dion MF, Kaplan T, Kim M et al (2007) Dynamics of replication-independent histone turnover in budding yeast. Science 315:1405–1408. https://doi.org/10.1126/science.1134053
32. Huang C, Zhang Z, Xu M et al (2013) H3.3-H4 tetramer splitting events feature cell-type specific enhancers. PLoS Genet 9:e1003558. https://doi.org/10.1371/journal.pgen.1003558
33. Katan-Khaykovich Y, Struhl K (2011) Splitting of H3-H4 tetramers at transcriptionally active genes undergoing dynamic histone exchange. Proc Natl Acad Sci 108:1296–1301. https://doi.org/10.1073/pnas.1018308108
34. Deal RB, Henikoff JG, Henikoff S (2010) Genome-wide kinetics of nucleosome turnover determined by metabolic labeling of histones. Science 328:1161–1164. https://doi.org/10.1126/science.1186777
35. Mito Y, Henikoff JG, Henikoff S (2005) Genome-scale profiling of histone H3.3 replacement patterns. Nat Genet 37:1090–1097. https://doi.org/10.1038/ng1637
36. Mito Y, Henikoff JG, Henikoff S (2007) Histone replacement marks the boundaries of cis-regulatory domains. Science 315:1408–1411. https://doi.org/10.1126/science.1134004
37. Verzijlbergen KFKF, Menendez-Benito V, Van Welsem T et al (2010) Recombination-induced tag exchange to track old and new proteins. Proc Natl Acad Sci 107:64–68. https://doi.org/10.1073/pnas.0911164107
38. Radman-Livaja M, Verzijlbergen KF, Weiner A et al (2011) Patterns and mechanisms of ancestral histone protein inheritance in budding yeast. PLoS Biol 9:e1001075. https://doi.org/10.1371/journal.pbio.1001075
39. Corrêa IR, Baker B, Zhang A et al (2013) Substrates for improved live-cell fluorescence labeling of SNAP-tag. Curr Pharm Des 19:5414–5420
40. Sun X, Zhang A, Baker B et al (2011) Development of SNAP-tag Fluorogenic probes for wash-free fluorescence imaging. Chembiochem 12:2217–2226. https://doi.org/10.1002/cbic.201100173
41. Corrêa IR (2015) Considerations and protocols for the synthesis of custom protein labeling probes. In: Site-specific protein labeling, Methods and protocols. Humana Press, New York, pp 55–79
42. Jasencakova Z, Scharf AND, Ask K et al (2010) Replication stress interferes with histone recycling and predeposition marking of new histones. Mol Cell 37:736–743. https://doi.org/10.1016/j.molcel.2010.01.033
43. Deaton AM, Gómez-Rodríguez M, Mieczkowski J et al (2016) Enhancer regions show high histone H3.3 turnover that changes during differentiation. elife 5:e15316. https://doi.org/10.7554/eLife.15316
44. Martin M (2011) Cutadapt removes adapter sequences from high-throughput sequencing reads. EMBnetjournal 17:10. https://doi.org/10.14806/ej.17.1.200
45. Langmead B, Salzberg SL (2012) Fast gapped-read alignment with bowtie 2. Nat Methods 9:357–359. https://doi.org/10.1038/nmeth.1923
46. Li H, Handsaker B, Wysoker A et al (2009) The sequence alignment/map format and SAMtools. Bioinformatics 25:2078–2079. https://doi.org/10.1093/bioinformatics/btp352
47. Ramirez F, Dundar F, Diehl S et al (2014) deepTools: a flexible platform for exploring deep-sequencing data. Nucleic Acids Res 42:W187–W191. https://doi.org/10.1093/nar/gku365
48. Dunham I et al (2012) An integrated encyclopedia of DNA elements in the human genome. Nature 489:57–74. https://doi.org/10.1038/nature11247
49. Robinson JT, Thorvaldsdóttir H, Winckler W et al (2011) Integrative genomics viewer. Nat Biotechnol 29:24–26. https://doi.org/10.1038/nbt.1754

Chapter 8

MINCE-Seq: Mapping In Vivo Nascent Chromatin with EdU and Sequencing

Srinivas Ramachandran and Steven Henikoff

Abstract

The epigenome has been mapped in different cell types to understand the relationship between the chromatin landscape and the control of gene expression. Most mapping studies profile a large population of cells in various stages of the cell cycle, which results in an average snapshot of the chromatin landscape. However, chromatin is highly dynamic, undergoing rapid changes during active processes such as replication, transcription, repair, and remodeling. Hence, we need methods to map chromatin as a function of time. To address this problem in the context of replication, we developed the method MINCE-seq (Mapping In vivo Nascent Chromatin with EdU and sequencing). MINCE-seq is a genome-wide method that uses the passage of replication fork as a starting point to map the chromatin landscape as a function of time. MINCE-seq can measure chromatin dynamics in a time scale of minutes and at the resolution of individual nucleosome positions and transcription factor-binding sites genome-wide.

Key words Replication, Chromatin dynamics, Transcription factors, Genomics, Click chemistry

1 Introduction

Chromatin states are defined by the position and identity of proteins bound to the genome, including transcription factors (TFs) and nucleosomes. Chromatin states reflect genome function as evidenced by recent epigenome mapping efforts in many cell types using short-read sequencing [1]. Epigenome mapping is usually performed on a population of cells that are undergoing various states of activity. Hence, the chromatin states mapped are ensemble averages. However, chromatin is disrupted and reassembled constantly on time-scales of seconds to minutes due to active processes such as remodeling, replication, repair, and transcription.

Every protein-DNA contact in the genome is broken at least once during the cell cycle when the replicative helicase unwinds double-stranded DNA to supply single-stranded DNA to the DNA polymerase for replication. Thus, replication represents a starting point for the establishment of TF and nucleosome landscapes

Guillermo A. Orsi and Geneviève Almouzni (eds.), *Histone Variants: Methods and Protocols*, Methods in Molecular Biology, vol. 1832, https://doi.org/10.1007/978-1-4939-8663-7_8, © Springer Science+Business Media, LLC, part of Springer Nature 2018

genome-wide on newly replicated naked DNA. Furthermore, because replication disrupts the pre-existing chromatin landscape, replication represents a challenge to the cell to maintain specific chromatin landscapes through successive cell divisions. To understand how distinctive chromatin landscapes are maintained through replication, we need to be able to map the nascent chromatin landscape established right after replication and then track its maturation over time.

Epigenome mapping has been performed using a variety of DNA enzymes including methyl transferases [2], transposases [3, 4], and nucleases [5, 6]. One enzyme that has been used for decades for mapping nucleosomes, and more recently for TFs, is Micrococcal Nuclease (MNase), an endo-exo nuclease from *Staphyloccoccus aureus*. MNase, when added to intact nuclei in the presence of Ca^{2+}, digests away unbound DNA to generate minimal protections of protein-DNA complexes. Subjecting DNA fragments obtained after MNase treatment of nuclei to paired-end short-read sequencing enables determination of both the length of the protected fragment and its position in the genome. Because TFs usually protect short DNA fragments (<50 bp) and nucleosomes protect ~150 bp, computationally filtering the sequence data based on fragment lengths yields genome-wide maps of both TFs and nucleosomes from a single experiment. These maps show distinct patterns of nucleosomes and TFs at promoters, enhancers, and replication origins (Fig. 1).

To map the newly replicated chromatin landscape, we developed a novel method, "Mapping *In vivo* Nascent Chromatin with EdU and sequencing" (MINCE-seq) [7]. In MINCE-seq, we combine high-resolution mapping of the chromatin landscape using MNase with metabolic labeling with Ethynyl-deoxyUridine (EdU) [8]. EdU has an alkyne group that can be covalently attached to an azide group ex vivo using the bio-orthogonal click chemistry method. EdU is incorporated by active replication forks in place of thymidine, enabling high-stringency purification of newly replicated DNA genome-wide from asynchronous cultures. EdU and MNase can be used in most metazoan cell types, making MINCE-seq highly transferrable to different experimental systems. EdU labeling can be followed by a thymidine chase for various time intervals, thus enabling us to track the chromatin landscape as a function of time after replication genome-wide.

Applying MINCE-seq to Drosophila S2 and BG3 cell lines, we observed that nucleosomes replace TFs at active promoters and enhancers genome-wide due to the action of the replication-coupled chromatin assembly machinery. This finding implies that the steady-state chromatin landscape at active regions is established by a competition between transcription factors and nucleosomes. We present here the detailed experimental methodology of MINCE-seq (schematic in Fig. 2) that should be applicable to any desired cell type.

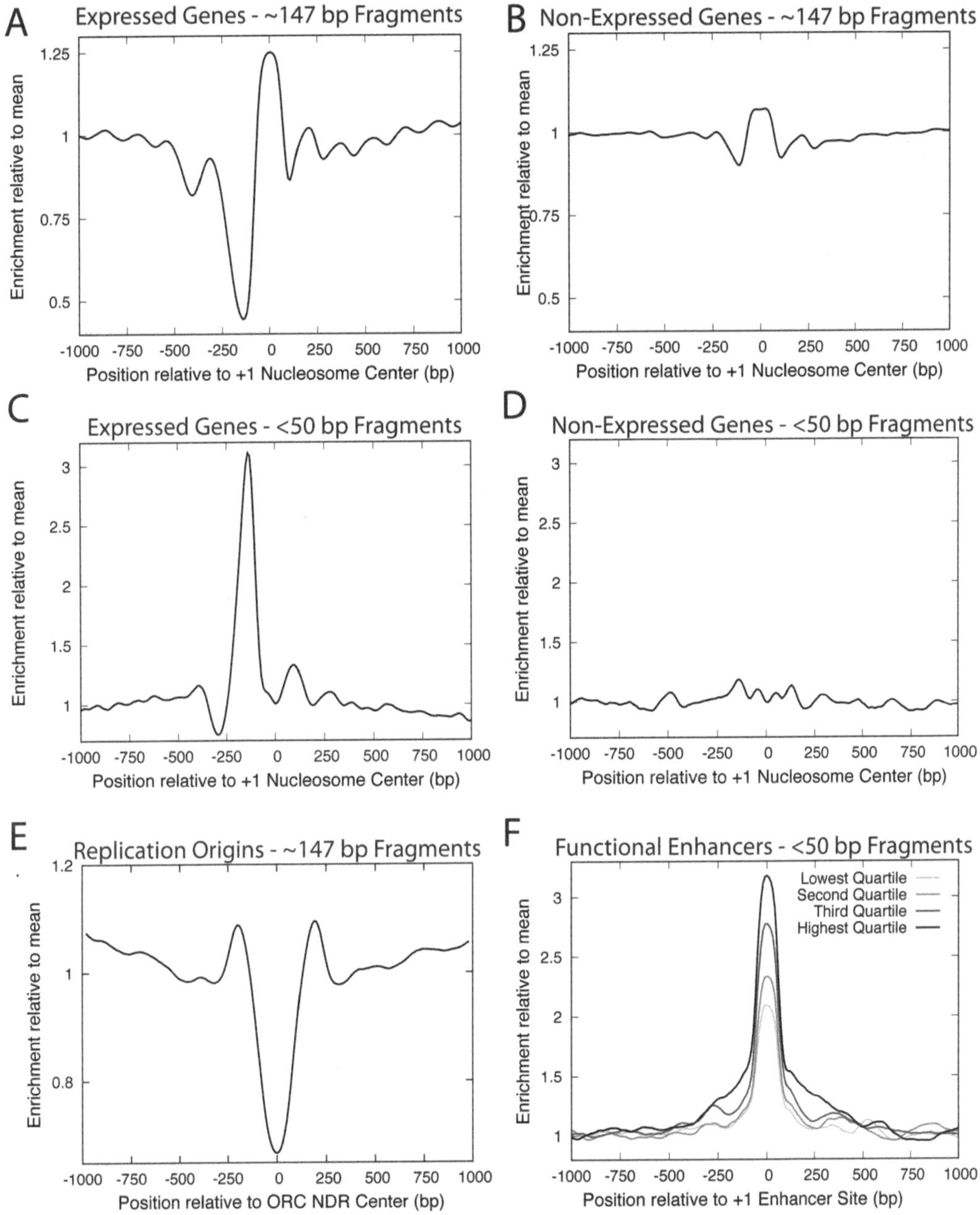

Fig. 1 Chromatin landscapes reflect genome function. (**a**) Average profile of 147 ± 5 bp fragments from MNase-seq data from ***Drosophila*** S2 cells (combined data from GEO accessions GSM1974516 and GSM1974518) over promoters of active genes show an array of nucleosomes upstream and downstream of the promoters, and nucleosome depletion at the promoters. (**b**) Average profile of 147 ± 5 bp fragments from MNase-seq data over promoters of inactive genes shows no nucleosome depletion at promoters and no nucleosome arrays upstream or downstream of the promoters. (**c**) Average profile of 25–50 bp fragments from MNase-seq data over promoters of active genes shows a peak with high enrichment coinciding with the nucleosome-depleted region. (**d**) Average profile of 25–50 bp fragments from MNase-seq data over promoters Fig. 1 (continued) of inactive genes shows no peak at the promoter. (**e**) Average profile of 147 ± 5 bp fragments from MNase-seq data over origin recognition complex (ORC) binding sites shows nucleosome depletion. ORC binding sites correspond to origins of replication. (**f**) Average profile of 25–50 bp fragments from MNase-seq data over functional enhancers (as defined by STARR-seq) shows a peak whose enrichment correlates with quartiles of enhancer strength; the highest quartile of enhancer strength has the peak with highest enrichment at the enhancer site followed by second, third, and then the lowest quartiles

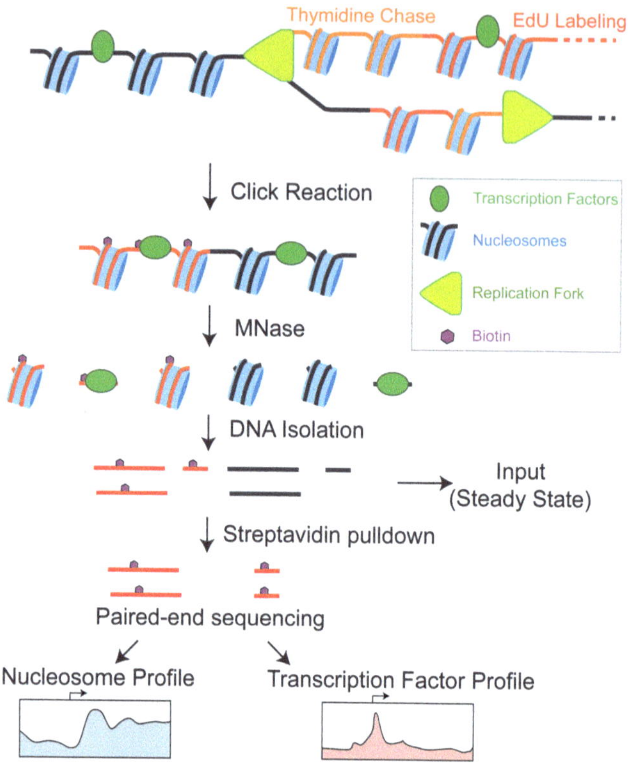

Fig. 2 Schematic of the MINCE-seq protocol

2 Materials

2.1 EdU Labeling and Click Reaction

1. Mid-log-phase cultured cells (*see* **Note 1**).
2. 10 mM 5-ethynyl-2′-deoxyuridine solution in dimethyl sulfoxide (DMSO). Can be stored at −20 °C.
3. 20 mM thymidine solution in DMSO.
4. Phosphate-buffered saline (PBS): 11.9 mM phosphate pH 7.4, 137 mM NaCl, 2.7 mM KCl. Keep ~500 mL of PBS in ice before starting the experiment for cold washes. For formaldehyde fixation and permeabilization with Triton X-100, keep some PBS at room temperature.
5. Formaldehyde—37% or 16% solutions—dilute to 1% in PBS just before cross-linking.
6. 1.25 M Glycine.
7. Triton X-100.
8. 30% BSA (w/v).
9. 100 μM Biotin TEG azide solution in DMSO.

10. Sodium ascorbate—freshly prepared solution of 10 mM in water.
11. Copper sulfate—freshly prepared solution of 2 mM in water.

2.2 Chromatin Preparation and DNA Isolation

1. Lysis buffer: 1% sodium dodecyl sulfate (SDS) w/v, 0.5 M sodium chloride, 100 mM Tris-HCl pH 8, 2 mM ethylenediaminetetraacetic acid (EDTA) in water.
2. Dilution buffer: 1% Triton X-100 (v/v), 3 mM calcium chloride, 2 mM EDTA in water.
3. Micrococcal nuclease, 0.2 Units/μL stock solution in water.
4. 0.2 M ethylene glycol-bis(β-aminoethyl ether)-tetraacetic acid (EGTA).
5. 10% SDS (w/v).
6. Proteinase K (Thermo Fisher Scientific, cat. no. EO0492).
7. Phenol-chloroform-isoamyl alcohol 25:24:1 (v/v).
8. RNAse A (DNAse- and Protease-free) 10 mg/mL solution.
9. 300 mM sodium acetate.
10. 20 mg/mL glycogen from mussels (Roche).
11. Isopropanol.
12. TE buffer: 10 mM Tris-HCl pH 8, 1 mM EDTA in water.
13. 200 proof ethanol.

2.3 Isolation of Biotinylated DNA

1. 2× Wash and Bind (2× WB) buffer in water (dilute this buffer 1:1 with water to obtain 1× WB buffer): 10 mM Tris-HCl pH 7.5, 1 mM EDTA, 2 M NaCl, 0.2% Tween 20.
2. Streptavidin M-280 dynabeads (Invitrogen).

2.4 General Equipment

1. Refrigerated centrifuges that can accommodate 15 mL tubes at speeds up to 1100 rcf and tabletop refrigerated centrifuges that can accommodate 1.5 mL tubes at speeds up to 16,000 rcf.
2. Nutating and end-over-end mixers.
3. Probe sonicator.
4. Heating blocks/water baths at 37 °C and 65 °C.

3 Method

1. Culture cells to mid-log-phase at the start of the experiment (For S2 cells, 4–6 × 10^6 cells/mL of media). The scale of the experiment with S2 cells is 2 × 150 cm^2 flasks with 30 mL of media in each, for a total of 240–360 × 10^6 cells.

2. Start the experiment by adding EdU to the cells in the culture flask to a final concentration of 10 μM (1:1000 dilution from the stock).
3. Replace the cells in the incubator for the duration of EdU labeling. EdU labeling for *Drosophila melanogaster* S2 and BG3 cells is performed for 10 min (*see* **Note 2**).
4. After EdU labeling, if performing thymidine chase, quickly change the media (*see* **Note 3**). For S2 and BG3 cells, which are semi-adherent, the media collected from the flask is spun-down in an aseptic polypropylene tube at 600 rcf for 3 min. The supernatant is discarded and the pelleted cells are resuspended in fresh media and added back to the flask. Thymidine is then added to a final concentration of 20 μM (1:1000 dilution from the stock).
5. After the desired time of chase (or if not performing chase, immediately after EdU labeling), harvest cells and pellet them in a polypropylene tube at 600 rcf for 3 min.
6. Wash cells once with cold PBS.
7. Resuspend the cells with 10 mL of 1% formaldehyde in PBS in a 15 mL polypropylene conical tube. Cross-link cells for 15 min in a nutating mixer at room temperature.
8. Quench with glycine: Add glycine to a final concentration of 250 mM, 1:5 dilution from the stock and mix in a nutating mixer for 5 min.
9. Pellet the cross-linked cells at 1100 rcf and wash three times with PBS.
10. Permeabilize cells by resuspending the cell pellet in 4 mL of PBS + 0.25% Triton X-100 and placing in a nutating mixer for 30 min at room temperature.
11. Pellet cells at 1100 rcf for 4 min and wash once with 4 mL of cold 0.5% BSA in PBS.
12. Assemble the click reaction by resuspending cells in PBS and adding the click reagents in the following order (final volume of reaction is 450 μL): Biotin TEG azide, sodium ascorbate, and finally copper sulfate. Make the sodium ascorbate and copper sulfate stock solutions fresh just before setting up the click reaction. The stock and final concentrations of click reagents are presented in Table 1. Place on a nutating mixer for 1 h.
13. Pellet the cells at 1100 rcf for 4 min and discard the supernatant.
14. Wash the cells three times with 5 mL of cold PBS to thoroughly remove copper.

Table 1
Click reaction recipe

Reagent	Stock concentration	Required concentration	Volume of stock to be added
PBS	N/A	N/A	414 μL
Biotin azide	2 mM	100 μM	22.5 μL
Sodium ascorbate	1 M	10 mM	4.5 μL
Copper sulfate	100 mM	2 mM	9 μL
Total volume			*450 μL*

15. Resuspend the cells gently in 200 μL of cold lysis buffer and incubate for 10 min on ice.
16. Add 1.8 mL of cold dilution buffer to the cells and mix well (total 2 mL).
17. Sonicate the lysed cells to solubilize the chromatin in ice using a probe sonicator (we have used a Branson digital 200 W sonifier) at 30% intensity, 2.5 s ON and 5 s OFF repeating for 40 s total.
18. Split into four 0.5 mL samples in 1.5 mL tubes and incubate at 37 °C for 5 min.
19. To each of the four samples add 1, 2, 4, and 8 μL of Micrococcal Nuclease (concentration of 0.2 U/μL) respectively and incubate for 10 min at 37 °C.
20. Quench the nuclease reaction by adding EGTA to a final concentration of 2 mM (1:100 dilution of 0.2 M stock).
21. Add Proteinase K (20 μg), SDS to a final concentration of 1% (1:10 dilution of stock) and mix well.
22. Incubate overnight at 65 °C to reverse the crosslinks and degrade the protein.
23. The next day, add an equal volume of Phenol:Chloroform:Isoamyl alcohol mix (25:24:1 ratio) and spin at 16,000 rcf in a tabletop centrifuge for 10 min.
24. Transfer the aqueous phase into a new tube.
25. Add 2 μL of RNase A. Incubate at 37 °C for 15 min to degrade the RNA molecules in the sample.
26. Add an equal volume of phenol:chloroform:isoamyl alcohol mix (25:24:1 ratio). Mix well by pipetting and then spin at 16,000 rcf in a tabletop centrifuge for 10 min.
27. Separate the aqueous phase and add sodium acetate to a final concentration of 30 mM (dilute from 300 mM stock solution), 2 μL of glycogen, and isopropanol to a final concentration of 70%. Mix well by inverting the tube several times.

28. Precipitate the DNA by spinning at 16,000 rcf at 4 °C for 30 min in a tabletop centrifuge.
29. Discard the supernatant and wash the pellet with 75% ethanol (v/v) in TE buffer: to the DNA pellet, first add 750 μL of 200-proof ethanol and then add 250 μL TE. Mix well by inverting the tube several times and precipitate by spinning at 16,000 rcf at 4 °C for 15 min.
30. Discard the supernatant and air-dry the DNA pellet. Resuspend the dry pellet in 400 μL TE; take a 25 μL aliquot to serve as input, which can be diluted to a final volume of 100 μL (*see* **Note 4**).
31. Analyze the size of the input DNA by 2% TBE agarose gel electrophoresis to determine the MNase concentrations at which di- and tri-nucleosomes are preserved. These samples can be used for streptavidin pull-down (*see* **Note 5**).
32. To the remaining DNA (in 375 μL) add an equal volume of 2× WB buffer.
33. Use 25 μL of streptavidin magnetic beads for each sample. Wash the beads in 1 mL of 1× WB buffer and resuspend to original volume in 1× WB buffer.
34. Add 25 μL of washed streptavidin magnetic beads to each sample of DNA.
35. Rotate at room temperature for 30 min and place the tubes on the magnet.
36. Save the unbound fraction and wash the beads three times. Resuspend the beads with 1 mL of 1× WB buffer and rotate at room temperature for 5 min for each wash.
37. Resuspend the beads in 20 μL of TE. Add SDS to a final concentration of 1% and 20 mg of Proteinase K per mL of the reaction and incubate at 65 °C for 15 min.
38. Add an equal volume of phenol:chloroform:isoamyl alcohol mix (25:24:1 ratio) and spin at 16,000 rcf in a tabletop centrifuge for 10 min.
39. Transfer the aqueous phase into a new tube and add sodium acetate to a final concentration of 30 mM (dilute from 300 mM stock solution), 2 μL of glycogen and isopropanol to a final concentration of 70%. Mix well by inverting the tube several times.
40. Precipitate the DNA by spinning at 16,000 rcf at 4 °C for 30 min in a tabletop centrifuge.
41. Discard the supernatant and air-dry the DNA pellet. Resuspend the dried pellet in 20 μL of nuclease-free water and proceed for library preparation using a published protocol [5] (*see* **Notes 6** and **7**).

4 Notes

1. We have performed MINCE-seq on *Drosophila* cells (S2, BG3) and mammalian cells (K562). Any cells that are actively dividing and can incorporate EdU can be used.
2. Optimize the EdU labeling time for the chosen cell type: Sufficient EdU labeling for robust isolation of newly replicated DNA depends on the fraction of cells in S-phase, the number of active replication forks, and the length of EdU labeling. Early S-phase has a higher number of active replication forks but would lead to poor sampling of inactive regions of the genome that replicate in late S-phase. Longer EdU labeling would result in a higher amount of DNA labeled, but would miss the early events post-replication. The replication fork moves at ~2 kb/min [9]. So for 10 min of labeling, we capture events that occur within ~20 kb from the replication fork.
3. To determine kinetics of chromatin maturation post-replication, we can perform the thymidine chase for different durations after EdU labeling (for example, 1 h, 2 h, etc.).
4. Before proceeding to the biotin pull-down, perform a dot blot analysis using 1–3 μL of input DNA to confirm efficient biotin labeling. Biotin labeling could be inefficient either due to an insufficient fraction of cells in S-phase or due to an inefficient click reaction. Use biotinylated DNA as positive control.
5. Perform agarose gel electrophoresis to elucidate the degree of MNase digestion to choose a sample with optimal digestion. Nascent chromatin is more sensitive to nuclease digestion [10], so a lower level of digestion of bulk chromatin (as indicated by agarose gel electrophoresis of input DNA) would be preferable for MINCE-seq. We recommend using MNase amounts that would result in 30–40% mononucleosomes (as a fraction of mono-, di-, tri-nucleosome, etc. bands) on a 2% agarose gel after electrophoresis.
6. Strategies to increase the proportion of cells in S-phase include treatment with hydroxy urea (HU) to block cells in S-phase then release into media without HU, a single thymidine block, and for S2 cells, letting them reach stationary phase before splitting into fresh media [11].
7. The steady state and newly replicated chromatin landscapes are obtained by performing paired-end sequencing of DNA from input and biotin pull-down samples. Alignment of paired-end sequencing data to the reference genome gives information on both length and the genomic location of all the sequenced fragments. Short fragments (<50 bp) and longer fragments (~147 bp) obtained after MNase treatment represent TFs and

nucleosomes respectively. Thus, the distributions of nucleosomes and TFs around genomic landmarks can be determined at steady state and post-replication using the sequencing data from MINCE-seq.

References

1. Ho JW, Jung YL, Liu T et al (2014) Comparative analysis of metazoan chromatin organization. Nature 512:449–452
2. Krebs AR, Imanci D, Hoerner L et al (2017) Genome-wide single-molecule footprinting reveals high RNA polymerase II turnover at paused promoters. Mol Cell 67:411
3. Buenrostro JD, Wu B, Chang HY, Greenleaf WJ (2015) ATAC-seq: a method for assaying chromatin accessibility genome-wide. Curr Protoc Mol Biol 109:21.29.1–21.29.9
4. Wang H, Mayhew D, Chen X et al (2012) "calling cards" for DNA-binding proteins in mammalian cells. Genetics 190:941–949
5. Henikoff JG, Belsky JA, Krassovsky K et al (2011) Epigenome characterization at single base-pair resolution. Proc Natl Acad Sci U S A 108:18318–18323
6. Hesselberth JR, Chen X, Zhang Z et al (2009) Global mapping of protein-DNA interactions in vivo by digital genomic footprinting. Nat Methods 6:283–289
7. Ramachandran S, Henikoff S (2016) Transcriptional regulators compete with nucleosomes post-replication. Cell 165:580–592
8. Salic A, Mitchison TJ (2008) A chemical method for fast and sensitive detection of DNA synthesis in vivo. Proc Natl Acad Sci U S A 105:2415–2420
9. Blumenthal AB, Kriegstein HJ, Hogness DS (1974) The units of DNA replication in Drosophila melanogaster chromosomes. Cold Spring Harb Symp Quant Biol 38:205–223
10. Seale RL (1978) Nucleosomes associated with newly replicated DNA have an altered conformation. Proc Natl Acad Sci U S A 75:2717–2721
11. Rizzino A, Blumenthal AB (1978) Synchronization of Drosophila cells in culture. In Vitro 14:437–442

Chapter 9

RChIP-Seq: Chromatin-Associated RNA Sequencing in Developmentally Staged Mouse Testes

Tatiana A. Soboleva and David J. Tremethick

Abstract

Chromatin is a dynamic macromolecular structure comprised of histones and a wealth of non-histone proteins. Recently, it has become clear that RNA is also an integral component of chromatin playing an important role in regulating its structure and function. Central to the understanding of RNA function is the ability to identify and genomically map interactions between chromatin components and RNA.

Here, we describe a new method, RChIP-seq (RNA-associated-Chromatin-Immuno Precipitation followed by next-generation sequencing) that allows the identification of RNA species that are directly bound to specific components of chromatin in the mouse testis.

Key words Chromatin, RNA-binding proteins, histone variant, CLIP, next-generation sequencing

1 Introduction

RChIP-seq is based on the well-known iCLIP method developed by Jernej Ule and colleagues [1, 2] but it also includes a chromatin immunoprecipitation (ChIP) step to immunoprecipitate the chromatin protein-RNA complex of interest. In addition, the method has been improved by modifying the RNase digestion conditions, which has increased the sensitivity of the approach. These modifications enabled us to discover that the testis (and brain-specific) histone variant, H2A.B.3, is a RNA-binding protein [3].

The major steps of RChIP-seq include UV-crosslinking of germ cells (to capture RNA-protein interactions) followed by formalin crosslinking of whole cells to retain, potentially unstable, protein-DNA interactions. Subsequently, nuclei are isolated and soluble chromatin is released by sonication. Immunoprecipitation of RNA-associated chromatin is performed using antibodies specific to the chromatin protein of interest, which are covalently bound to magnetic epoxy resin. This helps to minimize IgG leakage and release of contaminating nonspecific RNA species.

Guillermo A. Orsi and Geneviève Almouzni (eds.), *Histone Variants: Methods and Protocols*, Methods in Molecular Biology, vol. 1832, https://doi.org/10.1007/978-1-4939-8663-7_9,

Following stringent washing steps with 1 M NaCl-containing buffer, RNA-bound to the protein of interest is radioactively labeled and this complex, resolved on the SDS protein gel, is transferred to a nitrocellulose membrane to purify the protein-bound RNA. Following this, the RNA is converted into cDNA for next-generation sequencing. The original iCLIP idea of intramolecular cDNA circularization allows for the identification of the cross-linked site at nucleotide resolution. For scientists working in the field of spermatogenesis, who are particularly interested in performing RChIP-seq in round spermatids, we describe the best ages of adolescent mice when round spermatids are enriched [4–6] and therefore, an additional purification of this subtype of male germ cells may not be required.

2 Materials

2.1 Equipment and Plastics

1. UV-irradiation oven, Stratalinker 1800, (Stratagene).
2. Bioruptor Sonicator (Diagenode).
3. Nanodrop (ThermoFisher Scientific).
4. Microtube rotation wheel.
5. Microtube magnet (e.g., DynaMag-5, ThermoFisher).
6. 37 °C incubator.
7. Electrophoresis chamber.
8. Western transfer apparatus.
9. Thermo shaker.
10. Inverted microscope.
11. Dounce glass homogeniser, 7 ml.
12. Refrigerated microcentrifuge.
13. Fine tip scissors.
14. Curved tweezers.
15. Petri dishes (3 cm and 10 cm in diameter).
16. 50 ml and 15 ml centrifuge tubes.
17. 1.5 ml microtubes.
18. 1 ml, 200 μl, 20 μl, 10 μl RNase/DNase-free filtered tips.
19. 70 μm cell strainer.
20. 5 ml- and 1 ml-syringes.
21. Nitrocellulose membrane protran BA85.
22. X-ray film.
23. Scalpel blades.
24. 2 ml Phase Lock Gel Heavy tube (e.g., 713–2536, VWR).

25. 1 cm glass pre-filters (Whatman).
26. Costar SpinX column (Corning Inc.).
27. Swing-bucket refrigerated centrifuge that can accommodate 15- and 50-mL tubes (e.g. Beckman Coulter Allegra-X-22R).

2.2 Reagents

1. 8% (v/v) formaldehyde solution, EM grade.
2. 2.5 M glycine.
3. RNase I (*see* main Method for unit requirements).
4. T4 Polynucleotide Kinase 10,000μ/ml (NEB).
5. RNasin (Promega).
6. RNA ligase (NEB).
7. PEG400.
8. ^{32}P-γ-ATP (Perkin Elmer).
9. Dynabeads M-270 epoxy resin (ThermoFisher Scientific).
10. Antibody of choice. The ChIP-grade antibody will generally work in RChIP-seq. To validate antibody, one can perform immunoprecipitation followed by western blotting to *see* if the antibody detects a single band corresponding to the size of the protein of interest. Immunofluorescence examination can also be done to ensure that the antibody detects protein in its native conformation.
11. Slide-A-Lyser dialysis cassettes, 0.5–3 ml, 3.5 K MWCO or similar dialysis tubing.
12. Precast NuPAGE 4–12% Bis-Tris gels (ThermoFisher Scientific) or equivalent.
13. Precast 6% TBE-urea gel or equivalent.
14. Precast 6% TBE gel or equivalent.
15. 4×NuPAGE loading buffer or equivalent.
16. 2×TBE-urea loading buffer.
17. 5× TBE loading buffer.
18. 20×MES Buffer (ThermoFisher Scientific).
19. 19 20×MOPS buffer (ThermoFisher Scientific).
20. PAGE ruler plus (Fermentas).
21. Pre-stained protein size marker (Fermentas).
22. Proteinase K 20 mg/ml.
23. Phenol/chloroform, pH 6.7.
24. Glycoblue or Linear acrylamide.
25. 3 M sodium acetate, pH 5.5.
26. 100% and 80% Ethanol (Molecular Biology grade).
27. 10 mM dNTP mix.

28. Superscript III reverse transcriptase.
29. Circligase II (Epicentre).
30. 50 mM $MnCl_2$.
31. BamHI (Fast Digest, Fermentas).
32. Q5 High Fedelity 2× Mix (NEB).
33. Sybr Green II.
34. AEBSF. Prepare 200 mM stock, store at −20 °C, avoid freeze-thaw cycles.
35. DTT. Prepare 500 mM stock, store at −20 °C, avoid freeze-thaw cycles.
36. Protease inhibitor cocktail, EDTA-free (Roche).

2.3 Buffers

1. PBS-GL: DPBS supplemented with 5.6 mM D-glucose, 5.4 mM DL-lactate, 0.5 mM DTT, 0.2 mM AEBSF, EDTA-free protease inhibitor cocktail (Roche), pH 7.1 [7].
2. Buffer I (swelling): 25 mM HEPES, 15 mM NaCl, 10 mM KCl, 2 mM $MgCl_2$, 0.2% (v/v) NP-40, 1 mM EDTA, 0.5 mM EGTA, 0.5 mM DTT, 0.2 mM AEBSF, supplemented with protease inhibitor cocktail, pH 7.6.
3. Buffer II: 0.6 M sucrose, 120 mM KCl, 15 mM HEPES, 15 mM NaCl, 2 mM $MgCl_2$, 0.2% (v/v) NP-40, 1 mM EDTA, 0.5 mM EGTA, 0.5 mM DTT, 0.2 mM AEBSF and protease inhibitor cocktail, pH 7.6.
4. Buffer III (sucrose cushion): 1.2 M sucrose, 60 mM KCL, 15 mM HEPES, 15 mM NaCl, 2 mM $MgCl_2$, 1 mM EDTA, 0.5 mM EGTA, 0.5 mM DTT, 0.2 mM AEBSF and EDTA-free protein inhibitor cocktail (Roche), pH 7.6.

Note that DTT, AEBSF and Roche protease inhibitor cocktail is added to each buffer just prior to use on the day of experiment.

5. Sonication buffer: 50 mM HEPES, 1 mM EDTA, 0.5 mM EGTA, 0.1%(w/v) SDS, 0.1%(w/v) Sodium Deoxycholate, pH 7.6.
6. Dialysis buffer: 10 mM Tris–HCl, 1 mM EDTA, 0.5 mM EGTA, 4% (v/v) glycerol, pH 7.6

Antibody coupling buffers:

7. Buffer A: 0.1 M sodium phosphate buffer, pH 7.4.
8. Buffer B: 3 M ammonium sulfate, 0.1 M sodium phosphate, pH 7.4.
9. Buffer C: 0.1 M citrate, pH 3.1.
10. Buffer E1: PBS, pH 7.4.
11. Buffer E3: PBS, 0.5–1%(v/v) Tween 20, pH 7.4.

12. 2×RNase treatment buffer: 100 mM Tris–HCl, 200 mM NaCl, 10 mM $MgCl_2$, pH 7.6.
13. 1×RNase treatment buffer: 50 mM Tris–HCl, 100 mM NaCl, 5 mM $MgCl_2$, pH 7.6.
14. High-salt buffer: 50 mM Tris–HCl, 1 M NaCl, 1 mM EDTA, 1%(v/v) NP-40, 0.1%(w/v) SDS, 0.5%(w/v) sodium deoxycholate, pH 7.4.
15. Wash buffer: 20 mM Tris–HCl, 10 mM $MgCl_2$, 0.2%(v/v) Tween-20, pH 7.4.
16. 5×PNK pH 6.5 buffer: 350 mM Tris–HCl,50 mM $MgCl_2$, 5 mM DDT, pH 6.5. Freeze aliquots of the buffer, *do not thaw and freeze again*.
17. 4×Ligation buffer: 200 mM Tris–HCl, 40 mM $MgCl_2$, 4 mM DTT, pH 7.8. Freeze aliquots of the buffer, *do not thaw and freeze again*.
18. PK buffer: 100 mM Tris–HCl, 50 mM NaCl,10 mM EDTA, pH 7.4.
19. PK – Urea buffer: 100 mM Tris–HCl, 50 mM NaCl, 10 mM EDTA, 7 M urea, pH 7.4.

3 Methods

All procedures are carried out on ice and all centrifugation steps are carried out at 4 °C unless otherwise specified.

3.1 Preparation of Total Germ Cell Suspension and UV-Crosslinking of RNA to Protein

1. Sacrifice 12 balb/c male mice, aged 28–30 day old (*see* **Note 1**), by cervical dislocation or other methods of choice.
2. Extract testis and place them in 4–5 ml of ice-cold PBS-GL buffer (*see* **Note 2**).
3. Cut a small incision in tunica albuginea and squeeze seminiferous tubules out into a small petri dish containing 5 ml PBS-GL.
4. Wash tubules in PBS-GL to remove blood and interstitial cells (e.g., Leydig cells). Process half of the testis at one time.
5. Combine tubules from 6 mice in 2 ml of PBS-GL and mince with fine scissors for 2 min to release germ cells (*see* **Note 3**).
6. Process the other half of testis and combine all cells plus remnants of seminiferous tubules in a 15 ml tube.
7. To facilitate further release of germ cells from the tubules, pipette the mixture 10 times with a 1 ml tip that was cut at the end to enlarge the tip opening.
8. Put the cell suspension through a 70 μm cell strainer, add PBS-GL, 1 ml at a time, to facilitate cell passage. Use a 5 ml

syringe plunger to gently push the remaining cells through the 70 μm strainer.

9. Wash pooled cells gently in 10 ml of PBS-GL twice and spin for 10 min at 300 g to pellet the cells.
10. Resuspend cells in 55 ml of PBS-GL (on average, the cell yield is 2–4 × 10^8 cells with a viability of 75–85%).
11. Add 6.5 ml of cell suspension into a 10 cm tissue culture dish, remove the lid, and place the dish on ice. Irradiate once with the appropriate UV dose in the Stratalinker (*see* **Note 4** and Fig. 1). Repeat the irradiation procedure for the rest of the cells, keeping some cells non-irradiated as a control.
12. Spin cells down to combine all irradiated cells together and all control cells together, proceed to the next step.

3.2 RChIP Sample Preparation

Fixation of cells with formalin and extraction of soluble chromatin with sonication.

1. Resuspend cells in 20 ml of PBS-GL.

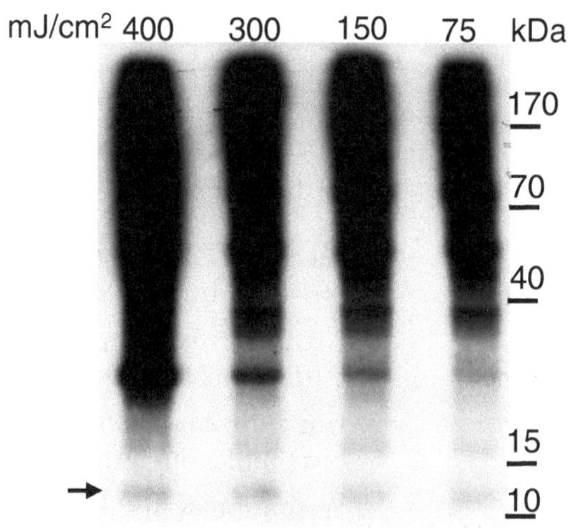

Fig. 1 Optimization of UV-irradiation. To avoid over-crosslinking, optimization of UV-irradiation dose at 254 nm is very important. Total male germ cell suspension from 28 to 30do testis was irradiated with different doses of UV, then cells were lysed, treated with 40μ of RNase I (Ambion) and cleared lysates were immunoprecipitated with anti-H2A.B.3 antibody. The immunoprecipitated samples were labeled on the beads with γ[^{32}P]-ATP and eluted samples were subjected to protein gel electrophoresis followed by a transfer to Nitrocellulose membrane. The signal was visualized by exposure of the membrane to X-ray film for 4 h at room temperature. The arrow shows the weak signal from labeled H2A.B.3. Note that the accumulation of the H2A.B.3 band was not observed as the lysis buffer contained 0.1% SDS and RNase I concentration were not optimized at this initial experiment

2. Add 20 ml of fresh 2.4% formalin in PBS-GL (1.2% final concentration of formalin) to the cells and incubate the cells on a wheel for exactly 10 min all at room temperature (RT).
3. Stop reaction by adding glycine to a final concentration of 0,125 mM (2 ml of 2.5 M stock to 40 ml volume) and continue the incubation on a wheel for an additional 3 min at RT.
4. Wash the cells twice with cold 15 ml of PBS and spin for 10 min at 300 × *g*.
5. Resuspend the cells in 2 ml of Swelling Buffer I, incubate on ice for 10 min.
6. Take 10 μl aliquot of cells and check under microscope with trypan blue—all nuclei should be stained blue.
7. Add 2 ml of Buffer II, mix and transfer into ice-cold glass Dounce homogenizer and gently apply 5 strokes with a "tight" pestle, avoid forming bubbles.
8. Check the cells under a microscope and apply 5–10 extra strokes if the cells still retain cytoplasm.
9. Prepare 2× 15 ml tubes with 8 ml of Buffer III (sucrose cushion) per each cell prep (if processing irradiated and non-irradiated samples on the same day, prepare 4 tubes with Buffer III).
10. Carefully layer 2 ml of cell suspension on the top of an 8 ml sucrose cushion.
11. Centrifuge at 3,900 × *g* for 38 min at 4 °C. The centrifugation will pellet the nuclei and the cell debris will stay on the top.
12. Very carefully remove the supernatant to make sure the top solution does not touch the nuclei. Use 1 ml pipette and change tips every time.
13. Resuspend nuclei in 1.8 ml of Sonication buffer and distribute between 6 x1.5 ml tubes, 300 μl in each for sonication (*see* **Note 5**).
14. Sonicate nuclei for 15 min at maximum voltage with 30 s on/off cycle at 4 °C using Bioruptor Sonicator (*see* **Note 6**).
15. Centrifuge sonicated chromatin in microcentrifuge at 10,000 × *g* for 15 min at 4 °C.
16. Collect and combine the supernatant, containing soluble RNA-associated chromatin.
17. Dialyze the supernatant overnight in 500 ml of Dialysis buffer at 4 °C using a Slide-A-Lyser dialysis cassette.
18. Centrifuge dialyzed chromatin in microcentrifuge at 10,000 × g for 5 min at 4 °C, discard any pellet.
19. Measure DNA concentration using a Nadodrop. Concentration usually ranges between 300 and 900 ng/μl.

20. Divide material into aliquots and store at −80 °C.

Covalent coupling of antibodies to magnetic beads (see **Notes 7** *and* **8***).* All steps are performed at RT except for incubation step that is performed at 37 °C.

21. Carefully weigh out 5 mg of Dynabeads M-270 epoxy resin (ThermoFisher Scientific) (3.3×10^8) and resuspend in 1 ml of Buffer A (*see* **Note 9**).
22. Vortex for 30 s and incubate with tilting and rotation for 10 min.
23. Place the tube in a magnet for 1 min and discard the supernatant.
24. Remove the tube from the magnet and resuspend the washed beads in 1 ml Buffer A. Vortex for 30 s.
25. Place the tube in a magnet for 1 min and discard the supernatant.
26. Resuspend the beads in the same volume of Buffer A as the antibody volume (e.g., 150 μl). Mix or vortex.
27. Per 5 mg of beads use 20–35 μg of antibody (*see* **Notes 10** and **11**). Using Buffer A, bring antibody to the same volume (150 μl) as the volume of resuspended beads.
28. Combine the equal volumes of bead slurry with the diluted antibody (150 μl each) and mix thoroughly.
29. Add the same volume of Buffer B (150 μl). The final concentration of ammonium sulfate should be 1 M for efficient coupling.
30. Incubate for 16–24 h at 37 °C with slow tilt rotation. Make sure that the fluid in the tube is mixing well.
31. Place the tube on the magnet for 2 min. Gently turn the magnet upside-down twice, to ensure collection of any remaining beads in the cap. Remove the supernatant.
32. Wash the coated beads a total of four times with 1 ml Buffer E1.
33. Resuspend the beads and apply to a magnet for 2 min for each wash.
34. Remove physically absorbed antibody by washing for 10 min in Buffer E3.
35. Resuspend the coated beads to the desired concentration in Buffer E1 and store at 4 °C.

RNA digestion (*see* **Note 12** and Fig. 2).

36. For High RNase treatment (*see* **Note 13**) add to each 250 μl aliquot of RNA-associated chromatin, 500 μl of 2×RNase

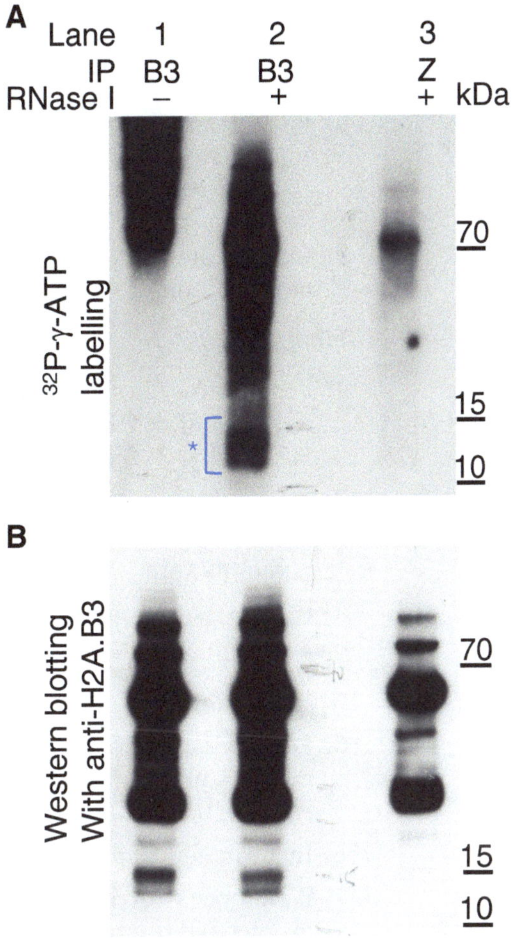

Fig. 2 Optimized RNase I treatment conditions lead to increased sensitivity and allowed to detect RNA-binding property of histone variant H2A.B.3. (**a**) The autoradiograph of RNA-associated chromatin treated (lane 2, 3) or not treated (lane 1) with RNase I (400 μ) in an SDS-free RNase I treatment buffer (50 mM Tris–HCl, 100 mM NaCl, 5 mM $MgCl_2$, pH 7.6) and immunoprecipitated with anti-H2A.B.3 antibody (lane 1,2) or anti-H2A.Z antibody (negative control, lane 3). The asterisk (*) indicates detection of H2A.B.3 that is directly bound to RNA (compare to Fig. 1. (**b**) The same membrane was subjected to western blotting with anti-H2A.B.3 antibodies to show that in both lanes, 1 and 2, H2A.B.3 can be detected (seen as a double band under 15 kDa). High molecular weight signal is mainly produced by the IgG leakage from the beads

treatment buffer, 2 μl of 500 mM DTT, Roche protease inhibitor cocktail and nuclease-free water to a final volume of 1 ml.

37. Mix well and divide into 2×500 μl aliquots. To each aliquot, add 4 μl of RNase I (Ambion, ThermoFisher Scientific), mix thoroughly, and incubate at 37C for 30 min, shaking at 1000 rpm.

38. For Low RNase treatment (*see* **Note 14**), repeat two steps above but use 5 μl of 1/250 dilution of RNase I per each 500 μl aliquot and incubate exactly for 3 (three) min at 37C with shaking at 1000 rpm.
39. Immediately after this incubation, transfer samples on ice to cool down.

3.3 RChIP, cDNA Preparation, and Amplification

The rest of the method follows the iCLIP protocol [1, 2] with minor modifications. It is reproduced here for convenience:

Immunoprecipitation

1. Per IP, use 1/5 of bead slurry (from Subheading 3.2. **step 35**), remove Buffer E1, and add RNA-associated chromatin (from Subheading 3.2, **step 39**).
2. Rotate the samples for 2 h at 4 °C.
3. Discard the supernatant and wash the beads 2× with 900 μl High-salt buffer, 10 min with rotation at 4 °C.
4. Wash 2× with 900 μl wash buffer, 10 min with rotation at 4 °C.

Dephosphorylation of RNA 3'ends

5. Discard the supernatant and resuspend the beads in 20 μl PNK mix (15 μl water 4 μl 5× PNK pH 6.5 buffer; 0.5 μl PNK enzyme; 0.5 μl RNasin).
6. Incubate for 20 min at 37 °C.
7. Wash once with 500 μl wash buffer for 1 min, then place on a magnet and discard the supernatant.
8. Repeat this wash once with High-salt buffer and then twice with wash buffer.

Linker ligation to RNA 3′ends

9. Carefully remove the supernatant and resuspend the beads in 20 μl ligation mix (9 μl water; 4 μl 4× ligation buffer; 1 μl RNA ligase; 0.5 μl RNasin; 1.5 μl 20 μM pre-adenylated linker L3; 4 μl PEG400).
10. Incubate overnight at 16 °C.
11. Repeat the washes as per **steps 6.3–6.4**.

RNA 5' end labeling

12. Remove the supernatant and resuspend the beads in 8 μl of hot PNK mix (0.4 μl PNK; 0.8 μl^{32}P-γ-ATP; 0.8 μl 10× PNK buffer; 6 μl water).
13. Incubate for 5 min at 37 °C.
14. Place tubes on a magnet and remove the hot PNK mix.
15. Wash the beads twice with 500 μl of wash buffer as per **step 6.3**. Dispose of radioactive supernatants in accordance with the guidance of your intuition.

16. Resuspend the beads in 20 μl 1× NuPAGE loading buffer.
17. Incubate on a thermomixer at 70 °C for 10 min.
18. Immediately place on a magnet to precipitate the empty beads and load the supernatant on the gel (*see* **step 8**).

SDS-PAGE and membrane transfer

19. Load the samples on a 4–12% NuPAGE Bis-Tris gel according to the manufacturer's instructions. Use 0.5 l of 1× MOPS running buffer. Also load 5 μl of a prestained protein size marker.
20. Run the gel for 50 min at 180 V.
21. Remove the gel front and discard as solid waste (contains free radioactive ATP).
22. Transfer the protein-RNA complexes from the gel to a nitrocellulose membrane using the Novex wet transfer apparatus according to the manufacturer's instructions (ThermoFisher Scientific, transfer 1 h at 30 V).
23. After the transfer, rinse the membrane in PBS buffer, then wrap it in saran wrap.
24. Expose it to a Fuji film at −80 °C (perform exposures for 30 min, 1 h and overnight).

RNA isolation

25. Isolate the protein-RNA complexes from the low-RNase experiment using the autoradiograph from **step 8.5** as a mask. Cut this piece of membrane into several small slices and place them into a 1.5 ml microtube.
26. Add 200 μl PK buffer and 10 μl proteinase K to the membrane pieces. Incubate shaking at 1100 rpm for 20 min at 37 °C.
27. Add 200 μl of PK-Urea buffer and incubate for 20 min at 37 °C.
28. Collect the solution and add it together with 400 μl of RNA phenol/chloroform, pH 6.7. to a 2 ml Phase Lock Gel Heavy tube.
29. Incubate for 5 min at 30 °C, shaking at 1100 rpm. Separate the phases by spinning for 5 min at 10,000 × *g* at room temperature.
30. Transfer the aqueous layer into a new tube (be careful not to touch the gel with the pipette). Add 0.5 μl glycoblue and 40 μl 3 M sodium acetate pH 5.5 and mix. Then add 1 ml 100% ethanol, mix again, and precipitate overnight at −20 °C.

Reverse transcription

31. Spin for 20 min at 12,000 × *g* and 4 °C. Remove the supernatant and wash the pellet with 0.5 ml 80% ethanol.

32. Resuspend the pellet in 7.25 μl RNA/primer mix (6.25 μl water; 0.5 μl Rclip primer 0.5 pmol/μl; 0.5 μl 10 mM dNTP mix). For each experiment or replicate, use a different Rclip primer containing individual barcode sequences.
33. Incubate for 5 min at 70 °C before cooling to 25 °C.
34. Add 2.75 μl RT mix (2 μl 5× RT buffer; 0.5 μl 0.1 M DTT; 0.25 μl Superscript III reverse transcriptase).
35. Incubate for 5 min at 25 °C, 20 min at 42 °C, 40 min at 50 °C and 5 min at 80 °C before cooling to 4 °C.
36. Add 90 μl TE buffer, 0.5 μl glycoblue and 10 μl 3 M sodium acetate pH 5.5 and mix. Then add 250 μl 100% ethanol, mix again and precipitate overnight at −20 °C.

Gel purification of cDNA

37. Spin down and wash the samples (*see* 11.1 for conditions), then resuspend the pellets in 6 μl of water.
38. Add 6 μl 2×TBE-urea loading buffer. Heat samples to 80 °C for 3 min directly before loading.
39. Load the samples on a precast 6% TBE-urea gel and run for 40 min at 180 V as described by the manufacturer. Also load a low molecular weight marker for subsequent cutting (*see* below).
40. Cut three bands at 120–200 nt (high), 85–120 nt (medium), and 70–85 nt (low). Use the upper dye and the marks on the plastic gel support to guide excision. Note that the Rclip primer and the L3 sequence together account for 52 nt of the CLIP sequence.
41. Using a 1 ml syringe plunger with removed rubber end, crush the gel into small pieces (until it resembles slushy). Add 400 μl TE Incubate shaking at 1100 rpm for 1 h at 37 °C (*see* **Note 15**).
42. Place two 1 cm glass pre-filters into a Costar SpinX column. Transfer the liquid portion of the sample to the column. Spin for 1 min at 10,000 × *g* into a 1.5 ml tube.
43. Add 0.5 μl glycoblue and 40 μl 3 M sodium acetate pH 5.5, then mix the sample. Add 1 ml 100% ethanol, mix again and precipitate overnight at −20 °C.

Ligation of primer to the 5'end of the cDNA

44. Spin down and wash the samples (*see* 11.1 for conditions), then resuspend the pellets in 8 μl ligation mix (6.5 μl water; 0.8 μl 10× CircLigase Buffer II; 0.4 μl 50 mM MnCl2; 0.3 μl; Circligase II) and incubate for 1 h at 60 °C.
45. Add 30 μl oligo annealing mix (26 μl water; 3 μl FastDigest Buffer; 1 μl 10 μM cut_oligo). Incubate for 1 min at

95 °C. Then decrease the temperature every 20 s by 1 °C until 25 °C is reached.

46. Add 2 μl BamHI (Fast Digest, Fermentas) and incubate for 30 min at 37 °C.
47. Add 50 μl TE and 0.5 μl glycoblue and mix.
48. Add 10 μl 3 M sodium acetate pH 5.5 and mix, then add 250 μl 100% ethanol.
49. Mix again and precipitate overnight at −20 °C.

PCR amplification

50. Spin down and wash the samples (*see* 11.1 for conditions), then resuspend the pellet in 19 μl water.
51. Prepare the PCR mix (19 μl cDNA; 1 μl primer mix P5/P3, 10 μM each; 20 μl Q5 High Fedelity 2× Mix).
52. Run the following PCR programme: 98 °C for 30 s, [98 °C for 10 s, 65 °C for 30 s, 72 °C for 30 s]$_{15\text{–}22\ \text{cycles}}$, 72 °C for 2 min, 4 °C forever (*see* **Note 16**).
53. Mix 8 μl PCR product with 2 μl of 5× TBE loading buffer and load on a precast 6% TBE gel. Stain the gel with Sybr Green II and analyze with a gel imager.
54. The barcode in the Rclip primers allow to multiplex different samples before submitting for high-throughput sequencing.
55. Submit 15 μl of the library for sequencing and store the rest.

4 Notes

1. Round spermatid enrichment at different ages of juvenile mice.

   We and others [5, 8, 9]observed that during the first round of spermatogenesis in juvenile balb/c mice, there are several stages of development when round spermatids become enriched in the testis, representing the major cell type (*see* Table 1). There is also a clear window of time, at the age of 24 days, when only early round spermatids are present (stages I-IV) while the late round spermatids form later (stages V-VIII). At the age between 26 and 30 days, round spermatids represent up to 80% of all germ cells in the testis while elongating and condensing spermatids are still absent and spermatocytes and spermatogonia represent a minority population (around 20%). Hence, if pure population of round spermatids is not required (for example a protein of interest is only expressed during round spermatid stage, which is a common feature of many post-meiosis expressed proteins), an additional purification step can be omitted and total germ cells from 26–30 day old mice can be used for further investigation. If a pure population of round spermatids is required, a number of techniques can be employed to purify round spermatids prior to RChIP-seq [4, 10, 11].

2. We found that PBS-GL buffer [7] keeps cells alive longer than PBS and it does not contain serum or other constituents of cell culture media (e.g., amino acids, vitamins) that potentially can interfere with further steps of the procedure and produce nonspecific signals in RChIP-seq.
3. We do not use enzymatic methods to release germ cells from seminiferous tubules, as we noticed that cell viability and yield is generally not as good as with the mechanical release of cells using fine-end scissors and a 70 μm strainer. With our methods, we mostly obtain 2–4 × 10^8 total germ cells with viability of 75–85%, measured by trypan blue incorporation.
4. UV-crosslinking dose must be adjusted for each new protein, we found that 75 mJ/cm^2 dose avoids over-crosslinking and contamination with nonspecific signal. Figure 1 shows that at higher UV dose there is a lot more nonspecific signal detected while specific signal intensity does not change.
5. Usually, for ChIP-seq procedures in formalin fixed cells, we use MNase I digest to release mono-nucleosomes, however, for RNA-ChIP-seq method, MNase I cannot be used, as it also digests RNA, therefore, sonication procedure is used.
6. It is important to optimize sonication conditions for your samples. Sonicate initially for 10, 15, 20, or 25 min to choose the best time. Spin samples down, take 20 μl of supernatant, and reverse formalin cross-linking by adding 2 μl of 4 M NaCl and incubating at 65C for 4h or overnight. To purify DNA, add 20 μl of Phenol:Chloroform:Isoamyl Alcohol (25:24:1, v/v) to each sample. Vortex each tube for 20 s and spin at top speed (at RT) for 5 min. Transfer exactly 10 μl of top aqueous phase to a new tube and add 2 μl of 80% glycerol. Run samples on 1.5% agarose in TAE buffer with 1Kb ladder. (In this gel, 80% glycerol is used instead of loading dye). Analyze samples. The best sonication condition will produce a smear between 200 and 500 bp with no high molecular weight DNA species present. If the DNA purification step is omitted, even well-sonicated DNA may stay in the wells. To avoid this, add SDS to your samples to a final concentration of 1% prior to electrophoresis.
7. We found that covalent coupling of primary antibodies to magnetic beads is far more superior to the non-covalent binding of antibodies to Protein A/G coupled magnetic beads. First, the amount of antibody that can be covalently linked to the beads is greater than the amount that binds to protein A/G; second, there is far less contamination with the RNA

molecules that nonspecifically interact with IgG heavy and light chains and co-elute with IgGs that are non-covalently bound to protein A/G beads.

8. Alternatively, antibody can be coupled to Dynabeads M-270 Epoxy using a Dynabeads Antibody Coupling Kit (ThermoFisher Scientific).
9. Due to the short stability of hydrophilic epoxy groups in aqueous buffers, rehydrated uncoupled beads cannot be stored in Buffer A and must be prepared fresh for each coupling. The beads can also be resuspended in organic solvents like diglyme or DMF and stored for a year at 4 °C (refer to Dynabeads M-270 epoxy manual for detail).
10. We usually conjugate 20–35 μg of antibody with 4–5 mg of beads in a total volume of 300–450 μl. After conjugation, antibodies bound to beads can be stored at least for a month at 4 °C without a noticeable loss of activity or degradation. This amount is sufficient for 4–5 IPs.
11. Antibody additives: It is important to note tha*t antibodies stored in glycerol cannot be used for coupling to dynabeads,* as the antibody function may be severely affected. Antibody storage in NaN_3-containing buffers can lead to a slight decrease (<10%) in antibody coupling efficiency. Additives like BSA or Gelatin will not interfere with coupling but will result in coupling to beads along with antibody, which can add a beneficial blocking effect but may also result in isolation of BSA or Gelatin-binding proteins. For more information refer to Dynabeads Antibody Coupling Kit manual (ThermoFisher Scientific).
12. RNase treatment: We found that buffers that are used for RNase treatment in CLIP/iCLIP procedures as well as some commonly used cell lysis buffers (e.g., RIPA buffer) contain high concentration of SDS (0.1%) [1, 2, 12]. 0.1% SDS completely and irreversibly inactivates RNase I [13]. We also found that for detecting the RNA-binding ability of proteins that only transiently and occasionally interact with RNA, RNase I amount needs to be at least 200–400 units (high RNase treatment) and the buffer should contain not more than 0.02% SDS (*see* Fig. 2).
13. High RNase treatment: This treatment is important if the protein has not been shown previously to interact directly with RNA in vivo, it is also an important control for antibody specificity.
14. Only Low RNase treatment samples used for consequent RNA extraction, cDNA synthesis and NGS.
15. This step is crucial, as efficient elution of cDNA from the gel is absolutely necessary. We found that crushing the gel with

the 1 ml syringe plunger with the removed black rubber stopper, (before adding the TE buffer) until the gel turns into very fine slushy helps to elute cDNA more efficiently.

16. Q5 High-fidelity 2× master mix contains new generation high-fidelity thermostable DNA polymerase that allows for robust DNA amplification. Using this particular enzyme allowed us to decrease the number of PCR cycles from 25–35 to only 15–22, which in turn minimizes PCR amplification bias in sequencing libraries.

References

1. Huppertz I, Attig J, D'Ambrogio A et al (2014) iCLIP: protein-RNA interactions at nucleotide resolution. Methods 65(3):274–287. https://doi.org/10.1016/j.ymeth.2013.10.011
2. Konig J, Zarnack K, Rot G et al (2011) iCLIP--transcriptome-wide mapping of protein-RNA interactions with individual nucleotide resolution. J Vis Exp (50). https://doi.org/10.3791/2638
3. Soboleva TA, Parker BJ, Nekrasov M et al (2017) A new link between transcriptional initiation and pre-mRNA splicing: the RNA binding histone variant H2A.B. PLoS Genet 13(2):e1006633. https://doi.org/10.1371/journal.pgen.1006633
4. Bellve AR (1993) Purification, culture, and fractionation of spermatogenic cells. Methods Enzymol 225:84–113
5. Bellve AR, Cavicchia JC, Millette CF et al (1977) Spermatogenic cells of the prepuberal mouse. Isolation and morphological characterization. J Cell Biol 74(1):68–85
6. Oakberg EF (1956) Duration of spermatogenesis in the mouse and timing of stages of the cycle of the seminiferous epithelium. Am J Anat 99(3):507–516. https://doi.org/10.1002/aja.1000990307
7. Pelengaris SA, Moore HD (1995) Separation of round spermatids from the rat using an immunoselection panning technique. Mol Reprod Dev 41(3):348–354. https://doi.org/10.1002/mrd.1080410310
8. Oakberg EF (1956) A description of spermiogenesis in the mouse and its use in analysis of the cycle of the seminiferous epithelium and germ cell renewal. Am J Anat 99(3):391–413. https://doi.org/10.1002/aja.1000990303
9. Soboleva TA, Nekrasov M, Pahwa A et al (2012) A unique H2A histone variant occupies the transcriptional start site of active genes. Nat Struct Mol Biol 19(1):25–30. https://doi.org/10.1038/nsmb.2161
10. Lassalle B, Ziyyat A, Testart J et al (1999) Flow cytometric method to isolate round spermatids from mouse testis. Hum Reprod 14(2):388–394
11. Pivot-Pajot C, Caron C, Govin J et al (2003) Acetylation-dependent chromatin reorganization by BRDT, a testis-specific bromodomain-containing protein. Mol Cell Biol 23(15):5354–5365
12. Peach M, Marsh N, Miskiewicz EI et al (2015) Solubilization of proteins: the importance of lysis buffer choice. Methods Mol Biol 1312:49–60. https://doi.org/10.1007/978-1-4939-2694-7_8
13. Ambion Rnase I (cloned)https://tools.thermofisher.com/content/sfs/manuals/sp_2294.pdf

Chapter 10

Bioinformatic Analysis of Nucleosome and Histone Variant Positioning

Jakub Mieczkowski and Michael Y. Tolstorukov

Abstract

Assays profiling nucleosome positioning and occupancy are often coupled with high-throughput sequencing, which results in generation of large data sets. These data sets require processing in specialized computational pipelines to yield useful information. Here, we describe main steps of such a pipeline, and discuss bioinformatic and statistical aspects of assessing data quality, as well as data visualization and further analysis.

Key words Nucleosome positioning, Histone variants, Next generation sequencing, MNase, Data processing and analysis

1 Introduction

Nucleosome composition, occupancy, and positioning are often examined in epigenetic research, as this information is important for understanding the regulatory mechanisms of the genome readout and the maintenance of the cell epigenetic state [1–10]. Most experimental approaches currently used for profiling nucleosome occupancy and composition involve three steps: (a) chromatin fragmentation, (b) an optional step that allows enrichment for the nucleosomes of a desired type, e.g., those bearing a specific histone variant or modification, and (c) sequencing of the released DNA fragments [8, 11, 12]. Additionally, the methods based on DNA methylation were proposed to analyze nucleosome positioning, e.g., NOMe-seq [13, 14], however, these methods are not discussed in this chapter.

Chromatin fragmentation can be achieved with enzymatic digestion, sonication, or with chemical (e.g., hydroxyl radical)

Electronic supplementary material: The online version of this chapter (https://doi.org/10.1007/978-1-4939-8663-7_10) contains supplementary material, which is available to authorized users.

Guillermo A. Orsi and Geneviève Almouzni (eds.), *Histone Variants: Methods and Protocols*, Methods in Molecular Biology, vol. 1832, https://doi.org/10.1007/978-1-4939-8663-7_10, © Springer Science+Business Media, LLC, part of Springer Nature 2018

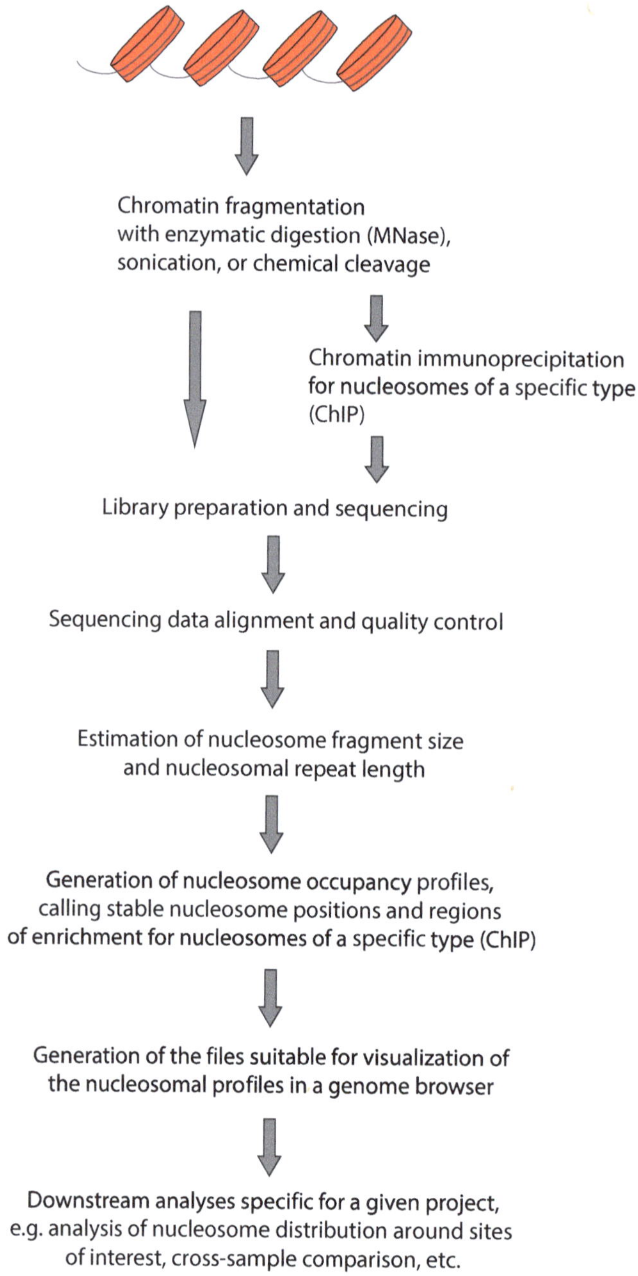

Fig. 1 A schematic illustration of the main steps of a nucleosome-profiling pipeline. The pipeline includes both experimental (top rows) and bioinformatics (middle and bottom rows) components

cleavage [15–17] (Fig. 1). Arguably, enzymatic digestion is applied in the majority of recent studies and it is most commonly performed with micrococcal nuclease (MNase) that cleaves DNA between nucleosomes, preferentially preserving nucleosome-protected genomic fragments ([18–21], *see* [11, 22] for further review). However, digestion can also be performed with other enzymes that cleave DNA [23–25]. Digestion is preferentially used for profiling nucleosome positioning with high resolution, and can be done both with and without histone enrichment step, while sonication is generally used in combination with the histone enrichment step when "base-pair" resolution of resulting profiles is not required. Enrichment for nucleosome of a specific type is typically performed with chromatin immunoprecipitation (ChIP) [4].

In the genome-wide studies, the released DNA fragments are subjected to massively parallel sequencing, which generates reads representing either one end (single-end sequencing) or both ends (paired-end sequencing) of the initial DNA fragments. The resulting data sets comprise hundreds of millions or even billions of the sequenced reads and require bioinformatic processing for further use. Such processing includes steps specific for nucleosomal data as compared to the data sets of other types. For instance, in contrast to transcription factor-binding profiles that typically contain tens of thousands punctate peaks, the nucleosomal profiles contain millions of less pronounced peaks of varying fuzziness as well as broader regions of enrichment or depletion. Each of these features of nucleosomal profiles provides biologically relevant information and requires a specialized bioinformatics tool for detection and analysis. For instance, a number of computational algorithms have been developed by now for calling and, in some cases, cross-sample comparison of the nucleosome positions (e.g., [26–29]).

Here, we describe in detail several algorithms and techniques that allow data quality control, processing, and visualization as applicable to nucleosomal assays. Specifically, we describe analyses of the nucleosome fragment sizes and nucleosomal repeat length, identification of the regions of enrichment and depletion in the nucleosomal profiles, calling stable nucleosome positions, and GC-content correction and cross-sample comparison of the nucleosome occupancy profiles. We also address technical issues that may arise during nucleosome data processing as well as briefly discuss downstream analyses and new assays that aim at more in-depth interrogation of the primary structure of chromatin.

2 Materials

2.1 Data

Computer files comprising sequences and quality information reported for the reads that represent nucleosomal DNA fragments. Most commonly these files are stored either in plain text format (FASTQ) or in a binary form specifically designed for short sequence reads (SRA). Each sample is typically represented by one file (single end reads) or by two related files (paired-end reads).

2.2 Software

The analyses described below can be performed using R environment for statistical computing, which is freely available from http://ww.r-project.org for Windows, Linux or MAC OS. However, these analyses depend neither on operating system nor on particular computational environment. Therefore, similarly to previously published computational protocols [30–32], the majority of the described analyses are illustrated by the "pseudo-code" examples rather than by the code fragments written in any specific programming language (unless we describe a specific software package). We also provide an example of the code generated in R environment, which includes all major steps of the pipeline described below, as Supplementary File 1. The computational procedures described here use software suite bedTools [33] and R packages SPP [34], nucleR [27], and DESeq [35].

3 Methods

3.1 Sequencing Data Alignment and Quality Control

1. The sequenced reads can be aligned to the genome of interest using short read aligners such as bowtie [36], BWA [37], and others. The aligners use the data files in FASTQ format as input, and produce the output files that specify the coordinates of the genomic alignments of the sequenced reads. Sequence Alignment Map format either in the plain text (SAM) or binary (BAM) form is most commonly used for the output files.
2. The aligned data should be examined to make sure that the alignment rates are satisfactory (at least above 50% *see* **Note 1**) and that the total number of reads available for the subsequent analysis is sufficient (such a number would depend on the size of the genome and type of nucleosomes under analysis). As a further control, the genomic positions with abnormally high numbers of mapped reads should be identified and removed from consideration [38], since they likely represent PCR artifacts. The resulting "clean" set of the genomic coordinates representing sequenced reads is taken for further analysis.

3.2 Estimation Nucleosome Protection Length

Information on the length of the DNA fragments protected by the nucleosomes from shearing can be obtained from the sequencing data. The computational procedures to achieve this goal differ for paired- and single-end sequencing data.

3.2.1 Paired-End Data

1. For paired-end data, the fragment size can be computed directly by finding difference between the genomic positions of the 5′-ends of the mate reads in each pair (also *see* **Note 2**).

The distribution of the fragment sizes (histogram) should have a maximum close to the expected nucleosome size, which can be estimated during the sequencing library preparation (e.g., ~150 bp for the MNase-digested chromatin) (Fig. 2).

3.2.2 Single-End Data

1. For single-end reads, only a representative or "characteristic" fragment size can be estimated using an approach based on computing cross-correlation of read frequencies on the DNA "+" and "-" strands [34] (Fig. 3). For each chromosome **c** compute vectors $n^s_c(x)$ representing frequencies (counts normalized by the library size) of the 5′-ends of the reads mapped to each genomic position, *x*, on the strand *s*.
2. Select a range of the lengths, ***Hc***, so that it is expected to cover most of the nucleosomal fragments in the library. Typically, the length range from 50 bp to 300 bp is sufficient (***Hc*** = [50,300]).
3. For each chromosome and each tested shift value *h* from the range ***Hc*** compute Pearson's correlation coefficients $P(a,b)$ between symmetrically shifted frequency vectors corresponding to opposing strands. $P(n_c^{s+}(x + h/2), n_c^{s-}(x-h/2))$.
4. The final cross-correlation value for each h, $X(h)$, is a weighted sum of the Pearson's correlation coefficients computed for this shift for each chromosome. Weighting is performed using fractions of sequenced reads aligned to a particular chromosome (N_c/N, where N_c is the number of reads aligned to the chromosome **c**, and N is the total number of aligned reads respectively).

Pseudocode:

```
> For each chromosome c compute two strand specific read frequency vectors n^s_c(x)
> For each h from Hc do {
>         X(h) = Σ_c (N_c/N) P(n_c^{s+}(x + h/2), n_c^{s-}(x - h/2))
> }
```

$$X(h) = \sum_c \frac{N_c}{N} \mathrm{P}\left(n_c^{s+}\left(\mathrm{x} + \frac{\mathrm{h}}{2}\right), n_c^{s-}\left(\mathrm{x} - \frac{\mathrm{h}}{2}\right)\right)$$

5. The function $X(h)$ can be further analyzed similarly to the histogram of the fragment length in the case of paired-end sequencing, e.g., the estimated characteristic fragment size is equal to $\max_h\{X(h)\}$ *see* **Note 3** for further discussion on the applications of such an analysis.

3.3 Identification of Nucleosomal Repeat

The periodicity in the translational positioning of nucleosomes (nucleosomal repeat) can be estimated from sequencing data using auto-correlation analysis [38]. Unlike strand cross-correlation analysis, the auto-correlation analysis focuses on the distribution of read frequencies on the same DNA strand (Fig. 3 bottom

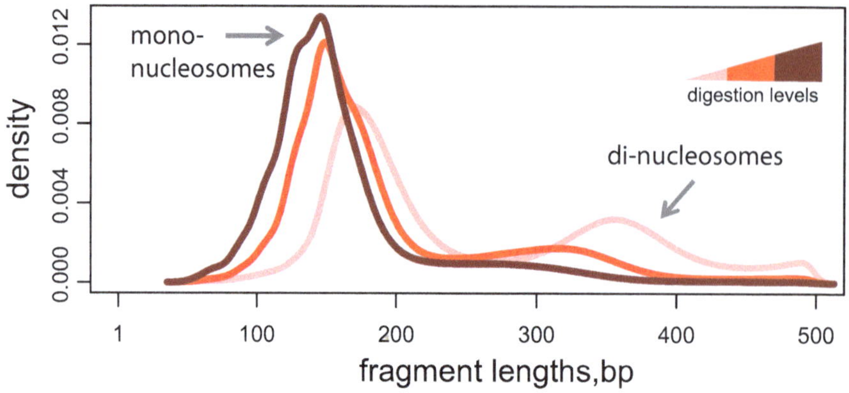

Fig. 2 Distributions of the digestion fragment lengths. The color scheme indicates digestion levels, with pink corresponding to under-digested sample and dark red to over-digested sample. The digested DNA was obtained from S2 Drosophila melanogaster cells. Peaks corresponding to mono- and di-nucleosomal fragments are indicated with arrows

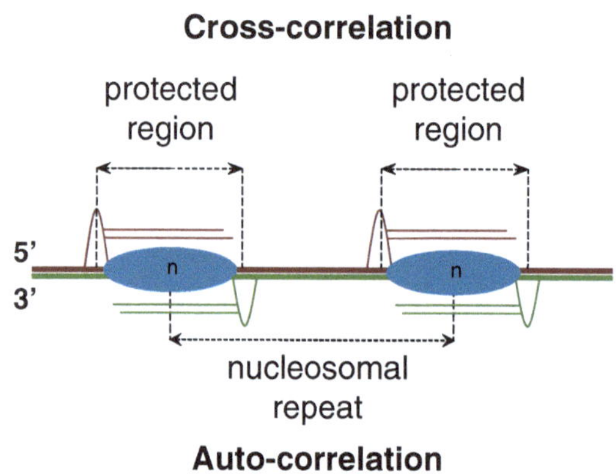

Fig. 3 An illustration of the algorithms used to compute cross- and auto-correlations of sequenced read frequencies. Shown at the top and bottom respectively. Nucleosomes are represented with blue ovals. DNA positive (red) and negative (green) strands are shown with thick solid lines. Mononucleosomal fragments released by chromatin fragmentation are shown with thin red and green solid lines. Distances between buildups in the frequencies of mapped tags (bell-shaped curves) were used to identify the length of DNA protected by the histone core (cross-correlation, top) or the nucleosomal repeat (auto-correlation, bottom)

see **Note 4**). The analyses of the single and paired-end data differ only by the first step as described below.

1. For the single-end data obtain read frequency vectors $n^s_c(x)$ as described in Subheading 3.2.2 (i.e., two vectors representing 5′-ends of the reads mapped to the positive and negative DNA strands). For the paired-end data obtain a vector $n_c(x)$ representing frequencies (counts normalized by the library size) of the midpoints between 5′-ends of the mate reads (i.e., digestion fragment centers) at each genomic position, x. There is

only one frequency vector in the case of paired-end data and it can be used instead of vector $n^s_c(x)$ corresponding to the positive strand in the subsequent analysis.

2. Similarly to the analysis of nucleosome protection length, select a range of the repeat lengths, ***Hc***, to be considered. This range it typically selected from 150 bp to 500 bp so that at least two periodically located nucleosomal peaks could be detected (***Hc*** = [150,500]).
3. Compute Pearson's correlation coefficients ***P(a,b)*** between symmetrically shifted read frequency vectors corresponding to the same strand for each shift value ***h*** from ***Hc*** and chromosome ***c***.
4. The final auto-correlation $A(h)$ is a sum of weighted Pearson's correlation coefficients:

$$A(h) = (N_c^{s+}/N)*P(n_c^{s+}(x+h/2),\ n_c^{s+}(x-h/2)) + (N_c^{s-}/N)*P(n_c^{s-}(x+h/2),\ n_c^{s-}(x-h/2))$$

Pseudocode:

```
> For each chromosome c compute two strand specific tag frequency vectors
  n^s_c(x)
> For each h from Hc do {
>         A(h) = Σ_c Σ_{s in {s+,s-}} (N_c/N) P( n^s_c(x + h/2), n^s_c(x - h/2) )
> }
```

3.4 Computation of Nucleosomal Occupancy and/or Enrichment Profiles

The occupancy profiles can be used for visual inspection of the data in a genome browser and for hypothesis generation (*see* **Note 5**).

1. Divide genome into **D** nonoverlapping bins of the selected size. The choice of bin size depends on the depth of sequencing and usually bins from 10 to 50 bp are used (*see* **Notes 6** and **7**).
2. Calculate the frequencies of fragment centers in each bin as the number of fragment centers per million of mapped reads or read pairs (i.e., counts normalized by the library size). In the case of the single-end data the fragment centers are estimated as the positions of 5′-ends of the sequenced reads shifted in the 3′-end direction by half of the characteristic fragment size (*see* Subheading 3.2.2). The fragment centers can be estimated directly from the mate read coordinates in the case of the paired-end data.
3. If the nucleosomal DNA fragments were obtained with chromatin immunoprecipitation step the fragment frequencies computed for ChIP sample should be normalized to the corresponding frequencies computed for non-immunoprecipitated control (input). The normalization can be done either by division of the immunoprecipitated reads frequencies by the input read frequencies (to compute "enrichment") or by subtraction

of the input read frequencies from the immunoprecipitated reads frequencies (to compute "abundance"). In the case of enrichment estimation, it is useful to add pseudo-counts to fragment numbers in each non-overlapping bin to avoid division by zero.

4. The fragment frequencies or enrichments (abundances) in bins can be used for further analysis as well as to produce files in one of the wiggle formats (e.g., "bigwig"), suitable for visualization in a genome browser such as UCSC browser, or others (*see* **Notes 8**).

Pseudocode:

```
>   Divide genome into D non-overlapping bins of equal size
>   For each bin d from D do {
    v[d] = ((number of reads aligned within bin d)) * (1,000,000/(library
>   size))
>   }
    # if the ChIP step was applied e.g. for profiling of a specific histone
    variant, perform the procedure described above for the ChIP and
    input samples, then:
  > enrichment = (v.ChIP+pseudocount)/(v.input+pseudocount)
  > abundance = v.ChIP-v.input
```

3.5 Identification of Stable Nucleosome Positions and Regions of Enrichment in the Nucleosomes of a Specific Type

3.5.1 Stable Nucleosome Positions

Detection of the stable nucleosome positions is methodologically similar to calling peaks in the case of transcription factor-binding profiles. There are multiple algorithms developed for this purpose by us and others [26, 27, 38]. Here, we describe the algorithm implemented in the R/Bioconductor package "nucleR" [27], which is effective for calling nucleosome positions from the paired-end sequencing data. This algorithm comprises the following major steps:

1. Data reading-in and "cleaning." The data can be read from the alignment output files in BAM format and the information on the positions of aligned reads is stored in GRanges object for further use. The coverage profiles are obtained from the sequencing data and they are additionally cleaned using the Fast Fourier Transform analysis (FFT). Briefly, the original profile is transformed into the "frequency space" with FFT, and then the inverse FFT is performed using only a subset of components (usually 2%). This allows obtaining profiles with more pronounced peaks and valleys than the original coverage profiles.
2. Calling dyad positions. Dyad positions are identyfied as local maxima.
3. Nucleosome scoring. Nucleosome positions corresponding to the local peaks are scored based on the peak height and fuzziness.

4. The results of nucleosome calling can be output to the bedGraph-format file for visualization in the genome browser or further analysis.
   R code:

```
# Package installation.
>    source("https://bioconductor.org/biocLite.R")
>    biocLite("nucleR") #try http:// if https:// URLs are not supported.
# Package loading
>    library("nucleR")
# Data reading-in and preparing data structure
>    data <- readBAM("data.file.bam", type = "paired")
>    data <- split(data, seqnames(data))
>    data <- lapply(data,function(d) {
>        strand(d) <- "*"
>        a<-as(d,"RangedData")
>        a[order(start(a)),]
>      })
>      reads <- lapply(data,function(c) processReads(c, type="paired"))
# Calculation of coverage
>      cover <- lapply(names(reads),function(nc)
>                 coverage.rpm(reads[[nc]])[[nc]])
# Smoothing and cleaning of the obtained signal
>      htseq_raw <-  lapply(cover,function(c) as.vector(c))
>      htseq_fft <- lapply(htseq_raw,filterFFT)
# Identification of nucleosomal positions
>      peaks <- lapply(htseq_fft,peakDetection,width=147)
>      names(peaks) <- names(reads)
# Saving identified peaks to bedGraph file
>      peaks2bed <- lapply(names(peaks),function(a){
>        cbind(rep(a,length(peaks[[a]])),start(peaks[[a]]),end(peaks[
[a]]),
>        round(score(peaks[[a]]),4))
>      })
>       peaks2bed <- do.call(rbind,peaks2bed)
# Saving positions of the identified nucleosomes.
> write(paste('track type=bedGraph name="Stable nucleosomes"
>            visibility=dense color=215,16,68 priority=20',sep="")
>             ,file="StableNucleosomes.bedGraph")
>  write.table(as.matrix(peaks2bed),file="StableNucleosomes.bedGraph",
>             row.names=FALSE, col.names=FALSE, quote=FALSE, sep=" "
>            ,append=TRUE)
```

Please *see* nucleR documentation for more details.

3.5.2 Regions of Enrichment

In the case of data generated with ChIP step, a procedure to call broad regions of significant enrichment can be used in addition to calling stable nucleosome positions. The corresponding algorithms compare ChIP and input samples to identify enrichment and its statistical significance; alternatively these estimates can be based on the randomization of ChIP data set. One of such algorithms is implemented in the R package SPP [34] as follows:

1. Data reading-in and initial processing. The alignment output in BAM format is read-in and the genomic positions with abnormally high read counts are filtered out. For data reading-in, one can use internal SPP function (*see* example code below) or function "read.bam.file" from R package MACC [39], which was specifically tested for paired-end data.
2. Estimation of regions where ChIP read frequencies are significantly higher than the input or randomized read frequencies, e.g., using the SPP function "get.broad.enrichment.clusters." This function evaluates the read enrichments and finds the largest region meeting the select significance and enrichment thresholds. An important parameter in this algorithm is the minimal size of the region of enrichment. With the currently used depths of sequencing it is recommended to use sizes around 500 bp or larger.
3. As an additional step, the identified regions of enrichment that are located closer to each other than a selected distance threshold can be joined. The levels of enrichment in such joint regions should be re-estimated.
4. Regions of enrichment can be saved in the ENCODE broadPeak format using the SPP function "write.broadpeak.info."

R code:

```
# Package installation.
>    require(devtools)
>    devtools::install_github('hms-dbmi/spp',build_vignettes=FALSE)
# Package loading
>    library("spp")
# Definition of function for joining identified regions located close to
each other
>    join.close.clusters <- function(cls,min.dist=100){
>           cl <- lapply((1:length(cls)),function(ic){
>               c <- cls[[ic]];o <- order(c$s);
>            s <- c$s[o];e <- c$e[o];
>            l <- length(c$s)
>               i=1;eind <- numeric();sind <-1
>               while(i<l){
>                   d <- s[i+1]-e[i]-min.dist
>                   if(d>0){
>                     eind <- c(eind,i);eind
>                      sind <- c(sind,i+1);sind
>                      i <- i+1
>                }else{
>                      i <- i+1}
>                   }
>                   cat(ic,names(cls)[ic],"\n")
>            return(list(s=s[sind],e=e[c(eind,l)]))
>        })
>        names(cl) <- names(cls)
>          return(cl)
```

```
>    }
# Data reading-in and cleaning
>       chip.data  <- read.bam.tags("chip.bam")  #ChIP sample
>       input.data <- read.bam.tags("input.bam") #input or control sam-
ple
>       binding.characteristics <- get.binding.characteristics(chip.
data,
>                  srange=c(50,500), bin=5,cluster=cluster);
>       chip.data  <- chip.data$tags
>       input.data <- input.data$tags
>       chip.data  <- remove.local.tag.anomalies(chip.data);
>       input.data <- remove.local.tag.anomalies(input.data);
# Identification of broad regions with ChIP signal higher than input
signal
>       regions <-get.broad.enrichment.clusters(chip.data,input.data,
>             window.size=1e3, z.thr=3,
>             tag.shift=binding.characteristics$peak$x/2)
# Joining regions that are located closer to each other than m
>       regions <- join.close.clusters(regions,min.dist=m)
>       regions <- lapply(chrn,function(chr){
>                    d  <- regions[[chr]]
>                    sc <- points.within(abs(chip.data[[chr]] +
>                        binding.characteristics$peak$x/2),
>                           d$s, d$e, return.point.counts = T)
>                    cc <- points.within(abs(input.data[[chr]] +
>                        binding.characteristics$peak$x/2),
>                          d$s, d$e, return.point.counts = T)
>                d$rv <- log2((sc + 1/input.in.bins$lib.size)/(cc +
>                        1/input.in.bins$lib.size))
>                    return(d)
>             })
# Saving positions of the identified regions.
>       write.broadpeak.info(regions,"clusters.broadPeak")
```

3.6 GC-Content Normalization of the Nucleosome Occupancy Profiles and Cross-Sample Comparison

Comparison of nucleosome occupancy profiles is commonly required in the projects that involve different cell types or conditions *see* **Note 9**. The different levels of digestion in the samples can hinder such a comparison even for the replicates of the same cell type. To address this issue we have developed a normalization procedure based on the comparison of the GC-content in each set of nucleosome fragments to a desired target GC-content distribution, and subsequent regularization of the nucleosome occupancy values [40].

1. Compute the GC content of the regions ±100 bp around the genomic location of centers of the sequenced DNA fragment.
2. Produce histogram of the fragment GC-contents with 1% increment. Normalize the histogram, so that it represents fractions (or frequencies) of fragments with particular GC content (sum of all fractions is equal to 1).

3. Compute GC-correction coefficients as the ratios between the fractions from the fragment GC-content histogram and the fractions from the 'target' GC-content histogram. The target histogram may represent the GC-content distribution in a selected data set, in the entire genome, or it can be generated for a subset of the genome or a random model. Previously, for mammalian genomes, we used normal (Gaussian) distribution of GC-contents with mean equal to 50% or 48% and variance equal to 7.5%.
4. For each genomic position where at least one center of the nucleosomal fragment was mapped, normalize the total count of fragments using the GC-correction coefficients corresponding to the GC-content of this position ±100 bp.
Pseudocode:

```
>    For each aligned fragment i from the data set to be corrected do {
>       cen_i = center of fragment i
>       gc[i] = GC.content in [cen_i-100bp, cen_i-100bp]
>    }
>    For k in [0,1,2,..,100] do {
fraction[k] = ( Σ_{g in gc} 1_{g equal to k})/(length of gc)
>    }
>    For each fragment i from the 'target' set do {
>       cen_i = center of fragment i
>       gcT[i] = GC.content in [cen_i-100bp, cen_i-100bp] #alternatively
gcT can be set using Gaussian distribution
>    }
>    For k in [0,1,2,..,100] do {

>     TargetFraction[k] = ( Σ_{g in gcT} 1_{g equal to k} )/(length of gcT)
>    }
>    For k in [0,1,2,..,100] do {
>       GC.CorrCoeffs[k] = fraction[k]/TargetFraction[k]
>    }
>    For pos in genomic positions with mapped fragment centers do {
>       GC.pos_k = GC.content in [pos-100bp,pos+100bp]
>       FragmentCounts[pos] = FragmentCounts[pos]/GC.CorrCoeffs[GC.pos_k]
>    }
```

Here, the GC-content values are expressed as percentages (0,..,100) and $\mathbf{1}_{\{..\}}$ stands for Dirichlet function.

The following steps allow calling the regions of difference in the GC-corrected read frequency profiles corresponding to two samples. The GC-correction procedure may be omitted if the GC-content distributions of the two samples are similar or if there are other reasons to believe that such a correction is not required.

1. Summarize the read frequencies in bins, similar to the steps described in Subheading 3.4. Such frequencies are to be further normalized for the library size. The recommended bin size is 150–200 bp.

2. Identify the bins that have significantly different read frequencies using an appropriate statistical test. Algorithms similar to those developed for the estimation of differentially expressed genes can be used to this end. For instance, DESeq package can be used to estimate the significance of the difference between read frequencies in the bins that correspond to the same genomic locations in the two samples. Since DESeq uses negative binomial distribution for significance calculations, the non-integer per-bin frequencies should be rounded to the nearest integer value. It is recommended to apply this procedure for 25–100 kb segments of the genome independently to facilitate statistical evaluation of the differences.
3. Join the adjacent bins that are significantly different in the two samples to identify the regions of gain or loss of the nucleosome occupancy.

Pseudocode:

```
  >    library(DESeq)
  > Create a matrix "M" with read frequencies in non-overlapping bins.
Let each row correspond to a currently analyzed genomic segment and col-
umns to the samples to be compared. If the frequencies are not integer,
they should be rounded to the nearest integer values.
  >    Create a factor "conditions", corresponding to considered condi-
tions, e.g. "group1" and "group2". The length of the created factor should
be the same as the number of columns.
  >    cds = newCountDataSet(M,conditions)
  >    cds = estimateSizeFactors(cds)
  >    cds = estimateDispersions(cds)
  >    results = nbinomTest(cds,"group1","group2")
```

The object "results" contains the results of the differential occupancy analysis including P-values and fold-changes. The differentially occupied bins located next to each other can be joined using a procedure similar to that described in Subheading 3.5.2.

3.7 Nucleosome Occupancy Profiles around Transcription Start Sites

A visual inspection of nucleosome occupancy profiles around a set of genomic sites of interest is a powerful tool routinely used for data interpretation (Fig. 4) *see* **Note 10**. The sites of interest can represent binding sites of a transcription factor, transcription start, or end sites, etc. Here, we provide an example of such an analysis for the transcription start sites (TSS):

1. Select set of genes of interest.
2. For each of the selected genes identify position of the corresponding TSS.
3. For the selected genes compute number of fragment centers aligned to each genomic position in a region around TSS (e.g., [TSS - 2 kb, TSS + 2 kb]).

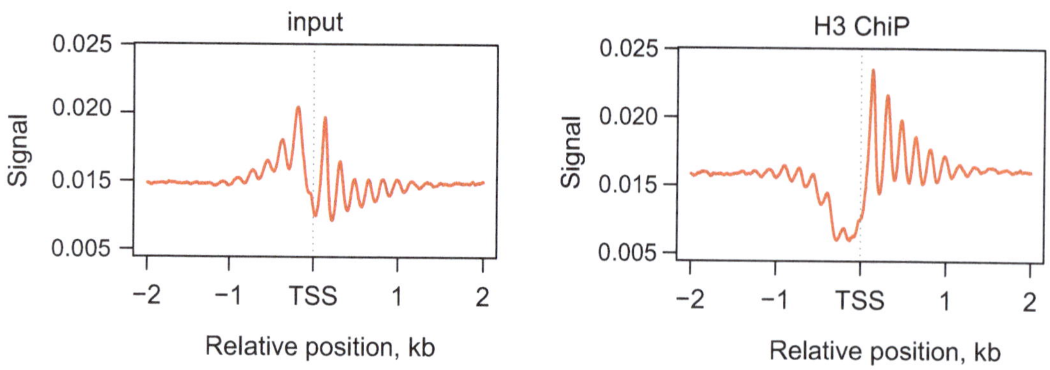

Fig. 4 Typical MNase-seq profiles around transcription start sites of the expressed genes. Pronounced peaks corresponding to nucleosome positions "-1", "+1", "+2", etc. are expected for bulk nucleosomes for moderate levels of digestion. Results are shown for the MNase digestion the chromatin from S2 Drosophila melanogaster cells without (left) and with (right) enrichment for H3 histone

4. Compute the average profile and normalize it to the library size. Usually, the average profiles require additional smoothing in a sliding window or by using a spline approximation.
5. Additionally, the matrix containing the normalized numbers of fragments at each genomic position for each TSS-proximal region may be used for heat map generation.

   Pseudocode:

```
>     G = a set of N genes of interest
>      For each gene g from G do {
>         tss = transcription start site of gene g
>         For i from -w to w do {
>             gn[w+1+i] = number of reads at tss+i position
>         }
>     }
>      For i in 1 to 2*w+1 do {
>         average.profile[i] = (g1[i] + g2[i] + ... + gN[i])/N
>     }
> smoothed.average.profile = smooth average.profile
> To plot heatmap initialize a matrix M with N rows and 2*w+1 columns
> For i in [1,..,N] do {
>         For j in [1,..,2*w+1] do {
>             M[i,j] = gi[j]
>         }
> }
```

4 Notes

1. Nucleosomes cover most of the genome, and as a result profiling "bulk" nucleosomes requires substantial sequencing depth. E.g., in the analysis of nucleosome positioning in the GM12878 cells about four billion single-end reads were pro-

duced [41]. Such sequencing depth, however, is not achieved in most studies, which usually report from several hundred million to one billion reads for mammalian genomes.

2. The reads representing nucleosomal fragments can align to either "+" or "-" DNA strand. E.g., in the case of paired-end data one mate read aligns to the "+" strand and another mate read in a pair aligns to the "-" strand. Short read aligners usually report the most left coordinate of the alignment for any read in the data set. This corresponds to the 5′-end of the reads aligned to "+" stand and to the 3′-end of the reads aligned to the "-" strand. The computational procedures described here often require estimation of the 5′-ends of all reads, because they represent the beginning and the end of nucleosomal fragments. To obtain the 5′-end coordinate of the reads aligned to the "-" strand the reported coordinate should be shifted right by the read length.
3. The estimation of nucleosome protection length, which can be done with cross-correlation analysis for single-end sequencing data and with direct analysis of the fragment lengths for paired-end data, is important for two reasons. First, in the case of enzymatic fragmentation of chromatin with MNase, this analysis allows evaluation of the digestion levels (Fig. 2). The bulk nucleosome samples comprising DNA fragments that are substantially shorter than 150 bp on average are over-digested and the samples with the average fragment length substantially longer than 150 bp are under-digested. Second, it has been reported that nucleosomes bearing some histone variants, e.g., H2A.B, protect shorter DNA fragments than bulk nucleosomes [42, 43], which should be accounted for in the analysis. Further, less stable nucleosomes, e.g., those located at chromatin "hotspots," can be expected to protect shorter DNA fragments than nucleosomes in "bulk" at the same level of digestion.
4. Nucleosomal repeat length can be estimated with the auto-correlation analysis both genome wide and in the subset of specific regions. Comparison of repeat length between genomic regions, cell types, and conditions may provide useful information on chromatin organization and regulation. Normally, and especially for "bulk" nucleosomes that are not enriched for a specific histone variant or modification, the auto-correlation function $A(h)$ exhibits several peaks. The location of the first peak corresponds to the nucleosomal repeat length, i.e., the most prevalent distance between the two adjacent nucleosomes in the genome, and the subsequent peaks represent further nucleosomes.
5. Currently, three metrics are used to represent nucleosome profile along the genomic coordinate. These metrics comprise

(a) frequency of nucleosomal fragments in bins, (b) coverage, i.e., number of nucleosomal fragments spanning over a given genomic location, and (c) enrichment (or abundance), a value usually used for the nucleosomes of a specific type, e.g., bearing a specific histone variant or modification. The procedures to compute nucleosome frequency and enrichment are described in detail here (Subheading 3.4); coverage values can be computed using standard tools for manipulations with sequenced reads, e.g., bedTools [33] or by other means, e.g., using packages such as nucleR (Subheading 3.5.1). Nucleosome frequency and coverage are both used in the literature to describe nucleosome occupancy.

6. The bin size is an important parameter for the computation of nucleosome occupancy (fragment frequencies). Using smaller bins increases the resolution of the profile, but if the depth of sequencing is not sufficient, the fluctuations in read frequencies in bins may result in less accurate estimation of occupancy. For most studies bins of 15–20 bp in sizes provide sufficient resolution for the generation of genome-wide occupancy profiles. If enrichment is computed, larger bin sizes are often selected. Nucleosome coverage is estimated without binning.
7. Time considerations. Processing large nucleosome positioning data sets may take considerable time (on the scale of hours) even when modern computer servers/clusters are used. To reduce processing times, parallelization (i.e., multi-threaded computations) is recommended. Aligners, such as bowtie, allow parallelization by simply including a corresponding parameter in the command line (*see* specific aligner documentation for more details). If computations are performed in R environment, parallelization can be achieved using R core tools (package "parallel").
8. Absolute values of the observed nucleosome occupancy are sensitive to the level of chromatin digestion used to generate the sequencing data [40, 44–46]. This complicates the cross-sample comparison, and to elevate this effect the experimental approaches based on using several digestions were proposed, e.g., refs [40, 47]. The GC-correction procedure described in Subheading 3.6 provides an additional way to address this experimental variability at the bioinformatic level. It is noteworthy that the comparison of the nucleosomal profiles obtained under different digestion levels may provide useful information, e.g., for the identification of "fragile" or "regulated" nucleosomes [48–52] (*see* also comment 9 for more discussion).
9. Recently, a number of new approaches based on MNase digestion were proposed to extend the information that can be obtained from a single assay beyond nucleosome occupancy and positioning. For instance, using salt fractionation of

chromatin and accounting for the digestion fragments of subnucleosomal sizes was shown to provide valuable information on physical stability of nucleosomes and non-histone protein binding [53, 54]. Another recently suggested approach, MACC, is based on the integrative analysis of several chromatin digestions of increasing depths, which allows profiling of nucleosome occupancy and chromatin accessibility in the same assay [40, 55]. When supplemented with histone enrichment step this approach also allows distinguishing non-histone protein binding from nucleosomal protection of genomic DNA. The pipeline described in this chapter will be helpful for these approaches as well, but it should be supplemented with additional steps specific for each assay.

10. Further analyses that can be performed using nucleosome occupancy profiles depending on the aims of a specific project. One example, described in Subheading 3.7, is the analysis of nucleosome occupancy around sites of interest, such as gene starts or protein-binding sites. Similar analysis can be performed for the distribution of stable nucleosome positions. Other examples include (a) correlation of enrichment in the nucleosomes of a specified type with other metrics, e.g., gene expression or regulatory protein binding at a subset of genes or enhancers, and (b) incorporation of nucleosomal metrics in data mining approaches, such as calling "chromatin states." In most of these analyses integration of nucleosomal metrics with other data types plays an important role, and application of such tools as regression analysis, data clustering, and machine learning algorithms is often required.

References

1. Felsenfeld G (1992) Chromatin as an essential part of the transcriptional mechanism. Nature 355:219–224. https://doi.org/10.1038/355219a0
2. Li B, Carey M, Workman JL (2007) The role of chromatin during transcription. Cell 128:707–719. https://doi.org/10.1016/j.cell.2007.01.015
3. Yuan G-C, Liu Y-J, Dion MF et al (2005) Genome-scale identification of nucleosome positions in S. cerevisiae. Science 309:626–630. https://doi.org/10.1126/science.1112178
4. Barski A, Cuddapah S, Cui K et al (2007) High-resolution profiling of histone methylations in the human genome. Cell 129:823–837. https://doi.org/10.1016/j.cell.2007.05.009
5. Schones DE, Cui K, Cuddapah S et al (2008) Dynamic regulation of nucleosome positioning in the human genome. Cell 132:887–898. https://doi.org/10.1016/j.cell.2008.02.022
6. Mavrich TN, Jiang C, Ioshikhes IP et al (2008) Nucleosome organization in the Drosophila genome. Nature 453:358–362. https://doi.org/10.1038/nature06929
7. Valouev A, Ichikawa J, Tonthat T et al (2008) A high-resolution, nucleosome position map of C. elegans reveals a lack of universal sequence-dictated positioning. Genome Res 18:1051–1063. https://doi.org/10.1101/gr.076463.108
8. Tolstorukov MY, Kharchenko PV, Park PJ (2010) Analysis of the primary structure of chromatin with next-generation sequencing. Epigenomics 2:187–197. https://doi.org/10.2217/epi.09.48
9. Almouzni G, Cedar H (2016) Maintenance of epigenetic information. Cold Spring Harb Perspect Biol 8:a019372. https://doi.org/10.1101/cshperspect.a019372
10. Zhang Z, Pugh BF (2011) High-resolution genome-wide mapping of the primary struc-

ture of chromatin. Cell 144:175–186. https://doi.org/10.1016/j.cell.2011.01.003
11. Lai WKM, Pugh BF (2017) Understanding nucleosome dynamics and their links to gene expression and DNA replication. Nat Rev Mol Cell Biol 18:548–562. https://doi.org/10.1038/nrm.2017.47
12. Cui K, Zhao K (2012) Genome-wide approaches to determining nucleosome occupancy in metazoans using MNase-Seq. Methods Mol Biol 833:413–419. https://doi.org/10.1007/978-1-61779-477-3_24
13. Han H, Cortez CC, Yang X et al (2011) DNA methylation directly silences genes with non-CpG island promoters and establishes a nucleosome occupied promoter. Hum Mol Genet 20:4299–4310. https://doi.org/10.1093/hmg/ddr356
14. Kelly TK, Liu Y, Lay FD et al (2012) Genome-wide mapping of nucleosome positioning and DNA methylation within individual DNA molecules. Genome Res 22:2497–2506. https://doi.org/10.1101/gr.143008.112.To
15. Brogaard K, Xi L, Wang J-P, Widom J (2012) A map of nucleosome positions in yeast at base-pair resolution. Nature 486:496–501. https://doi.org/10.1038/nature11142
16. Henikoff S, Ramachandran S, Krassovsky K et al (2014) The budding yeast centromere DNA element II wraps a stable Cse4 hemisome in either orientation in vivo. elife 3:e01861. https://doi.org/10.7554/eLife.01861
17. Voong LN, Xi L, Sebeson AC et al (2016) Insights into nucleosome organization in mouse embryonic stem cells through chemical mapping. Cell 167:1555–1570.e15. https://doi.org/10.1016/j.cell.2016.10.049
18. Noll M (1974) Subunit structure of chromatin. Nature 251:249–251
19. Lohr D, Van Holde KE (1975) Yeast chromatin subunit structure. Science 188:165–166
20. Kornberg RD, LaPointe JW, Lorch Y (1989) Preparation of nucleosomes and chromatin. Methods Enzymol 170:3–14
21. Zaret K (2005) Micrococcal nuclease analysis of chromatin structure. Curr Protoc Mol Biol: Unit 21.1
22. Tsompana M, Buck MJ (2014) Chromatin accessibility: a window into the genome. Epigenetics Chromatin 7:33. https://doi.org/10.1186/1756-8935-7-33
23. Buenrostro JJD, Giresi PPG, Zaba LCL et al (2013) Transposition of native chromatin for fast and sensitive epigenomic profiling of open chromatin, DNA-binding proteins and nucleosome position. Nat Methods 10:1213–1218. https://doi.org/10.1038/nmeth.2688
24. Ishii H, Kadonaga JT, Ren B (2015) MPE-seq, a new method for the genome-wide analysis of chromatin structure. Proc Natl Acad Sci U S A 112:E3457–E3465. https://doi.org/10.1073/pnas.1424804112
25. Chen PB, Zhu LJ, Hainer SJ et al (2014) Unbiased chromatin accessibility profiling by RED-seq uncovers unique features of nucleosome variants in vivo. BMC Genomics 15:1104. https://doi.org/10.1186/1471-2164-15-1104
26. Chen K, Xi Y, Pan X et al (2013) DANPOS: dynamic analysis of nucleosome position and occupancy by sequencing. Genome Res 23:341–351. https://doi.org/10.1101/gr.142067.112
27. Flores O, Orozco M (2011) nucleR: a package for non-parametric nucleosome positioning. Bioinformatics 27:2149–2150. https://doi.org/10.1093/bioinformatics/btr345
28. Woo S, Zhang X, Sauteraud R et al (2013) PING 2.0: an R/Bioconductor package for nucleosome positioning using next-generation sequencing data. Bioinformatics 29:2049–2050. https://doi.org/10.1093/bioinformatics/btt348
29. Vainshtein Y, Rippe K, Teif VB (2017) NucTools: analysis of chromatin feature occupancy profiles from high-throughput sequencing data. BMC Genomics 18:158. https://doi.org/10.1186/s12864-017-3580-2
30. Liou D-M, Chang W-P (2015) Applying data mining for the analysis of breast cancer data. Methods Mol Biol:175–189. https://doi.org/10.1007/978-1-4939-1985-7_12
31. Wang J (2017) Logical reasoning (inferencing) on microRNA data. Methods Mol Biol:197–209. https://doi.org/10.1007/978-1-4939-7046-9_15
32. Thimmaiah T, Voje WE, Carothers JM (2015) Computational design of RNA parts, devices, and transcripts with kinetic folding algorithms implemented on multiprocessor clusters. Methods Mol Biol:45–61. https://doi.org/10.1007/978-1-4939-1878-2_3
33. Quinlan AR, Hall IM (2010) BEDTools: a flexible suite of utilities for comparing genomic features. Bioinformatics 26:841–842. https://doi.org/10.1093/bioinformatics/btq033
34. Kharchenko PV, Tolstorukov MY, Park PJ (2008) Design and analysis of ChIP-seq experiments for DNA-binding proteins. Nat Biotechnol 26:1351–1359. https://doi.org/10.1038/nbt.1508

35. Anders S, Huber W (2010) Differential expression analysis for sequence count data. Genome Biol 11:R106. https://doi.org/10.1186/gb-2010-11-10-r106
36. Langmead B, Trapnell C, Pop M, Salzberg SL (2009) Ultrafast and memory-efficient alignment of short DNA sequences to the human genome. Genome Biol 10:R25. https://doi.org/10.1186/gb-2009-10-3-r25
37. Li H, Durbin R (2009) Fast and accurate short read alignment with burrows-wheeler transform. Bioinformatics 25:1754–1760. https://doi.org/10.1093/bioinformatics/btp324
38. Tolstorukov MMY, Kharchenko PVP, Goldman JAJ et al (2009) Comparative analysis of H2A.Z nucleosome organization in the human and yeast genomes. Genome Res 19:967–977. https://doi.org/10.1101/gr.084830.108
39. Cook A, Mieczkowski J, Tolstorukov MY et al (2017) Single-assay profiling of nucleosome occupancy and chromatin accessibility. Curr Protoc Mol Biol:21.34.1–21.34.18. https://doi.org/10.1002/cpmb.45
40. West JA, Cook A, Alver BH et al (2014) Nucleosomal occupancy changes locally over key regulatory regions during cell differentiation and reprogramming. Nat Commun 5:4719. https://doi.org/10.1038/ncomms5719
41. Kundaje A, Kyriazopoulou-Panagiotopoulou S, Libbrecht M et al (2012) Ubiquitous heterogeneity and asymmetry of the chromatin environment at regulatory elements. Genome Res 22:1735–1747. https://doi.org/10.1101/gr.136366.111
42. Tolstorukov MY, Goldman JA, Gilbert C et al (2012) Histone variant H2A.Bbd is associated with active transcription and mRNA processing in human cells. Mol Cell 47:596–607. https://doi.org/10.1016/j.molcel.2012.06.011
43. Bao Y, Konesky K, Park Y-J et al (2004) Nucleosomes containing the histone variant H2A.Bbd organize only 118 base pairs of DNA. EMBO J 23:3314–3324. https://doi.org/10.1038/sj.emboj.7600316
44. Chung H-R, Dunkel I, Heise F et al (2010) The effect of micrococcal nuclease digestion on nucleosome positioning data. PLoS One 5:e15754. https://doi.org/10.1371/journal.pone.0015754
45. Dingwall C, Lomonossoff GP, Laskey RA (1981) High sequence specificity of micrococcal nuclease. Nucleic Acids Res 9:2659–2673
46. Weiner A, Hughes A, Yassour M et al (2010) High-resolution nucleosome mapping reveals transcription-dependent promoter packaging. Genome Res 20:90–100. https://doi.org/10.1101/gr.098509.109
47. Rizzo JM, Bard JE, Buck MJ (2012) Standardized collection of MNase-seq experiments enables unbiased dataset comparisons. BMC Mol Biol 13:15. https://doi.org/10.1186/1471-2199-13-15
48. Chereji RV, Kan T-W, Grudniewska MK et al (2016) Genome-wide profiling of nucleosome sensitivity and chromatin accessibility in *Drosophila melanogaster*. Nucleic Acids Res 44:1036–1051. https://doi.org/10.1093/nar/gkv978
49. Iwafuchi-Doi M, Donahue G, Kakumanu A et al (2016) The Pioneer transcription factor FoxA maintains an accessible nucleosome configuration at enhancers for tissue-specific gene activation. Mol Cell 62:79–91. https://doi.org/10.1016/j.molcel.2016.03.001
50. Kubik S, Bruzzone MJ, Jacquet P et al (2015) Nucleosome stability distinguishes two different promoter types at all protein-coding genes in yeast. Mol Cell 60:422–434. https://doi.org/10.1016/j.molcel.2015.10.002
51. Teif VB, Beshnova D a, Vainshtein Y et al (2014) Nucleosome repositioning links DNA (de)methylation and differential CTCF binding during stem cell development. Genome Res 24:1285–1295. https://doi.org/10.1101/gr.164418.113
52. Xi Y, Yao J, Chen R et al (2011) Nucleosome fragility reveals novel functional states of chromatin and poises genes for activation. Genome Res 21:718–724. https://doi.org/10.1101/gr.117101.110
53. Henikoff JG, Belsky JA, Krassovsky K et al (2011) Epigenome characterization at single base-pair resolution. Proc Natl Acad Sci U S A 108:18318–18323. https://doi.org/10.1073/pnas.1110731108
54. Mieczkowski J, Cook A, Bowman SK et al (2016) MNase titration reveals differences between nucleosome occupancy and chromatin accessibility. Nat Commun 7:11485. https://doi.org/10.1038/ncomms11485
55. Henikoff S, Henikoff JG, Sakai A et al (2009) Genome-wide profiling of salt fractions maps physical properties of chromatin. Genome Res 19:460–469. https://doi.org/10.1101/gr.087619.108

Part III

Microscopy and Cell Biology

Chapter 11

Imaging Newly Synthesized and Old Histone Variant Dynamics Dependent on Chaperones Using the SNAP-Tag System

Júlia Torné, Guillermo A. Orsi, Dominique Ray-Gallet, and Geneviève Almouzni

Abstract

Distinct histone variants mark chromatin domains in the nucleus. To understand how these marks are established and maintained, one has to decipher how the dynamic distribution of these variants is orchestrated. These dynamics are associated with all DNA-based processes such as DNA replication, repair, transcription, heterochromatin formation and chromosome segregation. Key factors, known as histone chaperones, have been involved in escorting histones, thereby contributing to the chromatin landscape of given cell types. SNAP-tag-based imaging system enables the distinction between old and newly deposited histones, and has proved to be a powerful method for the visualization of histone variant dynamics on a cell-by-cell basis. This approach enables the tracking of specific variants in vivo and defining their timing and mode of deposition throughout the cell cycle and in different nuclear territories. Here, we provide a detailed protocol to exploit the SNAP-tag technology to assess the dynamics of newly synthesized and old histones. We then show that combining the SNAP-tagging of histones with the knockdown of candidate factors, represents an effective approach to decipher the role of key actors in guiding histone dynamics. Here, we specifically illustrate how this strategy was used to identify the essential role of the chaperone HIRA in deposition of newly synthesized histone variant H3.3.

Key words SNAP-Tag, Histone variants, Histone dynamics, Histone chaperones, H3.1, H3.3, HIRA

1 Introduction

Eukaryotic DNA is organized in the nucleus as a nucleoprotein complex known as chromatin. The most basic unit of chromatin is the nucleosome, which contains a core particle composed of an octamer of two copies of each of the four core histones H3, H4, H2A and H2B, wrapped by 147 bp of DNA [1, 2]. This particle

Electronic supplementary material: The online version of this chapter (https://doi.org/10.1007/978-1-4939-8663-7_11) contains supplementary material, which is available to authorized users.

Guillermo A. Orsi and Geneviève Almouzni (eds.), *Histone Variants: Methods and Protocols*, Methods in Molecular Biology, vol. 1832, https://doi.org/10.1007/978-1-4939-8663-7_11, © Springer Science+Business Media, LLC, part of Springer Nature 2018

displays a variety of forms, defined by the choice of histones that exist as distinct variants, and can be further diversified via harboring a collection of different post-translational modifications. This versatile module thus provides a means to define different chromosomal domains enriched in particular variants and modifications [3]. Variants of the histone H3 include the centromeric CENP-A variant, which marks the mammalian centromere, a structure that is essential for kinetochore formation and chromosome segregation [4–7], and two other major non-centromeric variants: H3.1 and H3.3. Histone H3.1 contributes to the bulk of chromatin and is deposited in a DNA synthesis coupled (DSC) manner during DNA replication and repair [8–10]. New H3.1 deposition in the nucleus is thus detected at sites corresponding to newly-replicated DNA during S-phase [9]. In contrast, H3.3 is deposited in a DNA synthesis independent (DSI) manner and at any time during interphase. It is found enriched at genomic regions undergoing active nucleosome turnover and in a subset of heterochromatic regions including telomeres [9–14].

Histone chaperones selectively escort distinct histone variants and are critical for their deposition at specific territories [10, 15, 16]. For example, the chaperone HJURP (holliday junction recognition protein) mediates CENP-A deposition at the centromere in late mitosis/early G1 [17]. The CAF-1 complex (chromatin assembly factor 1) couples H3.1 deposition with DNA synthesis [8, 10]. The HIRA (histone regulator A) complex ensures H3.3 deposition in transcriptionally active regions and at sites of DNA repair [8, 10], while DAXX/ATRX (death domain associated protein/ alpha-thalassemia X-linked mental retardation) has been involved in H3.3 enrichment in heterochromatic regions [13].

To characterize chaperone function with respect to individual histone variants, and the mechanisms by which histone variants contribute to chromosome organization, it is critical to measure histone dynamics. At the single-cell level, this can be globally achieved with fluorescent-recovery after photobleaching (FRAP), as previously described [18, 19]. However, distinguishing old from newly-synthesized histones is essential to explore how deposition of new histones is achieved and to determine the levels of recycling of parental material. In yeast, this can be accomplished using recombination-induced tag switching (RITS) through the exchange of a fluorescent tag to distinguish old and recent proteins. However, this system is dependent on DNA recombination and thus achieves relatively low temporal resolution and its optimization for use in organisms other than yeast is not trivial.

The pioneering work from Lars Jansen using the SNAP-tag method to explore CENP-A deposition [20], inspired us to extend this approach to other variants [10, 20, 21]. SNAP-tag, a mutant form of the human DNA repair protein O6-alkylguanine-DNA alkyltransferase, was engineered to covalently and irreversibly react with a cell-permeable guanine derivative, O6-benzylguanine (BG) (Fig. 1a). The BG moiety can in turn be combined with

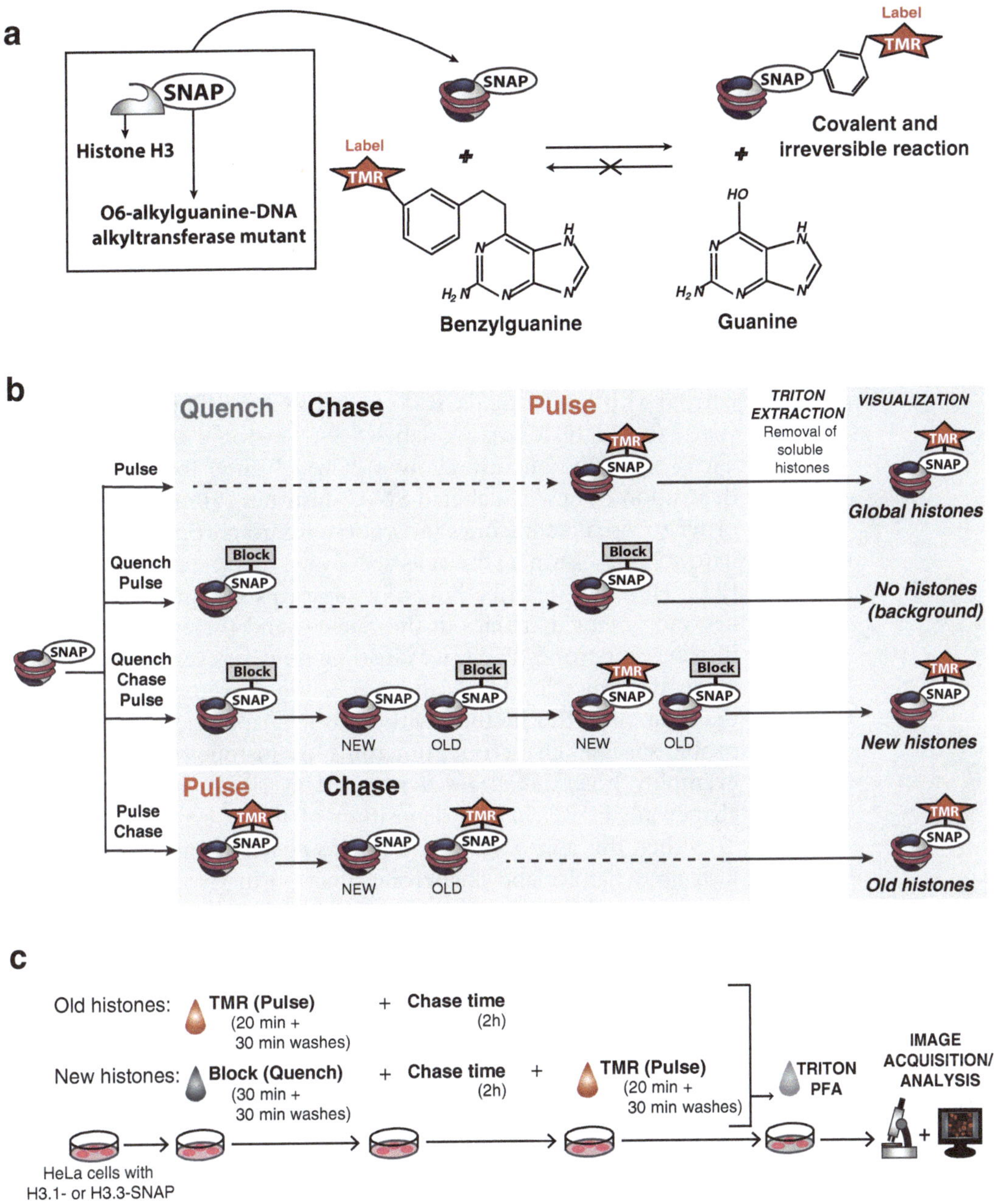

Fig. 1 Histone dynamics imaging strategy. (**a**) H3.1 or H3.3 histone variants are fused to the SNAP-tag, a 20-kDa mutant of the DNA repair protein O6-alkylguanine-DNA alkyltransferase that reacts specifically and covalently with benzylguanine derivatives in vivo. H3.1- or H3.3-SNAP histone variants incorporated into the nucleosome particle are irreversibly labeled with a synthetic probe such as the fluorescent TMR (exemplified in this figure) or the optically inert "Block". (**b**) Protocol scheme for labeling of global histones ("Pulse"), no-labeling control ("Quench-Pulse"), new histones ("Quench/Chase/Pulse") and old histones ("Pulse/Chase"). Triton extraction is performed to eliminate the soluble fraction and visualize chromatin associated/incorporated histones. (**c**) Experimental pipeline for cell preparation, staining, image acquisition and analysis for new and old histones

fluorophores such as tetramethylrhodamine (TMR), enabling visualization of SNAP-tag proteins by fluorescence microscopy, or with an optically inert group bromothenylpteridine (BTP), referred to as "Block".

Here, we describe the use of H3.1 and H3.3 histone variants fused to the SNAP-tag under different in vivo labeling protocols to selectively follow global, newly-synthesized or old proteins (Fig. 1b, c). To visualize global histones, we label all SNAP-histones with a pulse of TMR ("Pulse"). Alternatively, to track the fate of newly-synthesized histones, we first quench pre-existing SNAP-histones with Block, then allow a "Chase" time for synthesis and deposition of new proteins, and these new SNAP-histones are labeled with a pulse of TMR ("Quench/Chase/Pulse"). Finally, to visualize old histones, we label SNAP-histones with a pulse of TMR and subsequently allow a "Chase" time for synthesis and deposition of new, unlabeled SNAP-histones ("Pulse/Chase"). In order to visualize histones incorporated into chromatin and eliminate soluble histones, the cells are Triton extracted before fixation [22]. Here, we describe how to exploit this system to characterize histone variant dynamics in the nucleus and their dependence on histone chaperones. First, we illustrate how to reveal distinct deposition dynamics for H3.1 and H3.3. We then present a strategy that combines this method with histone chaperone knockdown to probe specific chaperone functions in histone deposition. We exemplify how this strategy revealed a critical function for the chaperone HIRA in the deposition of newly synthesized H3.3. Together, this approach proves to be a powerful method to dissect histone dynamics and chaperone function in vivo, at the scale of individual cells.

2 Materials

2.1 Cell Lines

We use HeLa cell lines with stable ectopic expression of H3.1-SNAP-3XHA or H3.3-SNAP-3XHA [10, 20] (*see* **Notes 1 and 2**). In selected clones we verified that the expression levels of tagged H3.1 and H3.3 compared to the endogenous counterparts were compatible to use as tracers with minimal perturbation of the total levels (*see* **Note 3**) [9].

2.2 Cell Culture Reagents

1. Complete medium: Dulbecco's Modified Eagle Medium (D-MEM) (1×), liquid (high glucose) with L-Glutamine, 4500 mg/L D-Glucose, 110 mg/L Sodium Pyruvate. Supplement medium with 10% newborn calf serum (add 50 mL to a 500 mL DMEM bottle) and 100 U/mL penicillin and 100 μg/mL streptomycin.
2. Phosphate Buffered Saline 1× (PBS).

3. 24- and 6-well cell culture plates.
4. Glass coverslips (12 mm diameter, 0.17 mm thickness).

2.3 SNAP Labeling Reagents

1. Block: SNAP-Cell Block (S9106S, New England Biolabs). Dissolve one vial of 100 nmol by adding 50 μL of sterile dimethyl sulfoxide (DMSO) for a stock concentration of 2 mM. Aliquot and store at −20 °C (*see* **Note 4**).
2. TMR: SNAP-Cell TMR-Star (tetramethylrhodamine), (S9105S, New England Biolabs). Dissolve one vial containing 30 nmoles by adding 150 μL DMSO for stock concentration of 200 μM. Store at −20 °C, always protected from light (*see* **Note 5**).

2.4 Triton Extraction and Cell Fixation

1. Cytoskeletal (CSK) buffer: 10 mM Pipes-KOH, pH 7.0, 100 mM NaCl, 300 mM sucrose, 3 mM $MgCl_2$. Use stock solutions of 5 M NaCl, 300 mM MgCl2, 0.5 M Pipes-KOH. Prepare this buffer with fresh sucrose. Store at 4 °C for up to 2 weeks or aliquot it at −20 °C for longer term storage.
2. CSK buffer containing 0.5% Triton X-100.
3. PFA: 4% Paraformaldehyde (stock solution). Dilute 16% solution, Methanol free, RNase free in PBS, aliquot and store at −20 °C.

2.5 SiRNA for Knockdown Treatment

1. ON-TARGETplus siHIRA 3'UTR: gacctaagacctatgtaaa (Dhamarcon).
2. ON-TARGETplus Non-targeting siRNA #1 (Dhamarcon).
3. Lipofectamine RNAiMAX (Invitrogen).
4. OptiMEM® medium (Invitrogen).

2.6 Cell Imaging

1. DAPI (4′, 6-diaminido-2-phenylindole) 1 mg/mL (4000× stock) to label DNA.
2. Vectashield mounting medium (Vector).
3. Microscope slides.
4. Nail Polish.

2.7 Epifluorescence Microscopy Imaging: Acquisition and Analysis

1. Zeiss Axiovert Z1 equipped with a 63× objective, and a Hamamatsu digital camera (C11440).
2. Metamorph software is used for image acquisition (Downloaded in https://www.moleculardevices.com/systems/metamorph-research-imaging/metamorph-microscopy-automation-and-image-analysis-software).
3. FIJI software for fluorescence quantification. (Downloaded at https://fiji.sc).

3 Methods

All volumes indicated below are adapted to treatment and staining of one 12 mm coverslip in a 24-well plate well. Medium and PBS need to be pre-warmed at 37 °C before protocol procedure. All steps before cell fixation are carried out in a cell culture hood under sterile conditions to avoid contamination.

3.1 Newly Synthesized Histone Labeling: « Quench/Chase/Pulse » Strategy

During the "Quench" step all pre-existing histones are coupled to the Block. The "Chase" step allows new histones to be synthesized and deposited onto DNA. Finally, during the "Pulse" step, only the newly deposited histones are labeled (Fig. 1b and c) (*see* **Notes 6, 7** and **8**).

In each well seed 0.1 × 10^6 cells expressing the SNAP-tagged histones on glass coverslips in complete medium (*see* **Note 9**). Incubate cells in a humidified 37 °C incubator with 5% CO_2 overnight or until they reach the desired confluence (70–80%) (*see* **Note 10**).

3.1.1 Quench

1. Aspirate complete medium and add 200 μL of complete medium with 10 μM Block.
2. Incubate at 37 °C, 5% CO_2 for 30 min.
3. Wash cells twice with PBS: aspirate complete medium with block, add 1 mL of PBS to each well, aspirate off and add another 1 mL PBS.
4. Replace PBS with 1 mL of complete medium and incubate at 37 °C, 5% CO_2 for 30 min (excess Block diffuses out).
5. Wash cells twice with PBS (as in **step 3**).

3.1.2 Chase

This step can be omitted for a "Quench/Pulse" control, (Fig. 1b) (*see* **Note 8**).

1. Replace PBS with 1 mL of complete medium.
2. Incubate at 37 °C, 5% CO_2 for 2 h (this step allows synthesis and deposition of new SNAP-tagged histones into chromatin to replace pre-existing histones) (*see* **Note 11**).

3.1.3 Pulse

The "Pulse" step can be performed alone to label global histones (Fig. 1b) (*see* **Note 7**).

1. Aspirate medium and add 200 μL of complete medium with 2 μM fluorescent TMR (newly synthesized histones are labeled by the fluorophore at this stage) (*see* **Note 12**).
2. Incubate at 37 °C, 5% CO_2 for 20 min. From now, all incubation steps must be performed protected as much as possible from light.
3. Wash cells twice with 1 mL of PBS.
4. Replace PBS with 1 mL of complete medium.

5. Incubate at 37 °C, 5% CO_2 for 30 min (excess TMR fluorophore diffuses out).
6. Wash cells twice with 1 mL of PBS (*see* **Note 13**).

3.1.4 Triton Extraction and Fixation

Triton extraction eliminates soluble proteins and preserves the chromatin fraction, thus enriching for chromatin-incorporated histones [22] (*see* **Note 14**).

1. Wash cells with 1 mL of CSK buffer.
2. Replace the buffer with 250 μL of CSK buffer containing 0.5% Triton.
3. Incubate at room temperature for 5 min (*see* **Note 15**).
4. Rinse cells gently twice with 250 μL of CSK with no Triton.
5. Replace CSK by 250 μL 2% PFA to fix the cells (do this under a chemical aspirating hood according to applicable safety procedures).
6. Incubate at room temperature in the dark for 20 min.
7. Discard PFA from the wells.
8. Wash cells three times with 1 mL of PBS (*see* **Note 16**).

3.1.5 DAPI Staining and Coverslip Mounting

1. Dilute 4000× stock DAPI to 1× in PBS.
2. Remove PBS and add 250 μL/well of 1× DAPI.
3. Incubate at room temperature for 5 min.
4. Remove DAPI (discard according to applicable safety procedures) and wash twice with 1 mL of PBS.
5. Leave each coverslip in 1 mL of PBS per well and proceed to mounting of the slides.
6. Put one drop (~20 μL) of vectashield medium on a slide.
7. Pick up a single coverslip with tweezers, use precision wipes to remove excess PBS and lay it on the vectashield drop cell-side down, avoiding formation of bubbles.
8. Remove excess of vectashield with tissue paper and seal borders with nail polish.
9. Protect the slides from light (*see* **Note 17**).

3.2 Old Histones Labeling: « Pulse/Chase » Strategy

When performing a "Pulse" without previous blocking of the preexisting SNAP-histones, global histones will be labeled and visualized. However, a "Chase" step following the "Pulse" will allow new unlabeled histones to be synthesized and deposited. Visualization after "Pulse/Chase" will reveal old histones that were retained in chromatin (Fig. 1b, c). Prepare cells as described above (*see* Subheading 3.1).

3.2.1 Pulse

Proceed with "Pulse" as described above (*see* Subheading 3.1.3).

3.2.2 Chase

Continue with the Chase instructions (*see* Subheading 3.1.2).

3.2.3 Triton Extraction and Fixation

Perform Triton extraction and fixation as described in Subheading 3.1.4 (**steps 1–8**).

3.2.4 DAPI Staining and Coverslip Mounting

Proceed with DAPI staining and slide preparation as described in Subheading 3.1.5 (**steps 1–9**).

3.3 Newly Synthesized Histone Labeling Coupled to Chaperone Knockdown

Combined with the SNAP-tag technology, the function of selected chaperones can be assessed by depleting them using siRNA. Here, as an example, we show how to estimate the effect of HIRA knockdown on H3.1- and H3.3-SNAP deposition by performing a newly synthesized histone labeling assay ("Quench/Chase/Pulse") following siHIRA treatment (Fig. 3) (*see* **Notes 18** and **19**). Five days are required for completion of the entire procedure.

1. (**Day 1**) In a 6-well plate, seed cells in two wells, each containing 4 coverslips, using the same conditions as described above (Subheading 3.1) at 0.1 × 10^6 cells/well (10–20% confluency) (*see* **Note 20**).
2. Incubate overnight in a humidified 37 °C incubator with 5% CO_2 (this will allow plated cells to attach to the coverslip and begin cycling).
3. (**Day 2**) Prepare transfection solution for one well of a 6-well plate for siHIRA and one for siControl:
   (a) Add 9 μL of lipofectamine to 150 μL of OptiMEM medium.
   (b) In separate tubes, add 3 μL of 10 μM solution of siHIRA or siControl to 150 μL of OptiMEM medium.
   (c) Mix the 150 μL siRNA solution with the 150 μL lipofectamine solution.
   (d) Incubate at room temperature for 5 min.
   (e) Add 250 μL of solution to each corresponding well (*see* **Note 21**).
4. Incubate for 72 h in the humidified 37 °C incubator with 5% CO_2 (*see* **Note 22**).
5. (**Day 5**) Fill one well of a 24-well plate with 1 mL PBS for each desired labeling condition (*see* **Note 23**).
6. Transfer coverslips to a 24-well plate (one coverslip per well).
7. Proceed with "Quench/Chase/Pulse" strategy to label newly synthesized histones (*see* Subheadings 3.1.1, 3.1.2 and 3.1.3).
8. Perform Triton extraction, fixation, DAPI staining and coverslip mounting as described above (*see* Subheading 3.1.4).

3.4 Cell Imaging and Quantification

An epifluorescence microscope is used for image acquisition. Use a 63× objective to acquire single-plane images. The parameters (i.e., exposure time) (*see* **Note 24**) must be the same for all conditions

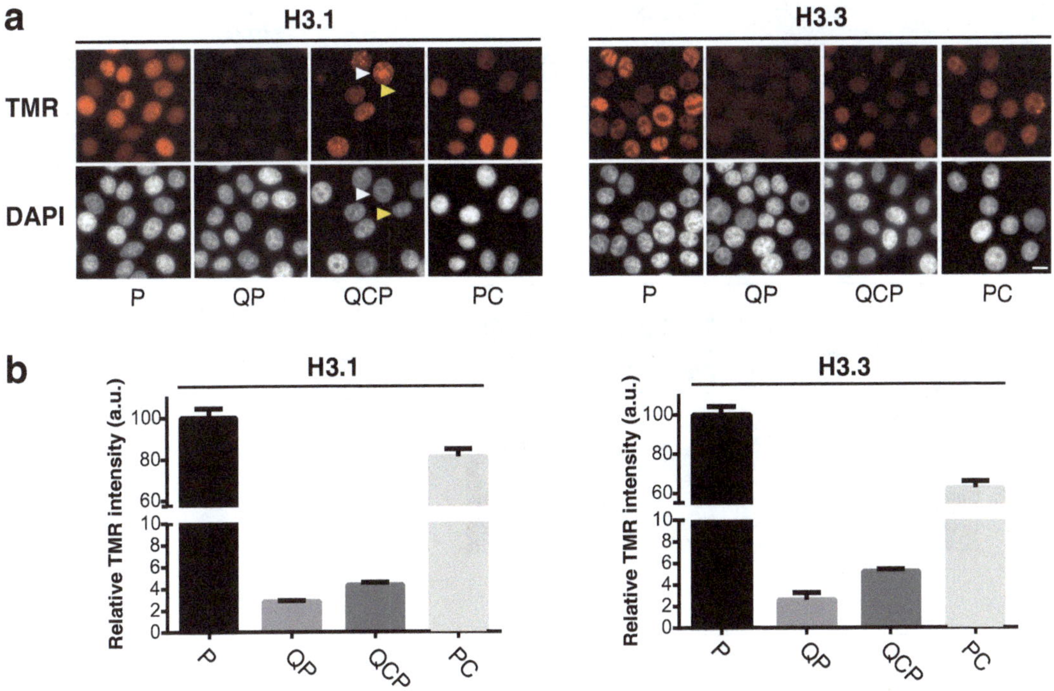

Fig. 2 Histone dynamics visualization and quantification. (**a**) Image visualization of global histones ("Pulse"), no-labeling control ("Quench-Pulse"), new histones ("Quench/Chase/Pulse") and old histones ("Pulse/Chase"). New H3.1-SNAP are detected at replication sites in S-phase cells, as this variant is deposited in a DSC manner (*see* yellow arrow). No new H3.1-SNAP are visualized in cells outside of S-phase (*see* white arrow). In contrast, new H3.3-SNAP are detected in all cells as this variant is deposited in a DSI manner at any time during interphase. Scale bar = 10 μm. (**b**) Cell-by-cell signal intensity is quantified and average is normalized to "Pulse" levels. "Quench/Pulse" leads to dramatically reduced TMR intensity as no histones should be available for labeling after the "Quench" step (this serves as a control for quenching efficiency). The addition of a "Chase" period before the "Pulse" ("Quench-Chase-Pulse") leads to increased TMR intensity, as newly synthesized histones are available for labeling. The "Pulse-Chase" exhibits reduced signal intensity compared to the "Pulse" as only old retained histones remain labeled

within a single experiment. Acquire images to analyze at least 200 nuclei per condition (Figs. 2a, 3b). Save single-color images in .tif format. For file nomenclature, every channel should be assigned a suffix number ("_1", "_2", etc.…).

FIJI (ImageJ) software is used to treat the images and quantify fluorescence signal within the nuclei area. This process is automated using two FIJI macros, one for the subtraction of the background and another one for fluorescence quantification within the nuclei (Supplementary Material Online).

1. Create a folder for every set of images (i.e., one folder per microscopy slide) and copy the images to be analyzed (here, DAPI channel and TMR channel) into the appropriate folder. Make sure each channel has a unique suffix number (*see* above - here, DAPI blue-channel files should be named "*_1.tif" and TMR, red-channel files should be named "*_2.tif").

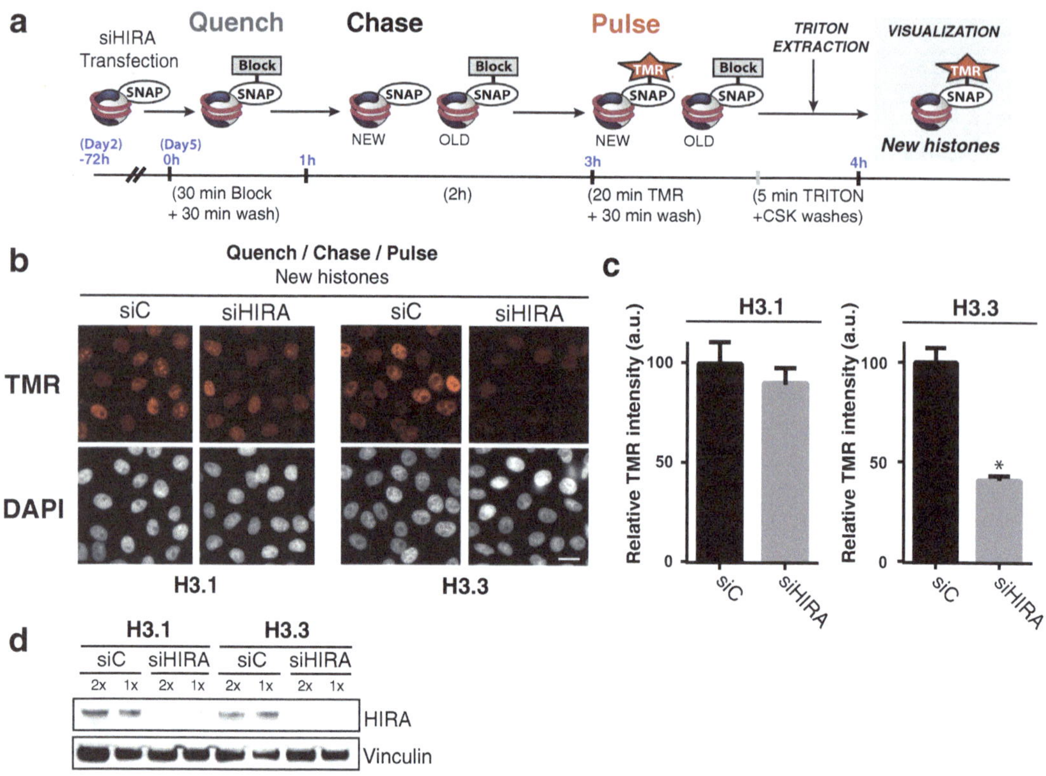

Fig. 3 New histones labeling assay coupled to chaperone depletion. (**a**) Protocol scheme for imaging new histones upon histone chaperone HIRA knockdown. (**b**) Visualization of newly synthesized H3.1- and H3.3-SNAP ("Quench/Chase/Pulse") under siHIRA and siControl conditions. Scale bar = 10 μm. (**c**) TMR intensity levels are quantified cell-by-cell and average is normalized to siControl. The intensity of new H3.3-SNAP is strongly decreased relative to siControl while the intensity of new H3.1-SNAP is only faintly reduced —Mann Whitney significance test. Scale bar = 10 μm. (**d**) Western blot analysis shows strong HIRA depletion upon siHIRA treatment. Note that biological results must be verified by using several different siRNAs (*see* **Note 23**)

2. Subtract background (for every image and both channels) (*see* **Note 25**):
   - Open an image.
   - Use the Oval Selection tool to manually select an area free of cells.
   - Run SubractBackground.ijm macro.
   - Save the background-substracted image.
3. Quantification (example for two-chanel DAPI/TMR images):
   - Run the Nucleus_2D_Quantification.ijm macro
   - Select the folder containing the images to be analyzed.
   - Select the number of channels to analyze (in this case select "2" - DAPI and TMR) (*see* **Note 26**).
   - Select the color of each channel, following their assigned number.

- For each channel/color, CHECK "Quantify Nucleus Fluorescence" if quantification needs to be reported (*see* **Note 27**).
- Select the channel that should be used for nuclei segmentation (in our example, DAPI—blue).
- Select minimum and maximum circularity and size parameters: set these parameters to include all nuclei and exclude debris (*see* **Note 28**).
- Click on "OK" to run the macro.

A semi-colon separated text file named "Results.txt" will be placed in the folder containing the files analyzed. In this file, every nucleus is represented by a line containing file of origin name, nucleus number, nucleus size, and measured fluorescence for all the selected channels. This data can be used to calculate how the signal intensity changes in the different conditions and to asses the variability of the signal from nucleus to nucleus in every condition (Figs. 2b, 3c).

4 Notes

1. In addition to the SNAP-tag system presented in this protocol, the CLIP-tag is a modified version of the SNAP-tag that provides an alternative. The CLIP-tag and the SNAP-tag systems can be combined for simultaneous labeling of two different SNAP and CLIP fusion proteins [23].
2. The H3.1 and H3.3 open reading frames (ORFs) were PCR amplified from complementary DNA and cloned between the KpnI/XhoI sites of pCENP-A-SNAP-3XHA [19]. replacing the CENP-A gene. The resulting H3.1-SNAP-3XHA or H3.3-SNAP-3XHA ORFs were subcloned into pBABE and used for retroviral production and delivery into HeLa cells as described previously [24]. Cells were selected by BlasticidinS.
3. The proportions of chromatin associated H3.1- and H3.3-SNAP-HA can be calculated and compared to the amount of total histone H3 by Western blot analysis using an anti-H3 antibody. Knowing that proportions of endogenous H3.1 and H3.3 variants in HeLa cells represent about 52% and 15% of total H3 respectively [25], we estimated that exogenous H3.1-SNAP-HA and H3.3-SNAP-HA in our cell lines represent about 10% of endogenous H3.1 and 30% of endogenous H3.3, respectively.
4. To completely resuspend SNAP-tag reagent, vortex at maximum speed for 10 min.

5. Aliquot the fluorescent substrate stock to avoid freeze-thaw cycles, which can alter its fluorescent properties. As an alternative to the fluorescent substrate, we also routinely use SNAP-Cell Oregon Green. Note that other fluorescent substrates for SNAP-tag are available.
6. To simultaneously visualize histones and replication sites use the EdU Imaging kit (Invitrogen) to label nascent DNA [10, 21].
7. To perform global histone labeling, a "Pulse" can be performed without previous "Quench" and "Chase" steps (Fig. 1b). The HeLa histone-SNAP-HA cells feature a Hemaglutinin (HA) tag, as described [9, 19]. This can be exploited to label global SNAP-tag histones by immunofluorescent detection of the HA epitope, independently of SNAP-tag chemistry, or to reveal SNAP-HA histones by Western blotting.
8. The no-labeling control "Quench/Pulse" (i.e., without the "Chase" step), must be performed in order to monitor quenching efficiency and background signal (Fig. 1b).
9. Use HeLa cells expressing SNAP-tag histones at 2–10 passages after thawing.
10. Overconfluence of seeded cells can lead to their detachment from the coverslip as this protocol includes several incubation and washing steps. Avoid working with cells over 80% confluence and monitor cell detachment after the washing steps.
11. "Chase" time can be adapted to the temporal scale of the expected biological process. Deposition of new proteins can be observed in fluorescence with Chase times as short as 30 min in our hands.
12. To label replication sites, 10 μM of EdU can be added together with TMR. The addition of EdU at the time of "Pulse" enables to detect cells in S-phase and study incorporation of newly synthesized histones at the replication sites [10, 21].
13. This washing step is important as insufficiently washed TMR results in strong non-specific fluorescence in the cell nucleus and may prevent precise quantification of the TMR signal. Insufficiently washed samples showing high background need to be discarded for fluorescence quantification.
14. Although the non-soluble histones retained after Triton extraction are likely mostly incorporated into chromatin, one cannot exclude remaining non-soluble non-nucleosomal histones associated with complexes or structures that are Triton resistant.
15. This timing can be different for every cell type. Insufficient CSK Triton washing results in TMR signal in the cell cytoplasm. Initially, we recommend 3–10 min (the ideal time of incubation is given when nuclear signal reaches a lower plateau, with minimal signal in the cytoplasm and minimal cell detachment).

16. If EdU has been added to visualize replication sites, perform EdU detection by following the detailed protocol provided with the Imaging EdU kit [21].
17. Store coverslips at 4 °C for short-term storage or at −20 °C for longer storage in the dark.
18. As a limitation of this technology, if a chaperone deposits histones at discrete regions, the amount of newly incorporated histones might not be sufficient to be detected. For example, this is the case of ATRX/DAXX, which deposits H3.3 at heterochromatin regions [9].
19. When probing histone dynamics upon induced perturbations (cell cycle arrest, histone chaperone knockdown, etc...) we strongly recommend Western blot analysis using an anti-HA antibody to check whether the levels of soluble and chromatin pools of SNAP-tag histones are affected.
20. In this experiment, one well (in a 6-well plate) is used for siControl and one well for siHIRA. Each well contains 4 coverslips which can be used to do "Quench/Chase/Pulse" in duplicates and to control for HIRA depletion efficiency by immunofluorescence.
21. Add solution drop by drop while carefully swirling the plate to progressively homogenize the incubation solution.
22. Incubation time can range from 24 to 72 h. OptiMEM containing siRNA can be replaced by complete medium 6 h after transfection.
23. Depletion efficiency has to be verified by Western blot analysis (Fig. 3d) and/or immunofluorescence. Verification of the results must be done by using several different siRNAs. We have also successfully tested ONTARGETplus J-013610-07 and J-013610-06 (Dhamarcon) [10].
24. Depending on the intensity of the fluorescence, choose an exposition time that is long enough for sufficient brightness but avoids over-saturation. In addition, short exposition will prevent photobleaching. To set the exposition parameters, the slide with the highest expected signal should be used.
25. Different levels of background can bias the quantification. We recommend discarding high background samples for fluorescence intensity quantification.
26. SNAP-tag labeling is compatible with subsequent antibody staining, not described in this protocol. This macro allows quantification of other color channels.
27. DAPI staining is used to mask the nuclei and allows segmentation of the area to be quantified. The nuclei mask is applied to the other channels, where the total and surface-normalized (mean) signal will be measured cell-by-cell.

28. The default parameters are for HeLa cells captured with a 63× objective. If either the size or morphology of the imaged cells is different, the parameters for minimum and maximum circularity of the cells can me modified as well as their minimum and maximum size.

Acknowledgements

We thank Shauna Katz for critical reading of the manuscript, Patricia Le Baccon and the PICT-IBiSA@Pasteur Imaging Facility, as well as all members of the Chromatin Dynamics team for stimulating discussions. This work was supported by la Ligue Nationale contre le Cancer (Equipe labellisée Ligue), ANR-11-LABX-0044_DEEP and ANR-10-IDEX-0001-02 PSL, ANR-12-BSV5-0022-02 "CHAPINHIB", ANR-14-CE16-0009 "Epicure", ANR-14-CE10-0013 "CELLECTCHIP", EU project 678563 "EPOCH28", ERC-2015-ADG- 694694 "ChromADICT", ANR-16-CE15-0018 "CHRODYT", ANR-16-CE12-0024 "CHIFT", ANR-16-CE11-0028 "REPLICAF", and Parisian Alliance of Cancer Research Institutes, PSL Aux Frontières Des Labex "TRACK".

References

1. Kornberg RD (1974) Chromatin structure: a repeating unit of histones and DNA. Science 184(4139):868–871
2. Oudet P, Gross-Bellard M, Chambon P (1975) Electron microscopic and biochemical evidence that chromatin structure is a repeating unit. Cell 4(4):281–300
3. Palmer DK, O'Day K, Trong HL et al (1991) Purification of the centromere-specific protein CENP-A and demonstration that it is a distinctive histone. PNAS 88:3734–3738
4. Gurard-Levin ZA, Almouzni G (2014) Histone modifications and a choice of variant: a language that helps the genome express itself. F1000Prime Reports 6:76
5. Malik H, Henikoff S (2003) Phylogenomics of the nucleosome. Nat Struct Biol 10:882–891. https://doi.org/10.1038/nsb996
6. Allshire RC, Karpen GH (2008) Epigenetic regulation of centromeric chromatin: old dogs, new tricks? Nat Rev Gen 9(12):923–937. https://doi.org/10.1038/nrg2466
7. Black BE, Bassett EA (2008) The histone variant CENP-A and centromere specification. Curr Opin Cell Bio 20(1):91–100. https://doi.org/10.1016/j.ceb.2007.11.007
8. Tagami H, Ray-Gallet D, Almouzni G et al. (2004) Histone H3.1 and H3.3 complexes mediate nucleosome assembly pathways dependent or independent of DNA synthesis. Cell 9;116(1):51–61, 116
9. Goldberg D, Banaszynski LA, Noh KM et al (2010) Distinct factors control histone variant H3.3 localization at specific genomic regions. Cell 140(5):678–691. https://doi.org/10.1016/j.cell.2010.01.003
10. Ray-Gallet D, Woolfe A, Vassias I et al (2011) Dynamics of histone H3 deposition in vivo reveal a nucleosome gap-filling mechanism for H3.3 to maintain chromatin integrity. Mol Cell Biol 22(3):874–885. https://doi.org/10.1128/MCB.22.3.874-885.2002
11. Mito Y, Henikoff JG, Henikoff S (2005) Genome-scale profiling of histone H3.3 replacement patterns. Nat Gen 37(10):1090–1097
12. Mito Y, Henikoff JG, Henikoff S (2007) Histone replacement marks the boundaries of cis-regulatory domains. Science 315(5817):1408–1411
13. Drané P, Ouararhni K, Depaux A et al (2010) The death-associated protein DAXX is a novel

histone chaperone involved in the replication-independent deposition of H3.3. Genes Dev 24(12):1253–1265. https://doi.org/10.1101/gad.566910

14. Wong LH, McGhie JD, Sim M et al (2010) ATRX interacts with H3.3 in maintaining telomere structural integrity in pluripotent embryonic stem cells. Genome Res 20(3):351–360. https://doi.org/10.1101/gr.101477.109
15. De Koning L, Corpet A, Haber JE, Almouzni G (2007) Histone chaperones: an escort network regulating histone traffic. Nat Struc Mol Biol 14(11):997–1007
16. Ray-Gallet D, Almouzni G (2010) Nucleosome dynamics and histone variants. Essays Biochem 48(1):75–87. https://doi.org/10.1042/bse0480075
17. Dunleavy EM, Roche D, Tagami H et al (2009) HJURP is a cell-cycle-dependent maintenance and deposition factor of CENP-A at centromeres. Cell 37(3):485–497. https://doi.org/10.1016/j.cell.2009.02.040
18. Kimura A, Umehara T, Horikoshi M (2002) Chromosomal gradient of histone acetylation established by Sas2p and Sir2p functions as a shield against gene silencing. Nat Genet 32:370–377. https://doi.org/10.1038/ng993
19. Meshorer E, Misteli T (2006) Chromatin in pluripotent embryonic stem cells and differentiation. Nat Rev Mol Cell Biol 7:540–546. https://doi.org/10.1038/nrm1938
20. Jansen JE, Black BE, Foltz DR, Cleveland DW (2007) Propagation of centromeric chromatin requires exit from mitosis. J Cell Biol 176(6):795–805
21. Clémént C, Vassias I, Ray-Gallet D, Almouzni G (2016) Functional characterization of histone chaperones using SNAP-tag-based imaging to asses de novo deposition. Methods Enzymol 573:97–117
22. Martini E, Roche DM, Marheineke K et al (1998) Recruitment of phosphorylated chromatin assembly factor 1 to chromatin after UV irradiation of human cells. J Cell Biol 143:563–575
23. Gautier A, Juillerat A, Heinis C et al (2008) An engineered protein tag for multiprotein labeling in living cells. Chem Biol 15(2):128–136. https://doi.org/10.1016/j.chembiol.2008.01.007
24. Shah JV, Botvinick E, Bonday Z et al (2004) Dynamics of centromere and kinetochore proteins; implications for checkpoint signaling and silencing. Curr Biol 14:942–952
25. Loyola A, Bonaldi T, Roche D et al (2006) PTMs on H3 variants before chromatin assembly potentiate their final epigenetic state. Mol Cell 24:309–316

Chapter 12

Real-Time De Novo Deposition of Centromeric Histone-Associated Proteins Using the Auxin-Inducible Degradation System

Sebastian Hoffmann and Daniele Fachinetti

Abstract

Measuring protein dynamics is essential to uncover protein function and to understand the formation of large protein complexes such as centromeres. Recently, genome engineering in human cells has improved our ability to study the function of endogenous proteins. By combining genome editing techniques with the auxin-inducible degradation (AID) system, we created a versatile tool to study protein dynamics. This system allows us to analyze both protein function and dynamics by enabling rapid protein depletion and reexpression in the same experimental setup. Here, we focus on the dynamics of the centromeric histone-associated protein CENP-C, responsible for the formation of the kinetochore complex. Following rapid removal and reactivation of a fluorescent version of CENP-C by auxin treatment and removal, we could follow CENP-C de novo deposition at centromeric regions during different stages of the cell cycle. In conclusion, the auxin degradation system is a powerful tool to assess and quantify protein dynamics in real time.

Key words Auxin-inducible protein degradation system, Centromere, Genome editing, Histones, CENP-C, Time-lapse imaging

1 Introduction

The ability to follow protein dynamics in vivo is extremely important to determine protein function and the role of specific proteins in complexes. This is particularly important for the assembly of large protein complexes, in which the association of specific proteins may rely on the presence of others. Therefore, the analysis of the dynamics of each complex component may yield essential information on the global function and the assembly of the complex itself.

Centromeres are DNA–protein structures necessary to maintain the balance in genetic information by controlling faithful chromosome segregation during cell division. They are the foundation for the assembly of the kinetochore that, in turn, is required

Guillermo A. Orsi and Geneviève Almouzni (eds.), *Histone Variants: Methods and Protocols*, Methods in Molecular Biology, vol. 1832, https://doi.org/10.1007/978-1-4939-8663-7_12,

to interact with spindle microtubules. Centromeres are epigenetically identified by the presence of a histone H3 variant named CENP-A. CENP-A is strongly enriched at centromeric regions and is the base for the assembly of the entire centromere–kinetochore complex. Altogether, the centromere/kinetochore can form a complex with over 100 proteins of about 150 nm in human cells [1]. Recently, it was proposed that the subunits of the constitutive centromere associated network (CCAN) interact between each other in an interdependent manner, and that this interdependency is necessary for the stability of the entire complex [2, 3]. Although most of the centromeric proteins are present along all stages of the cell cycle, their assembly is highly regulated and often restricted to a limited time window [4]. For example, CENP-A is deposited only once every cell cycle, at the exit of mitosis [5] via a tight regulatory mechanism [6].

Here we describe the auxin-inducible degradation (AID) system as a method to track protein dynamics throughout the cell cycle. The AID system is based on the transplantation in eukaryotes [7] of a ligand-induced degradation system found in plants [8, 9]. Addition of the plant hormone indole-3-acetic acid (IAA, auxin) mediates the interaction of the AID-tagged protein of interest with the ectopically expressed F-box from plant (Transport Inhibitor Response 1; TIR1), that can associate with Skp1 in eukaryotic cells and form an SCF-TIR1 complex to induce protein ubiquitination and degradation. AID size is about 25 kDa (similar to GFP), but can be reduced by ~1/3 without loss of functionality, as observed in budding yeast [10].

The auxin degradation system has already been successfully used in the past [2, 11–15] to study protein function following rapid and complete protein degradation at every cell cycle stage. As an example, using this system we recently demonstrated that CENP-A is dispensable for maintenance of an already assembled centromere–kinetochore complex [11].

In this chapter, we illustrate how to exploit one of the key characteristics of the AID system, its reversibility. We describe in detail how to (Fig. 1a) (1) generate a stable human cell line expressing the E3 ubiquitin ligase TIR1, (2) insert an AID tag coupled with a fluorescent protein (mRFP or EYFP) at an endogenous gene locus, (3) select and screen for correct AID integration, and

Fig. 1 (continued) PCR to distinguish normal (+/+), single tagged (+/AID-mRFP) or double tagged (AID-mRFP/ AID-mRFP) cells (**d**) Schematic illustration of the AID system to induce CENP-C degradation including the one incorporated at centromeric regions that interact with the CENP-A nucleosome (in red) and the CCAN complex (multicolor). (**e**) Representative immunofluorescence images of cells showing CENP-C depletion after 24 h treatment with IAA. ACA was used to mark centromere position. Scale bar = 5 μm. (**f**) Quantifications of the experiment in (**e**). Dots represent single centromere. Unpaired t test: ****$p < 0.0001$

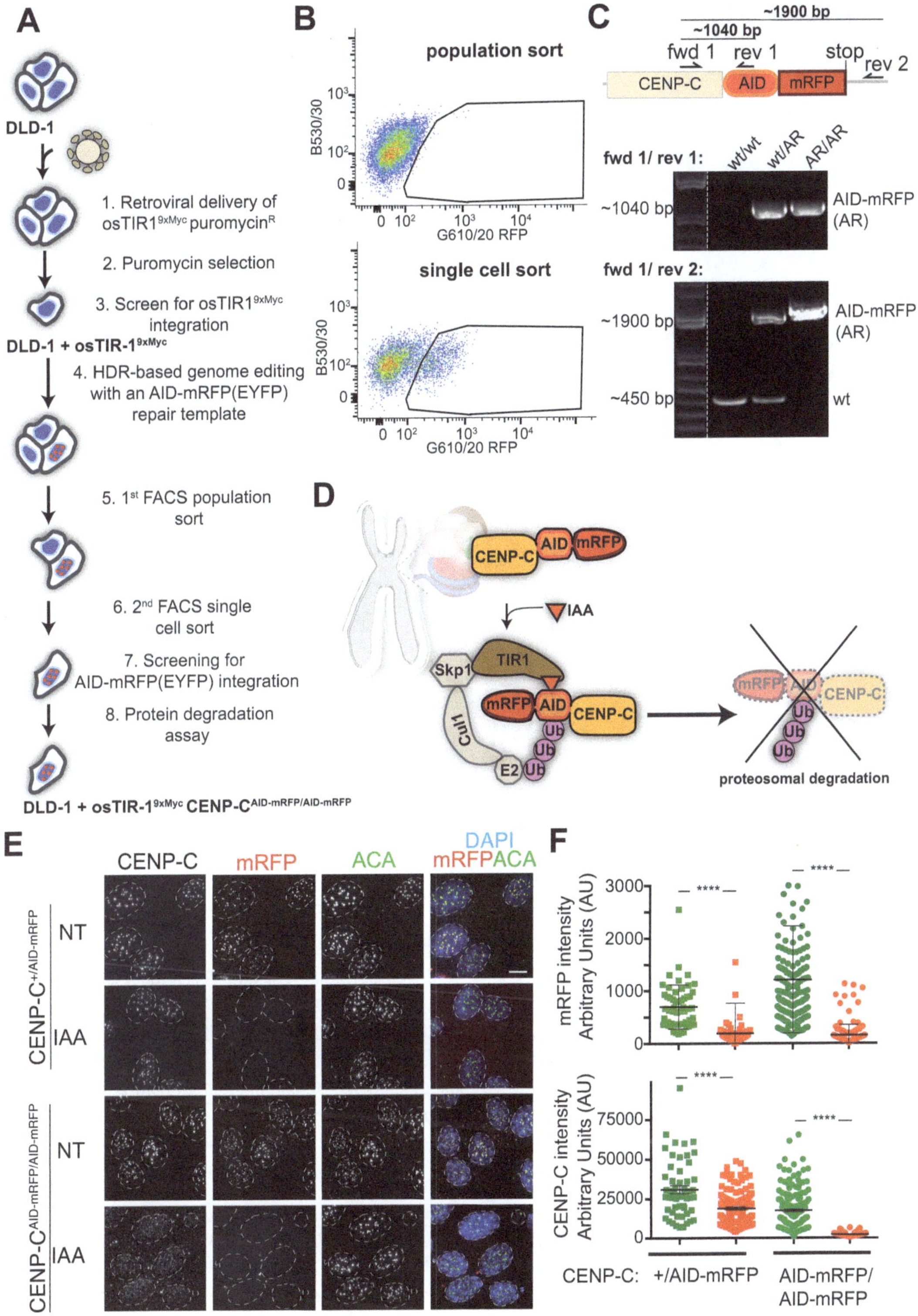

Fig. 1 CENP-C genome engineering. (**a**) Schematic of the TALEN-mediated genome editing strategy to endogenously tag CENP-C with AID and mRFP (or EYFP) in cells expressing osTIR1. (**b**) A two-step FACS selection procedure of cells in which CENP-C has been endogenously tagged with an AID and mRFP. In step one (upper plot) a population of red fluorescent cells was collected. In step two (lower plot) single red fluorescent cells were sorted into a 96-well plate. (**c**) CENP-C genotypes validated in the indicated cell lines using

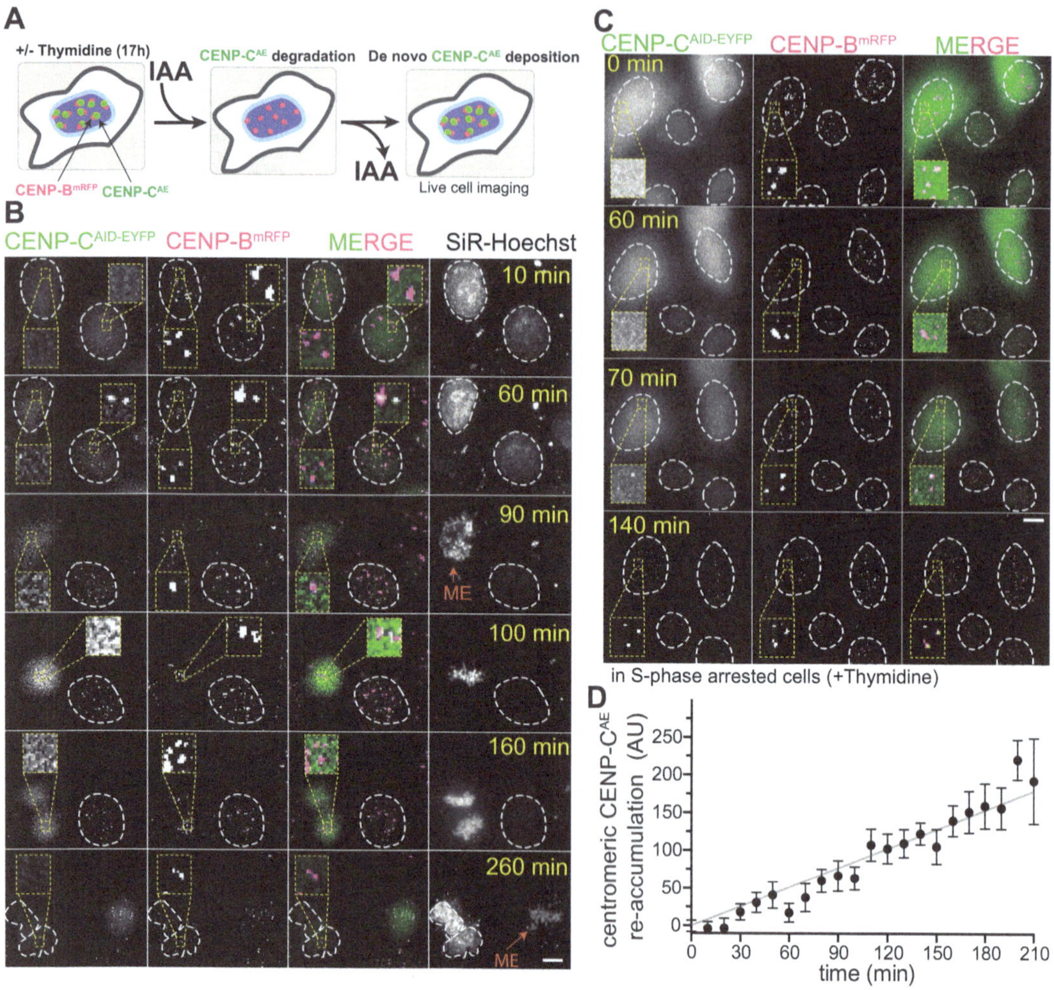

Fig. 2 Real time de novo deposition of CENP-C following rapid protein deletion and reactivation. (**a**) Experimental scheme to follow de novo deposition of CENP-C$^{AID-EYFP}$ in a DLD-1 cell line. (**b** and **c**) Representative live cell images of CENP-C$^{AID-EYFP}$ reloading in asynchronous DLD-1 cells (**b**) or in S-phase arrested cells (**c**) The inset highlighted with a yellow dashed square shows a magnification of a section of CENP-B-marked centromeres. Red arrows mark cell entry into mitosis (mitotic entry, ME). Images were acquired 1 h after IAA wash-out. (**d**) Graphs represent the mean from 10 cells of CENP-C reexpression and reaccumulation at centromeric regions marked by CENP-B-mRFP in S-phase arrested cells followed by live cell imaging. Error bars represent the SEM. AU = Arbitrary units. Scale bar = 5 μm

(4) measure protein dynamics by time lapse microscopy following rapid removal and reactivation of the AID-tagged protein pool (Fig. 2). Here, we focus on the histone CENP-A-associated protein CENP-C, the major component for kinetochore assembly in human cells. However, in principle, any protein of interest can be tagged and assayed for protein dynamics including histones, as we have previously shown for CENP-A [11].

2 Materials

2.1 Cell Culture Reagents

1. DMEM-GlutaMAX (Gibco or equivalent) supplemented with 10% fetal bovine serum (GE Healthcare), 100 U/ml penicillin, 100 U/ml streptomycin.
2. Trypsin (Gibco or equivalent).
3. Phosphate-buffered saline (PBS): 137 mM NaCl, 2.7 mM KCl, 10 mM Na_2HPO_4, 1.8 mM KH_2PO_4.
4. Cell culture incubator (5% CO2, 37 °C).
5. 500 mM auxin indole-3-acetic acid sodium salt (IAA) in water. Use at 500 μM final concentration.
6. 10 mg/ml puromycin. Use at 2 μg/ml final concentration.
7. 10 mg/ml polybrene. Use at 8 μg/ml final concentration.
8. 100 mM thymidine dissolved in water. Use at 2 mM final concentration.

2.2 Cell Line Generation

1. FuGENE HD (Promega) (or equivalent transfection system).
2. VSV-G plasmid.
3. Opti-MEM Reduced Serum Media (Gibco).
4. Cell Line Nucleofector Kit V, Lonza Nucleofector Device.
5. QIAGEN Plasmid Midi Kit (or equivalent plasmid preparation system).
6. SURVEYOR Mutation Detection Kit (Transgenomic or equivalent).

2.3 Cell Selection

1. Fluorescence-activated cell sorting (FACS) device.
2. FACS buffer: 1% BSA, 5 mM EDTA, 5 mM HEPES in 1× PBS.

2.4 Cell Screening

2.4.1 PCR

1. QuickExtract™ DNA Extraction Solution (Epicentre).
2. NanoDrop.
3. Q5 Hot Start High-Fidelity 2× Master Mix (NEB).
4. Oligomers flanking AID-mRFP at the C terminus of CENP-C:

   5′-GTTAGAGGAATCCACAGCAGT-3′ (forward)
   5′-TTACAAAGACAAATATTCCAACTA-3′ (reverse).
   5′-CTCATGAAAGGATCGGATGC-3′ (reverse, binds in the AID sequence).
5. Thermocycler.
6. Agarose gel electrophoresis system.
7. 1% agarose gels (w/v).
8. GelGreen stain (Interchim or equivalent).

2.4.2 Immunofluorescence Microscopy

1. Triton blocking buffer: 0.2 M glycine, 2.5% FBS, 0.1% Triton X-100 in 1× PBS.
2. Primary antibodies: CENP-C (1:1000 MBL International, guinea pig), CENP-B (1:1000 Abcam, rabbit).
3. Secondary antibodies: Fluorophore conjugated anti-guinea pig and anti-rabbit antibodies.
4. DAPI.
5. Antifading mounting reagent (Life technologies or equivalent).

2.4.3 Immunoblot

1. Sample buffer: 62.5 mM Tris, pH 6.8, 2.5% SDS (w/v), 10% glycerol (v/v), 5% β-mercaptoethanol (v/v), 0.002% bromophenol blue (v/v).
2. Running buffer: 25 mM tris base, 192 mM glycine, 0.1% SDS (w/v).
3. Blocking solution: 5% milk (w/v) in TBS with Tween (0.1%).
4. 4–15% Mini-PROTEAN TGX Precast Protein Gel (Bio-Rad).
5. Trans-Blot Turbo Transfer System (Bio-Rad).
6. Trans-Blot Turbo Midi Nitrocellulose Transfer Packs (Bio-Rad).
7. Tris-buffered saline (TBS)–Tween: 50 mM Tris–HCl, pH 7.5, 150 mM NaCl, 0.1% Tween (v/v).
8. Antibodies for immunoblot: Anti-Myc-tag, clone 4A6 (Merck chemicals, ref.: 05-724, used at 1 μg/ml in blocking solution), ECL™ Anti-mouse IgG, Horseradish Peroxidase linked whole antibody (from sheep, GE Healthcare UK, used 1:10,000 in blocking solution).
9. Pierce HCL (Thermo Scientific or equivalent).
10. Gel imaging system.

2.5 Live Cell Imaging

1. μ slide 8-well (IBIDI Cat. No. 80826 or equivalent).
2. Deltavision core system equipped with a CoolSNAP_HQ2 Camera and an Olympus UPlanSApo 100× oil-immersion objective (numerical aperture 1.4) for fixed cells and an Olympus 40× oil-immersion objective for live cell imaging (Applied Precision).
3. Software for data analysis: softWoRx (Applied Precision), ImageJ (open source), MetaMorph
4. CO_2-independent medium (Gibco) supplemented with 10% fetal bovine serum (GE Healthcare) 100 U/ml penicillin, 100 U/ml streptomycin.
5. SiR-DNA dye kit (Spirochrome).

3 Methods

3.1 Cell Line Generation

3.1.1 Retrovirus-Mediated osTIR1 Transgene Integration

Auxin mediated degradation of an AID-tagged protein is only achievable in the presence of the plant E3 ubiquitin ligase TIR1 (transport inhibitor response 1). To stably express TIR1 trans-gene from rice (*Oryza sativa*; *os*TIR1) into human cells (in our case colorectal adenocarcinoma DLD-1) this procedure exploits retrovirus-mediated integration. There are several methods to deliver DNA to mammalian cells in tissue culture. Here, we describe a method using Fugene HD (Promega) transfection reagent.

Retrovirus Production

1. Day 0: Seed 3×10^6 239-GP cells in a 10 cm cell culture plate (total volume: 10 ml).
2. Day 1: Transfection. Add 15 μl Fugene HD to 500 μl OptiMEM or free FBS/antibiotics DMEM medium and vortex for 1 s (use prewarmed medium).
3. Add 3 μg VSV-G and 5 μg osTir1-9xMyc construct in a pBABE vector backbone comprising a puromycin resistance cassette.
4. Incubate transfection mix for 15 min at RT and add the mix dropwise onto the cells. (Biosafety measures for working with the retroviruses need to be adopted at this point.)
5. Day 2: Replace culture medium with 6 mL fresh medium. Allow virus production for another 48 h.
6. Day 4: Collect 6 ml virus-containing medium and filter the medium through a 0.45 μm filter.
7. Aliquot the virus, snap-freeze in liquid nitrogen and store at −80 °C or use directly (see below).

Retrovirus Infection

Biosafety measures for working with the virus are required

8. Day 0: seed DLD-1 cells in a 6-well cell culture plate at 20% confluency (3 wells/transfection).
9. Day 1: replace medium with 2 ml of culture medium containing 8 μg/ml Polybrene.
10. Day 2: add 250 μl, 500 μl, and 750 μl of the virus to three separate wells and allow virus infection for 2 days.
11. Day 3: Wash out the virus three times with 2 ml culture medium. Wash cells once with 2 ml 1× PBS, remove PBS and add 300 μl trypsin to detach cells from the tissue culture dish. Incubate cells in trypsin for 5 min in the incubator. After about 2-5 min shake of the cells. Confirm cell detachment using a bright light microscope. When all cells are detached from the plate resuspend cells in 1.7 ml culture media. Pool and seed infected cells in a 15 cm cell culture plate. Add culture medium (total volume: 25 ml).

12. Day 4: add puromycin to select for cells with osTIR1 insertion (final puromycin concentration: 2 μg/ml for the DLD-1 cell line).

osTIR1 Screening

Two to seven days after puromycin addition refresh the cell culture medium in order to remove dead cells and to ensure puromycin efficacy. Two weeks after puromycin addition it is possible to observe the formation of puromycin resistant colonies. For colony isolation, proceed as follows:

13. Sterilize small (~1 cm^2) Whatman paper pieces.
14. Soak the paper in trypsin. Remove the excess of trypsin by squeezing the paper against an empty sterile surface (e.g., cell culture plate).
15. Wash the puromycin resistant colonies with 15 ml 1× PBS. Remove PBS.
16. Place the Whatman paper from step 14 directly onto a colony using sterile forceps.
17. After around 5 min in the cell culture incubator, cells start to detach. At this point, convey the colony into a 24-well plate containing 0.5 ml prewarmed cell culture medium by transferring the whole paper.
18. On the following day, remove the Whatman paper and replace the medium.

Immunoblot

Next, perform immunoblot using an anti-Myc antibody to screen all single cell-derived colonies for genomic integration and expression of osTIR1-9x-Myc construct.

19. Once the single cell clones are confluent collect cells by trypsinization as described in **step 11**. Here, use 100 μl of trypsin for each well of a 24-well plate and resuspend cells with 400 μl culture media. Seed 100 μl of cells into a 12-well plate in duplicate. Add culture media to the 12-well plate (total volume: 1 ml). Use the remaining 400 μl for western blot analysis.
20. Collect the 400 μl cells in a 1.5 ml Eppendorf tube. Harvest cells by centrifugation (2000 × g for 5 min).
21. Carefully remove the supernatant and wash the cell pellet once with 1× PBS. Resuspend cells in 150 μl sample buffer.
22. Boil samples at 100 °C for 10 min and separate proteins of the cell extracts by SDS–polyacrylamide gel electrophoresis (PAGE). Use a 10% Mini-PROTEAN® TGX™ precast protein gel and a biorad electrophoresis system or similar. Insert a SDS-PAGE gel into the designated electrophoresis system and add running buffer to the system. Load 30 μl protein samples and 10 μl of a protein ladder in separate gel pockets using a Hamilton syringe or gel loading tips for pipettes. Apply low voltage (70 V) for accurate protein separation.

23. After electrophoresis transfer proteins onto a nitrocellulose membrane with the Biorad Trans-Blot® Turbo™ Transfer System or equivalent.
24. After transfer, block the membrane in 25 ml blocking solution for 30 min at room temperature in a box that allows full coverage of the membrane.
25. Incubate the membrane with 10 ml primary antibody against the myc-tag in 0.5% milk TBS-Tween for 2 h at room temperature. Ensure that the membrane is completely covered.
26. Wash the membrane 3× with 25 ml TBS–Tween at room temperature (5 min per wash).
27. Incubate for 45 min with 10 ml secondary antibody linked to horseradish peroxidase at room temperature.
28. Wash 3× for 10 min with 25 ml TBS–Tween at room temperature.
29. Develop the membrane using Pierce HCl and a gel imaging system.

Based on the western blot result, pool single cell-derived colonies with similar high expression levels of osTIR1-9x-Myc (*see* **Note 1**).

Use this cell population for the subsequent genome editing process as described in the following sections.

3.1.2 AID-Tag Integration at the CENP-C Locus

In the next step, introduce an AID tag combined with a fluorescent protein [here: monomeric red fluorescent protein (mRFP) or enhanced yellow fluorescent protein (EYFP)] at the endogenous loci of a centromeric histone-associated protein (in this case CENP-C, but a similar strategy was used to target the histone H3 variant CENP-A). To this end, this procedure describes the use of a TALEN genome editing strategy (*T*ranscription *A*ctivator-*L*ike *E*ffector binding domain coupled with a FokI *N*uclease); however, it can easily be adapted for a CRISPR/Cas9 or any other genome editing strategy (*see* **Note 2**).

The AID-mRFP(EYFP)-tag is integrated at the target site via the homology-directed repair (HDR) (*see* **Note 3**).

In this case, genome targeting plasmids were delivered by electroporation using a Lonza Cell Line Nucleofector Kit and the Lonza Nucleofector electroporation device (but other transfection methods could be used).

Electroporation

1. One day before transfection, grow DLD-1 cells in a 10 cm cell culture dish in antibiotics-free cell culture medium (total volume: 10 ml).
2. Prepare a 1 μg plasmid mix containing 800 ng repair template plasmid, 100 ng left TALEN plasmid, and 100 ng right TALEN plasmid in a 1.5 ml Eppendorf tube (use Midi-Prep DNA quality for all plasmids). In addition to the transfection of the gene

targeting plasmids we also recommend to perform a separate control by transfecting 500 ng of a plasmid encoding soluble GFP.

3. Trypsinize cells as described previously (here: use 1.5 ml trypsin and resuspend cells with 8.5 ml culture medium). Collect cells in a 15 ml Falcon tube. From a fully confluent 10 cm cell culture dish three transfections can be performed.
4. Centrifuge cells for 5 min at 1000 × *g* at room temperature.
5. Aspirate medium completely and resuspend cells in 100 μl (for each transfection) of Lonza Cell Line Nucleofector kit V buffer with supplement (*see* **Note 4**).
6. Add resuspended cells to the plasmid cocktail and mix gently by pipetting up and down. Wait for 5 min.
7. Transfer the plasmid–cell mixture without generating bubbles into an electroporation cuvette provided in the nucleofector kit.
8. Place the cuvette in the Lonza Nucleofector device and apply program U-017 (*see* **Note 4**).
9. Transfer cells carefully with a plastic pipette (provided in the kit) to a 6-well cell culture plate containing 2 ml prewarmed cell culture medium without antibiotics. Shake the plate gently to dispense cells evenly.
10. One day post-transfection refresh the medium supplemented with antibiotics. Check the efficiency of transfection using a fluorescence microscope (e.g., EVOS FL Cell Imaging System). Transfection efficiencies of 70% to 90% with 50–90% cell viability are commonly reached.
11. Grow cells for around 3–5 days and then expand to obtain a bigger number of cells.

Clonal Selection

In order to select for DLD-1 clones in which the AID-mRFP/EYFP tag has been inserted a two-step FACS selection protocol was used (*see* **Note 5**).

12. Three to five days after transfection, harvest cells in 15 mL Falcon tube by trypsinization as described above.
13. Wash cells with 5 ml FACS buffer and resuspend cells in 0.5 to 1 ml FACS buffer. Also prepare a negative control with non-transfected DLD-1 cells in the same way.
14. Sort mRFP-positive cells in a tube and plate cells in a cell culture dish (Fig. 1b, upper panel) (*see* **Notes 6** and **7**).
15. Once cells are happily growing and a good confluency is achieved (normally after 2–5 days), perform a single-cell sorting into a 96-well plate using filtered conditional medium (50% fresh medium, 50% 3-day old medium) (*see* **Note 8**). At

this time, a strong enrichment of RFP-positive cells can be observed (Fig. 1b, lower panel).

Before performing single cells sorting, screening of cells for positive integration using PCR screening is recommended (see below).

3.1.3 Screening for clones

Following FACS selection there are multiple possibilities to screen for AID-mRFP/EYFP integration. Here we will present three of them (*see* **Note 9**).

Cell Viability Screening

The most straightforward way is to initially screen for cell viability if the protein of interest is essential as in the case of CENP-C. To this end:

16. Seed clones derived from single cell sorting into duplicate 24-well plates containing regular medium or regular medium + IAA (total volume: 0.5 ml).
17. Clones that die within 4 days of IAA treatment are selected.
18. Expand the corresponding untreated single cell clones.
19. Confirm the integration of AID-mRFP in these clones by PCR (as described below) and immunofluorescence microscopy.

PCR Examination for AID-mRFP Integration

Perform standard PCR to test for the integration of the AID-mRFP (EYFP) tag (*see* **Note 10**).

DNA extraction using QuickExtract™ DNA Extraction Solution:

20. Collect cells from a 1 ml fully confluent 12 cm cell culture dish by trypsinization and centrifugation in a 1.5 ml Eppendorf tube.
21. Wash cell pellets with PBS and resuspend in 50 μl QuickExtract™ DNA Extraction Solution (scaling of cell pellet/solution is possible) (Other DNA extraction procedures can be used).
22. Incubate for 5 min at RT.
23. Vortex the DNA extraction mix for 10 s.
24. Incubate for 6 min at 65 °C.
25. Vortex the DNA extraction mix for 15 s.
26. Incubate for 2 min at 98 °C.
27. Vortex the DNA extraction mix for 15 s.
28. Cool down DNA for PCR reaction (store at −20 °C).
29. Measure DNA concentration using a NanoDrop™ (use 300 ng/μl final DNA concentration for the PCR).
30. Prepare the PCR mix according to the table below.
31. Run PCR reaction according to the settings below.

20 μl PCR mix	PCR reaction settings
– 10 μl 2x Q5 polymerase mix – 1 μl 10 μM forward primer(5′-GTTAGAGGAATCCACAGCAGT-3′) – 1 μl 10 μM reverse primer(5′-TTACAAAGACAAATATTCCAACTA-3′) – 1 μl extracted DNA – 7 μl ddH_2O	1. 98 °C 3 min 2. 98 °C 30 s 3. 58 °C 30 s 4. 72 °C 2 min (repeat **steps 2** to **4**) 34 times 5. 72 °C 10 min 6. 12 °C hold

32. Analyze the PCR results by gel electrophoresis using a 1% (w/v) agarose gel containing GelGreen.
33. Excise the PCR products from the gel.
34. Purify the PCR product by using a standard DNA-agarose gel extraction kit and use a Sanger sequencing service to bidirectionally sequence the PCR product.
35. Verify the successful integration of the AID-mRFP/EYFP tag at the CENP-C locus with the sequencing results.

Immunofluorescence Microscopy Screen

CENP-$C^{AID\text{-}mRFP/AID\text{-}mRFP}$ clones identified by the viability screening and PCR analysis can be seeded for immunofluorescence microscopy analysis in duplicate on coverslips in a 24-well plate.

36. In order to test for the degradation of CENP-$C^{AID\text{-}mRFP}$ add IAA overnight to one of the slides prior to fixation (Fig. 1d, e) (*see* **Note 11**).
37. Preextract cells with 0.5 ml 1× PBS-T (0.1% Triton X-100) for 1 min at room temperature.
38. Fix cells by adding 200 μl 4% formaldehyde for 5–10 min at room temperature.
39. Wash out Formaldehyde with 0.5 ml 1× PBS-T (2 times).
40. Incubate for 30 min in 0.5 ml Triton blocking buffer at room temperature.
41. Incubate with 150 μl primary antibodies in Triton blocking buffer for 1 h (anti-CENP-C and anti-CENP-B) at room temperature.
42. Wash 3× with 0.5 ml PBS-T.
43. Incubate with 150 μl secondary antibodies in Triton blocking buffer for 45 min at room temperature.
44. Wash 3× with 0.5 ml PBS-T.
45. Incubate with 150 μl DAPI in PBS (1 μg/ml) in PBS for 10 min at room temperature.
46. Mount coverslips with antifading reagent.

Acquire fluorescence microscopy images using a fluorescent microscope such as the DeltaVision Core system (Applied Precision). We use a 100x Olympus UPlanSApo 100 oil-immersion objective (numerical aperture 1.4) and a 250 W Xenon light source. The system is equipped with a Photometrics CoolSNAP_HQ2 Camera. Acquire 4 μm z-stacks (step size: 0.2 μm). Quantification of CENP-C using an automated system [16] and ACA to mark centromeres reveals that CENP-C is completely depleted at the centromere after addition of IAA (Fig. 1f).

3.2 Auxin-Mediated Depletion, Reexpression, and Reloading of CENP-$C^{AID\text{-}EYFP}$ at the Centromere

The auxin degradation system is a useful tool to study consequences of rapid and complete protein depletion. Due to its reversibility, the system can also be used to study de novo reexpression and localization of AID-tagged proteins including histone variants. Here we use the auxin degradation system to follow de novo deposition of CENP-C following its complete degradation. To study the dynamics of CENP-C deposition along the cell cycle we used the CENP-$C^{AID\text{-}EYFP/AID\text{-}EYFP}$ cell line in which also one allele of CENP-B has been endogenously tagged with an mRFP-tag (the CENP-B tagging was carried out in a similar way as described in the previous section). The presence of mRFP on CENP-B allows us to mark centromere position and to follow de novo CENP-C deposition by live cell imaging. In order to completely deplete CENP-$C^{AID\text{-}EYFP}$ (henceforth referred to as CENP-C^{AE}) we added IAA to the culture medium. After CENP-C^{AE} depletion we washed-out auxin and allowed the reexpression of CENP-$C^{AID\text{-}EYFP}$. We followed CENP-C^{AE} reloading at the centromere by live-cell imaging in asynchronous cells or in cells arrested in S-phase (Fig. 2a).

3.2.1 Cell Preparation

1. Day 0: seed cells in five adjacent wells of an 8-well IBIDI slide (total volume: 300 μl).
2. Day 1: replace media with 300 μl 2 mM thymidine-containing cell culture medium overnight for the S-phase arrested condition.
3. Day 2: replace media with 200 μl IAA (500 μM final concentration) for 6 h (except for the Not Treated—NT condition) and also maintain thymidine for S-phase arrested cells.
4. Wash out IAA carefully three times with 300 μl culture medium.
5. Leave cells for 15 min in tissue culture incubator and repeat washes (*see* **Note 12**). During this time incubate cells with SiR-DNA dye kit (working solution 1 μM).
6. Replace media with CO_2 independent medium and start live-cell imaging immediately (for the S-phase arrested condition maintain thymidine in the cell culture medium).

3.2.2 Live Cell Imaging with Fluorescence Microscopy

1. Prior to image acquisition adjust temperature to 37 °C.
2. For imaging, we use the DeltaVision Core system (Applied Precision) with an Olympus 60X/1.42, Plan Apo N Objective. The microscope is equipped with a CoolSNAP_HQ2 camera.

3. Acquire 20 μm z-stacks (2 μm step size) with a 2 × 2 binning. Acquire images every 10 min (*see* **Note 13**).

3.2.3 Experimental Analysis

1. CENP-C^{AE} reloading can be qualitatively analyzed in different cell cycle phases using ImageJ (open source image processing software) after generating maximum intensity projections of the deconvolved z-stacks (*see* **Note 14**).
2. To quantify CENP-C^{AE} reexpression at the centromeres over time, save un-deconvolved 2D maximum intensity projections as un-scaled 16-bit TIFF images and signal intensities determined using MetaMorph (Molecular Devices).
3. Draw a 10 × 10 pixel circle around a centromere (marked by CENP-B^{mRFP}) and draw an identical circle adjacent to the centromere (background). The integrated signal intensity of each individual centromere is calculated by subtracting the fluorescence intensity of the background from the intensity of the adjacent centromere.
4. Average ten centromeres to provide the average fluorescence intensity for each individual cell and quantify about ten cells per experiment in S-phase arrested cell condition.

4 Conclusion

Several techniques exist to analyze protein dynamics such as Fluorescence Recovery After Photobleaching (FRAP), photoactivation of proteins, Recombination Induced Tag Exchange (RITE) and SNAP-based pulse labeling (for a detailed list *see* Table 1 in ref. 17). The AID system described here has several advantages over some of the aforementioned methods to analyze protein turn over. First and most important, it allows for the analysis of both protein function and dynamics in the same experimental setting by inducing rapid protein depletion (by IAA) and by achieving rapid protein reexpression (by IAA removal), both processes within minutes. Auxin is permeable (it passes through the cell membranes) and easy to wash out since it is dissolved in water (no need of DMSO), and it does not require the presence of any additional components (e.g., CRE recombinase in the RITE) to start monitoring protein dynamics. Since specific IAA-mediated protein degradation does not affect mRNA, the AID-tagged protein reaccumulates very rapidly, allowing live measurement of protein turnover at short timescales in every cell (in large numbers) and at every complex (in this case centromeres). Also, it does not require any particular laser or specialized equipment, so a standard fluorescence microscope can be used.

One disadvantage of the AID technique is that an "ad hoc" system is required for every protein of interest, involving extensive gene modification such as gene tagging that may disrupt protein function and the insertion of a transgene (TIR1). Also, target protein dynamics completely depend on protein expression levels, and there could be some photobleaching events during time-lapse experiments. However, at least one of these caveats is also found in all of the techniques mentioned above, highlighting again the unique advantage of using the AID system.

In summary, here we present the auxin degradation system as a unique tool to study the de novo deposition of centromeric proteins. This tool could be adapted to study a wide range of other proteins and protein complexes to gain better insight into their function and dynamics.

5 Notes

1. High level of osTIR1 expression is essential for rapid and complete degradation of AID-tagged target proteins. The majority of clones express the osTIR1-9x-Myc transgene due to the puromycin selection after virus infection; however, expression levels might change due to the different integration sites within the genome. In order to decide which clones to pick, band quantification on the western blot of osTIR1-9x-Myc expression level is critical. The usage of a positive reference for TIR1 expression level [13] is recommended. We commonly pool all clones that display expression levels that are at least half as strong as the level of the reference clone. An alternative strategy is to AID-tag the gene of interest before adding the TIR1. This strategy will help if different levels of TIR1 are required to degrade the protein of interest (e.g., too much TIR1 might lead to leaky protein degradation in the absence of IAA); however, it will preclude the usage of the "*Cell viability screening*" method described in this book chapter.
2. We used a TALEN-based genome editing approach, which allows for site-specific genomic modifications with a very low chance of off-target genome editing effects. The design of left and right TALEN DNA recognition FokI fusion constructs have been extensively described [18] and, due to space limitations, will not be addressed here. Briefly, TALENS are cloned into a modified version of pcDNA3.1 (purchased from Invitrogen) containing also a cDNA sequence for the Fok I endonuclease domain. TALENs CENP-C target sequences: GAGGAAAGTGTTCTTC and GGTTGATCTTTCATC [16]. TALENs cleavage efficiency is tested by using a surveyor assays as described in Ran et al. (2013) [19].
3. To exploit HDR in cells, a repair template plasmid (containing the *AID-mRFP or AID-EYFP* sequence flanked by a 0.3–1 kb

homologous DNA sequence on each site) is cotransfected along with plasmids expressing left and right TALEN DNA recognition proteins. The homologous DNA sequences are designed up- and downstream of the STOP genomic target site. Since HDR is a rare event we increased the repair template plasmid concentration relative to the left and right TALEN plasmids in the cotransfection mixture. Since the N-terminus of CENP-C is important for the interaction with Mis12, a crucial component for kinetochore formation [20], we decided to introduce the AID-mRFP tag at the C-terminus. Previously, we have also introduced the AID-tag at the N-terminus of other proteins such as CENP-A with a very similar approach. In our opinion, the amino terminal tagging might be advantageous since following double strand break formation by TALENs or Cas9, DNA will be more frequently repaired via the nonhomologous end-joining pathway. This error-prone repair can lead to rearrangements (deletion or insertion) and then, possibly, to deactivation of one of the alleles. In this case, the necessity to introduce the AID tag via HDR is reduced to one instead of two alleles.

4. The choice of the kit and the electroporation program are cell line dependent and need to be tested. Lonza provides a database for the correct choice of the kit and electroporation program but some troubleshooting might be required.
5. A one-step single cell sorting procedure after transfection commonly results in a high number of false-positive cells. This is due to low AID-mRFP/EYFP integration efficiency and low fluorescence intensity of cells expressing mRFP (or EYFP). This will highly depend on the endogenous promoter of the tagged gene of interest. We first enriched for a population of mRFP-positive cells. Next, after expansion of the sorted population, we performed single cell sorting for clone selection of integration of the desired tag.
6. Population sorting requires up to 1 h since the number of mRFP (EYFP)-positive cells might be very low (~0.1%). We sort cells in 200 μl CO_2-independent medium. The medium will preserve the correct pH value during the sorting procedure. Sorted cells can be centrifuged and resuspended in normal cell culture medium prior to seeding in a proper cell culture dish (depending on the number of cells sorted). However, centrifugation is not recommended when total number of cells is very low (<10,000).
7. If the integration of the AID-mRFP tag is very poor or not successful we recommend to use a recombinant adeno-associated virus system (rAAV) to deliver the repair template as described previously [21]. In this case, the efficiency of HDR

is expected to be higher since the repair template is delivered as a linear ssDNA instead of a circular dsDNA plasmid.

8. If a single cell sort resulted in poor survival rates we recommend to sort three cells in a 96-well plate instead of one cell per well. After selection of a three-cell population a limiting dilution can be performed to obtain a single clone. This will increase the chances of obtaining a positive clone.

9. At this step it is expected that at least one allele of the protein of interest (POI) is tagged. To achieve double heterozygous tagging more clones will likely need to be screened. Alternatively, a double selection strategy may be used (e.g., consecutive mRFP and EYFP tagging). If the protein is not tagged, we recommend to design different CRISPR/Cas9 guides and/or change the donor plasmid as described in **Note 6**, or to add to the construct a selectable cassette fused to a splicing site (e.g., P2A/T2A) or to a separate promoter. This latter strategy might be necessary if the POI is expressed at low levels and therefore not detectable by FACS. In the case of correct tagging but no protein depletion by IAA, a new target strategy needs to be generated. This might include moving the AID tag to a different protein extremity or adding a longer linker (in case the AID is sterically hidden by the protein itself).

10. We designed primers that bind upstream (forward) and downstream (reverse) of the donor template or inside the AID sequence (reverse). HDR-mediated integration of the tag will result in a shift of the PCR product size when using AID-mRFP flanking primers (e.g., with tag integration: ~1900 bp, without tag: ~450 bp; Fig. 1c). A PCR product with the reverse primer binding inside the AID sequence is only expected in the presence of the AID tag.

11. Duration of auxin treatment is protein and cell line-dependent. Despite that for CENP-C we can observe complete degradation in only 20 min [11], for screening of correct AID integration we prefer to use an overnight treatment to be sure to achieve complete protein degradation.

12. Multiple washing steps are absolutely necessary to remove IAA properly. Washing needs to be done carefully since, if performed too harshly, it can detach cells from the dish.

13. Endogenous protein levels of CENP-C^{AE} and also CENP-B^{mRFP} are low in cells. This represents a problem to perform live-cell imaging since the RFP and EYFP signals are bleached over time. By acquiring images every 10 min we found a good compromise between total duration of acquisition (limited by bleaching) and time resolution to monitor CENP-C^{AE} reloading.

14. CENP-C^{AE} reloading cannot be observed immediately before mitosis (G2), during mitosis or early after mitosis (early G1) (Fig. 2b) as observed by filming asynchronous cells (the phases of the cell cycle were estimated by counting the time that the cells need to reach mitosis). However, we have found that CENP-C^{AE} is reloaded in all S-phase arrested cells (Fig. 2c). This is in agreement with our previously reported findings that CENP-C can only be loaded at the centromere in mid G1-phase (only after CENP-A deposition that occurs immediately at mitotic exit [5]) and in S-phase [8]. We have found that CENP-C levels reaccumulate linearly at the centromere after about 30–60 min (Fig. 2d), however we started to acquire images approximately 1 h after IAA wash-out. Hence CENP-C^{AE} reexpression and reloading required a minimal time of about 90–120 min.

Acknowledgments

The authors would like to thank all the members of the D. Fachinetti laboratory and C. Bartocci (I. Curie) for helpful suggestions. We thank Solène Hervé for providing images for Fig. 1e. We also thank the FACS facility and the microscopy platform at Institut Curie. D.F. receives salary support from the CNRS. D.F. has received support by LabEx CelTisPhyBio, the Institut Curie and the ATIP-Avenir 2015 program. S.H. has received funding from the European Union's Horizon 2020 research and innovation programme under the Marie Skłodowska-Curie grant agreement No 666003. This work has also received support under the program Investissements d'Avenir launched by the French Government and implemented by ANR with the references ANR-10-LABX-0038 and ANR-10-IDEX-0001-02 PSL.

References

1. Varma D, Wan X, Cheerambathur D et al (2013) Spindle assembly checkpoint proteins are positioned close to core microtubule attachment sites at kinetochores. J Cell Biol 202(5):735–746. https://doi.org/10.1083/jcb.201304197
2. McKinley KL, Sekulic N, Guo LY et al (2015) The CENP-L-N complex forms a critical node in an integrated meshwork of interactions at the centromere-kinetochore Interface. Mol Cell 60:886–898. https://doi.org/10.1016/j.molcel.2015.10.027
3. Weir JR, Faesen AC, Klare K et al (2016) Insights from biochemical reconstitution into the architecture of human kinetochores. Nature 537:249–253. https://doi.org/10.1038/nature19333
4. Hoffmann S, Fachinetti D (2017) A time out for CENP-A. Mol Cell Oncol 4:e1293596. https://doi.org/10.1080/23723556.2017.1293596
5. Jansen LET, Black BE, Foltz DR, Cleveland DW (2007) Propagation of centromeric chromatin requires exit from mitosis. J Cell Biol 176:795–805. https://doi.org/10.1083/jcb.200701066
6. McKinley KL, Cheeseman IM (2015) The molecular basis for centromere identity and function. Nat Publ Group 17(1):16–29. https://doi.org/10.1038/nrm.2015.5

7. Nishimura K, Fukagawa T, Takisawa H et al (2009) An auxin-based degron system for the rapid depletion of proteins in nonplant cells. Nat Methods 6:917–922. https://doi.org/10.1038/nmeth.1401
8. Kepinski S, Leyser O (2005) The Arabidopsis F-box protein TIR1 is an auxin receptor. Nature 435:446–451. https://doi.org/10.1038/nature03542
9. Dharmasiri N, Dharmasiri S, Estelle M (2005) The F-box protein TIR1 is an auxin receptor. Nature 435:441–445. https://doi.org/10.1038/nature03543
10. Morawska M, Ulrich HD (2013) An expanded tool kit for the auxin-inducible degron system in budding yeast. Yeast 30:341–351. https://doi.org/10.1002/yea.2967
11. Hoffmann S, Dumont M, Barra V et al (2016) CENP-A is dispensable for mitotic centromere function after initial centromere/kinetochore assembly. Cell Rep 17:2394–2404. https://doi.org/10.1016/j.celrep.2016.10.084
12. Lambrus BG, Uetake Y, Clutario KM et al (2015) p53 protects against genome instability following centriole duplication failure. J Cell Biol 210:63–77. https://doi.org/10.1083/jcb.201502089
13. Holland AJ, Fachinetti D, Han JS, Cleveland DW (2012) Inducible, reversible system for the rapid and complete degradation of proteins in mammalian cells. Proc Natl Acad Sci U S A 109(49):E3350–E3357. https://doi.org/10.1073/pnas.1216880109/-/DCSupplemental
14. Brosh R, Hrynyk I, Shen J et al (2016) A dual molecular analogue tuner for dissecting protein function in mammalian cells. Nat Commun 7:11742. https://doi.org/10.1038/ncomms11742
15. Natsume T, Kiyomitsu T, Saga Y, Kanemaki MT (2016) Rapid protein depletion in human cells by Auxin-inducible Degron tagging with short homology donors. Cell Rep 15:210–218. https://doi.org/10.1016/j.celrep.2016.03.001
16. Fachinetti D, Han JS, McMahon MA et al (2015) DNA sequence-specific binding of CENP-B enhances the Fidelity of human centromere function. Dev Cell 33:314–327. https://doi.org/10.1016/j.devcel.2015.03.020
17. Bodor DL, Rodríguez MG, Moreno N, Jansen LET (2001) Analysis of protein turnover by quantitative SNAP-based pulse-chase imaging. Curr Protoc Cell Biol. Chapter 8; unit 8.8. https://doi.org/10.1002/0471143030.cb0808s55
18. Cermak T, Doyle EL, Christian M et al (2011) Efficient design and assembly of custom TALEN and other TAL effector-based constructs for DNA targeting. Nucleic Acids Res 39:e82–e82. https://doi.org/10.1093/nar/gkr218
19. Ran FA, Hsu PD, Wright J et al (2013) Genome engineering using the CRISPR-Cas9 system. Nat Protoc 8:2281–2308. https://doi.org/10.1038/nprot.2013.143
20. Kato H, Jiang J, Zhou BR et al (2013) A conserved mechanism for Centromeric nucleosome recognition by centromere protein CENP-C. Science 340:1110–1113. https://doi.org/10.1126/science.1235532
21. Kaulich M, Dowdy SF (2015) Combining CRISPR/Cas9 and rAAV templates for efficient gene editing. Nucleic Acid Therapeutics 25:287–296. https://doi.org/10.1089/nat.2015.0545

Chapter 13

Live Imaging of Parental Histone Variant Dynamics in UVC-Damaged Chromatin

Juliette Dabin, Anna Fortuny, Sandra Piquet, and Sophie E. Polo

Abstract

In eukaryotic cell nuclei, all DNA transactions, including DNA damage repair, take place on a chromatin substrate, the integrity of which is central to gene expression programs and cell identity. However, substantial chromatin rearrangements accompany the repair response, culminating in the deposition of new histones. How the original epigenetic information conveyed by chromatin may be preserved in this context is a burning question. Elucidating the fate of parental histones, which characterize the pre-damage chromatin state, is a key step forward in deciphering the mechanisms that safeguard epigenome stability. Here, we present an in vivo approach for tracking parental histone H3 variant dynamics in real time after UVC laser-induced damage in human cells.

Key words Live-cell imaging, Parental histone variants, Photoactivation, SNAP-tag technology, UVC laser damage

1 Introduction

In eukaryotic cells, the DNA associates with histone proteins in the form of chromatin, the basic unit of which is the nucleosome [1]. Chromatin organization is modulated at the nucleosomal level through the existence of histone variants [2] and their posttranslational modifications [3], and via the association of non-histone chromatin components, but also at the level of higher-order folding of the chromatin fiber into nuclear domains [4]. Altogether, these structural variations convey epigenetic information, which drives genome functions and cell identity [5]. Maintaining epigenome integrity is thus of critical importance. However, epigenome stability is challenged during all DNA transactions that take place on the chromatin substrate, including DNA damage repair

Juliette Dabin and Anna Fortuny equally contributed to this work.

Guillermo A. Orsi and Geneviève Almouzni (eds.), *Histone Variants: Methods and Protocols*, Methods in Molecular Biology, vol. 1832,
https://doi.org/10.1007/978-1-4939-8663-7_13,

[6]. Substantial chromatin rearrangements indeed accompany the repair response, as originally described in the Access-Repair-Restore model [7, 8]. This model postulates that damaged chromatin is first transiently disorganized to allow access to the repair factors, followed by restoration of chromatin structure and function. Our understanding of the molecular events underpinning such chromatin rearrangements has greatly improved in recent years, owing to the emergence of innovative methods that allow a more detailed examination of chromatin dynamics following DNA damage.

In particular, the development of novel imaging techniques, combining local induction of DNA damage and specific tagging of histone proteins, has been instrumental for visualizing a loss of histone density accompanied by chromatin decondensation at sites of UVC irradiation [9] and UVA laser micro-irradiation [10–12] in human cells. These approaches also proved invaluable for detecting the deposition of newly synthesized H2A and H3 histone variants at sites of DNA damage [13–17]. While new histones likely contribute to restoring chromatin structure following DNA damage, they also bring in new information [18], which raises the question of how the epigenetic landscape may be preserved. To address this issue, one needs to focus on parental histones, which characterize the pre-damage chromatin state and carry the original epigenetic information.

Here, we describe two complementary methods for real-time tracking of parental histone variants in response to local UVC irradiation in human cells [19]. The specific labeling of parental histones relies on histone protein tagging with a SNAP-tag [20], or a PhotoActivatable Green Fluorescent Protein (PA-GFP) [21]. We combine these technologies with cell micro-irradiation using a UVC laser implemented on a confocal microscope. Thus, we can follow parental histone redistribution and recovery during repair progression in UVC-damaged chromatin [19].

2 Materials

2.1 Cell Culture

1. Human U2OS cell lines stably expressing H3.3-SNAP or H3.3-PA-GFP (*see* **Note 1**). The *H3F3B* sequence encoding human H3.3 is cloned into plasmids encoding the SNAP-tag (New Englands Biolabs) or PA-GFP [22]. Stable cell lines are established by transfection followed by antibiotic selection of clones [19].
2. Dulbecco's Modified Eagle Medium with high glucose, GlutaMAX, sodium pyruvate, and phenol red (DMEM high glucose GlutaMAX™ Supplement, Gibco) supplemented with 10% fetal bovine serum, 100 U/ml penicillin and 100 μg/ml

streptomycin, 100 μg/ml G418 disulfate salt solution and 200 μg/ml hygromycin B. Store at 4 °C.

3. Dulbecco's Phosphate-Buffered Saline without calcium, magnesium, and phenol red (DPBS 1×).
4. Sterile round quartz coverslips 25 mm, thickness No. 1 (Neyco) (*see* **Note 2**).
5. Six-well cell culture plates (TPP).

2.2 SNAP Reagent

1. Fluorescent SNAP substrate (200 μM stock solution): dissolve 30 nmol SNAP-Cell® TMR-STAR (New England Biolabs) into 150 μl sterile dimethyl sulfoxide (DMSO). Store 30 μl aliquots at −20 °C (*see* **Note 3**).

2.3 Microscopy Equipment (see Note 4)

1. Zeiss fully motorized Axiovert microscope with a LSM700 confocal module (Carl Zeiss) adapted for UVC transmission with all-quartz optics, a temperature-controlled chamber and Zen software (Carl Zeiss).
2. UVC laser: 2 mW pulsed (7.8 kHz) diode-pumped solid-state laser emitting at 266 nm (Rapp OptoElectronics, Hamburg GmbH), directly coupled to the microscope stand. A neutral density filter OD1 (10% T) can be added to the light path. The UVC laser is fixed but the position of the damage spot can be precisely controlled by moving the motorized stage of the microscope via a custom macro on Zen software (*see* **Notes 5** and **6**).
3. Photoactivation laser: 405 nm laser diode, 5 mW power.
4. Imaging lasers: 405 nm, 5 mW power; 488 nm, 10 mW power; 555 nm, 10 mW power.
5. Objectives: Quartz 40x/0.6 Ultrafluar glycerol objective (Carl Zeiss) and LD LCI Plan-Apochromat 25x/0.8 multi-immersion objective for oil, water or glycerol immersion (Carl Zeiss).
6. Glycerol for immersion.

2.4 Live-Cell Imaging

1. Chamlide CMB 35 mm dish type 1-well magnetic chamber for round coverslip (Live Cell Instrument).
2. Thermoregulated microscope chamber with CO_2 entry (*see* **Note 7**).

2.5 Image Analysis and Processing

1. ImageJ software (U. S. National Institutes of Health, Bethesda, Maryland, USA, http://imagej.nih.gov/ij/).
2. Microsoft® Excel® or equivalent spreadsheet software.
3. Adobe® Photoshop® or equivalent image processing software.

3 Methods

Here, we describe a protocol for monitoring the in vivo dynamics of parental histone H3.3 variants at sites of DNA damage in human cells. This method combines either SNAP-tag-based labeling [20] or photoactivation [21] of parental histones (Fig. 1) with local UVC irradiation of cells, followed by live-cell imaging. First, parental SNAP-tagged H3.3 histones are labeled with a fluorescent SNAP substrate (**step 1**). Alternatively, PA-GFP-tagged H3.3 histones are photoactivated with a 405 nm laser diode (**step 2**). Next, cells are locally irradiated with a UVC laser (266 nm) coupled to a confocal microscope to induce damages in subnuclear regions of interest (**step 3**). Live-cell imaging following UVC irradiation allows visualization of parental histone dynamics during the repair response (**step 4**). Quantification of the fluorescence associated with parental histones (**step 5**) reveals their dynamics in damaged chromatin (Fig. 2, *see* also **Note 8**). The whole protocol can be completed in 4 to 6 days (*see* **Note 9**).

3.1 SNAP Labeling of Parental Histones (see Note 10)

1. Grow U2OS cells stably expressing H3.3-SNAP and GFP-XPC on quartz coverslips in 6-well plates with 2.5 ml culture medium per well so that they reach 70–80% confluency 3 days later. Incubate in a humidified 37 °C incubator with 5% CO_2 overnight.
2. The next day, dilute the fluorescent SNAP substrate in fresh culture medium to a final concentration of 2 μM (*see* **Note 11**). Pulse-label all SNAP-tagged histones by incubating cells with 1 ml of this solution per well for 20 min at 37 °C, 5% CO_2.
3. Wash out the excess of fluorescent substrate by rinsing cells twice with 3 ml PBS (*see* **Note 12**), then incubate cells with 2.5 ml culture medium per well for 30 min.
4. Chase time: rinse cells twice with 3 ml PBS and incubate cells in 2.5 ml culture medium for 48 h at 37 °C, 5% CO_2 (*see* **Notes 13** and **14**).

3.2 Photoactivation of Parental Histones (see Note 10)

1. Grow U2OS cells stably expressing H3.3-PA-GFP and RFP-XPC on quartz coverslips in 6-well plates with 2.5 ml culture medium per well so that they reach 70–80% confluency 3 days later. Incubate overnight in a humidified 37 °C incubator with 5% CO_2.
2. The next day, transfer each coverslip to a Chamlide magnetic chamber and place the chamber on the confocal microscope stage with controlled temperature and CO_2 conditions (37 °C, 5% CO_2).

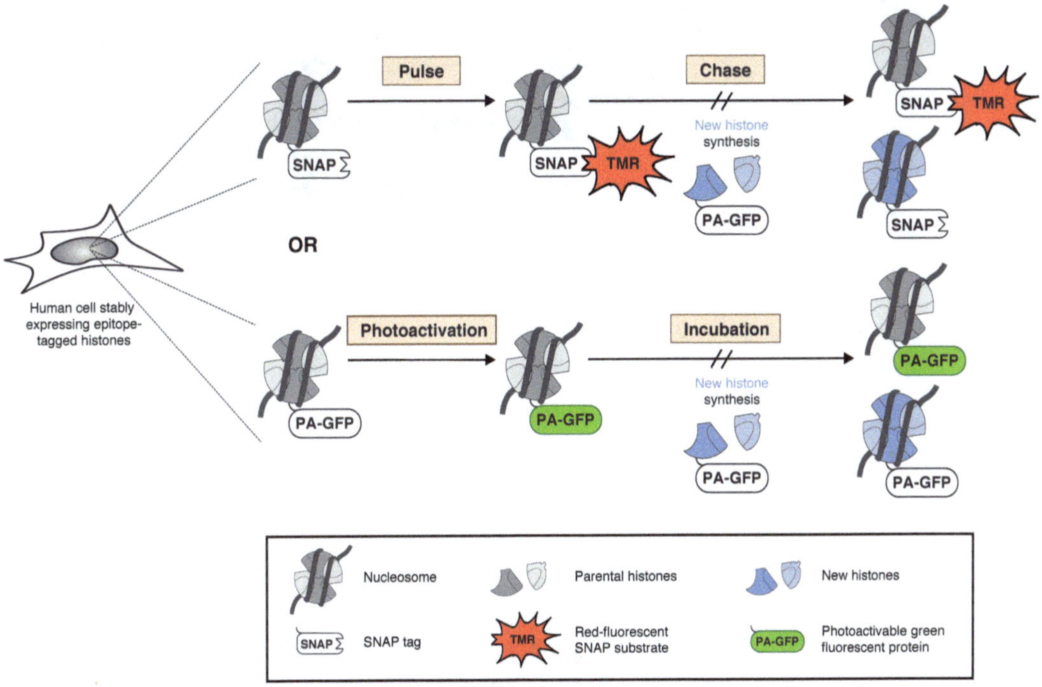

Fig. 1 Scheme of the assay for labeling parental histones. Two days prior to UVC irradiation, U2OS cells stably expressing SNAP- or PA-GFP-tagged histones are pulsed with a red fluorescent SNAP reagent (TMR, top) or photo-activated with a 405 nm laser (bottom), respectively. The subsequent chase/incubation time ensures that the pulse-labeled or photo-activated histones are properly incorporated into chromatin

3. Using the 40× glycerol immersion objective (*see* **Note 15**), photoactivate parental histones by bleaching a whole field of cells with the 405 nm laser diode using the following settings: maximum power, 5 iterations, 6.30μs/pixel scan speed. Cell nuclei can be visualized in transmitted light or via the fluorescently labeled repair factor. Repeat until 100 cell nuclei have been photoactivated (*see* **Note 16**).
4. Clean carefully the glycerol on the bottom of the coverslip before transferring it back to the cell culture plate and incubate for 48 h at 37 °C, 5% CO_2 (*see* **Note 13**).

3.3 Local UVC Irradiation of Cells

1. Two days after parental histone labeling or photoactivation, transfer each coverslip to a Chamlide magnetic chamber (*see* **Notes 17** and **18**) and place the chamber on the confocal microscope stage with controlled temperature and CO_2 conditions (37 °C, 5% CO_2).
2. Localize the fields of interest containing fluorescent nuclei (i.e., fluorescently labeled parental histones) and acquire an image of each field before UVC irradiation using the 25× multi-immersion objective with glycerol (*see* **Notes 15** and **19–21**). Use the appropriate imaging lasers depending on the

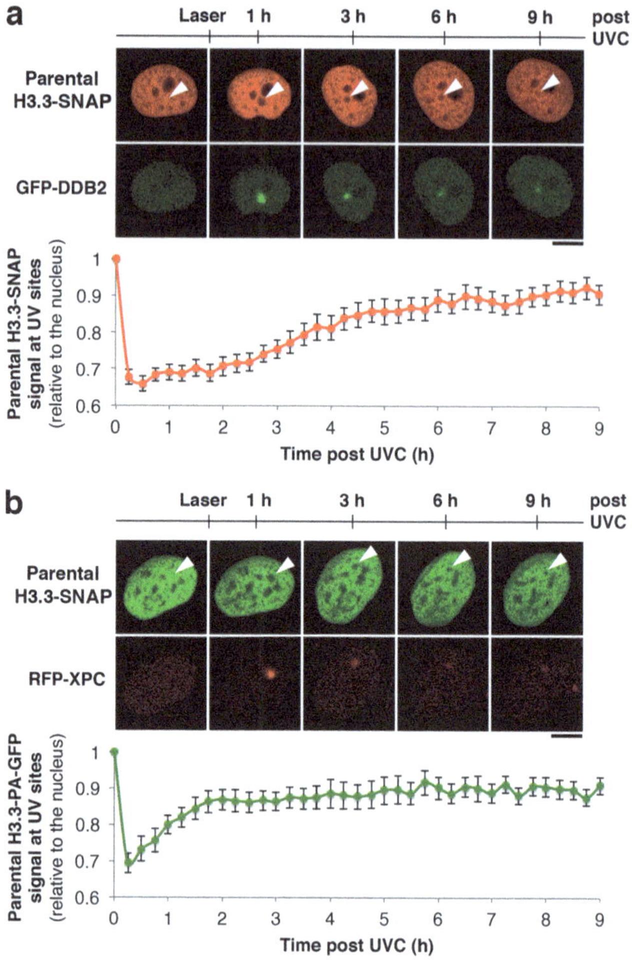

Fig. 2 Parental H3.3 redistribution and recovery at UVC damage sites. (**a**) Dynamics of parental histones H3.3 (red) at the indicated time points after UVC laser micro-irradiation in U2OS cells stably expressing H3.3-SNAP and GFP-DDB2. (**b**) Dynamics of parental histones H3.3 (green) at the indicated time points after UVC laser micro-irradiation in U2OS cells stably expressing H3.3-PA-GFP and RFP-XPC. White arrowheads point to the sites of laser impact. The graphs show quantification of red (**a**) and green (**b**) fluorescence in irradiated areas. Error bars represent SEM from n cells scored in two independent experiments. Scale bars: 10 μm. Adapted from ref. 19

fluorescence of parental histones (488 nm laser for H3.3-PA-GFP, 555 nm laser for H3.3-SNAP-tag-TMR). Save the position of each field of cells before moving to the next one to retrieve the cells more easily.

3. For UVC irradiation, switch to the 40× quartz objective with glycerol (*see* **Note 15**) and insert the neutral density filter (*see* **Note 22**).

4. Using the UV macro, mark the positions to be targeted by the UVC laser (266 nm) in as many cell nuclei as possible in a given field (one damage spot per nucleus, *see* also **Note 23**). Each cell nucleus is irradiated for 50 ms at maximum laser power (*see* **Notes 24–26**).

3.4 Live-Imaging of Parental Histones

1. Immediately following UVC irradiation, switch back to the 25× multi-immersion objective with glycerol to acquire images of the irradiated cells.
2. Capture one image every 15 min for several hours to overnight (*see* **Notes 19, 21**, and **27**). Use the appropriate imaging lasers depending on the fluorescence of parental histones, repair factors, and DNA if needed. Retrieve each field of cells using the positions saved previously. Cells should be kept at 37 °C, 5% CO_2 throughout the acquisition.
3. Once the acquisition is over, the quartz coverslip can be fixed with 2% paraformaldehyde if immunostaining is necessary or it can be recycled (*see* **Note 2**).
4. Clean the Chamlide magnetic chamber with distilled water and then 70% ethanol for reuse.

3.5 Image Analysis and Processing (see Note 28)

1. Open images with the ImageJ software.
2. Using the wand tool, select the UVC-damage region in an irradiated nucleus based on the fluorescence of the repair factor (XPC or DDB2) 15 min after irradiation. The selected area is then manually copied on each other image of the same cell nucleus before and after irradiation (*see* **Note 29**).
3. Measure the fluorescence signal (mean gray value) associated with parental histones within this region at each time point before and after UVC irradiation.
4. Also measure parental histone fluorescence intensity in the whole cell nucleus at each time point before and after UVC irradiation. The position of the nucleus is determined using the wand tool, based on the fluorescence of parental histones at each time point. Always measure background fluorescence in the same field.
5. Repeat **steps 2** to **4** for all damaged cells (*see* **Note 29**).
6. Export the results to an Excel spreadsheet.
7. Subtract background fluorescence from all measurements.
8. Divide the histone fluorescence intensity in the UVC-damaged area by the intensity measured in the corresponding nucleus. Normalize the results relative to before irradiation.
9. Selected images are mounted using Adobe Photoshop.

4 Notes

1. This protocol is optimized for U2OS cells stably expressing epitope-tagged H3.3, but it can be adapted to other cell types, stably or transiently expressing SNAP- or PA-GFP-tagged histones (H3 variants or other histone variants). Stable monoclonal cell lines are preferred to ensure minimal cell-to-cell variability in transgene expression. The co-expression of a fluorescently tagged repair factor, such as GFP- or RFP (Red Fluorescent Protein)-tagged XPC (Xeroderma Pigmentosum, complementation group C), is recommended to be able to visualize the damage site and assess repair progression. Other repair factors can also be used for the same purpose like DNA Damage Binding protein 2 (DDB2), as shown in Fig. 2.
2. Quartz coverslips are used to allow the transmission of UVC light through the coverslip to the sample. Cells normally do not detach from the coverslip but if this happens, it can be avoided by coating the coverslips with poly-L-lysine or collagen-fibronectin before seeding the cells. Quartz coverslips can be recycled after an experiment by performing the following washes (10 min each at room temperature): one wash in 1% SDS (Sodium Dodecyl Sulfate, MP Biomedicals), three washes in distilled water, two washes in 100% ethanol.
3. Centrifuge the fluorescent SNAP substrate before pipetting to avoid aggregates. The fluorescent SNAP substrate must be protected from light and is sensitive to repeated freeze-thaw cycles. It is thus advisable to prepare aliquots of the stock solution.
4. Any equivalent confocal microscope can be used as long as optics are changed to quartz and that it can be coupled to a 266 nm laser.
5. The UVC laser setup that we use for our experiments is available at the Cell and Tissue Imaging Facility of the Institut Curie (Paris, France). To date, the only other confocal microscope equipped with a 266 nm UVC laser and all-quartz optics is a Zeiss Axiovert 200 M with LSM 510 confocal module (Erasmus MC, Rotterdam, the Netherlands) [23].
6. Local UVC irradiation can also be performed on the bench with a UVC lamp (254 nm) through micropore filters [24].
7. Instead of CO_2, HEPES buffer can be added to the culture medium (25 mM final concentration) to maintain the pH.
8. For a deeper understanding of the mechanisms underlying histone dynamics at UV sites, this protocol can be combined with siRNA-mediated depletion of histone chaperones and repair factors or with inhibition of enzymatic activities involved in the UV damage response [19].

9. The total duration of the experiment may vary depending on the duration of live cell imaging (a few hours to overnight) and on the final number of cells to be analyzed.
10. Subheadings 3.1 and 3.2 are to be executed alternatively rather than sequentially.
11. In cells that do not express a GFP-tagged repair protein, the red-fluorescent substrate SNAP-Cell® TMR-star can be replaced by the green fluorescent SNAP-Cell® Oregon-Green® at a final concentration of 4 μM from a 1 mM stock solution in DMSO.
12. The washing procedures are critical to minimize background signal. The specificity of SNAP labeling can be controlled on the parent U2OS cell line, which does not express SNAP-tagged proteins and thus should not be stained.
13. The 48 h chase/incubation time allows incorporation of labeled parental histones into chromatin and can be adapted depending on the turnover of histone proteins.
14. Parental histone labeling can be combined with specific labeling of newly synthesized histones using distinct fluorescent SNAP substrates for new and parental histones (red-fluorescent SNAP-Cell® TMR-star and green-fluorescent SNAP-Cell® Oregon-Green®) [17, 19].
15. The same immersion medium, i.e., glycerol, should be used for photo-bleaching, imaging, and UVC irradiation.
16. Photoactivation efficiency is monitored by imaging with the 488 nm laser. The indicated settings should lead to maximal photoactivation of PA-GFP-tagged histones, but they may need to be adjusted depending on the objective and laser power available.
17. Following SNAP-tag labeling, it is preferable to rinse the cells in PBS before transferring the coverslip to the chamber to remove remaining aggregates of fluorescent SNAP substrate from the culture medium.
18. If DNA visualization is wanted during live-cell imaging, incubate cells before UVC irradiation with Hoechst 33258 nuclear staining dye at a final concentration of 10 μg/ml for 30 min.
19. Acquisitions can be performed in 2D or in 3D (z-stacks). You can save your images in any file format supported by the ImageJ software.
20. The 25× objective is preferred for image acquisitions due to a larger numerical aperture.
21. You can use the autofocus mode for image acquisitions (focusing on the fluorescent parental histones).
22. For safety reasons, it is important to close the microscope chamber during the UVC irradiation procedure and the lamp

shutter should also be closed to avoid transmission of the UVC light through the eyepieces.

23. As there is no scanning mode for the UVC laser, only damage spots can be obtained. It is not possible to draw lines.
24. The damage spot is 2 μm in diameter corresponding to ca. 2% of the nuclear volume. The UVC dose delivered at the site of laser micro-irradiation is estimated at 600 J/m^2 and does not cause major cytotoxicity: the mortality rate during an overnight imaging experiment is only around 10%, the damaged cells do repair and can go through mitosis.
25. Other laser wavelengths can be used for generating other types of DNA damage including DNA breaks [23].
26. We did not observe bleaching of TMR star, Oregon green, Hoechst, GFP, or RFP fluorescence by the UVC laser in the conditions used for local UVC irradiation.
27. To avoid phototoxicity during long-term imaging such as overnight experiments, it is recommended to adjust the acquisition settings in order to minimize the laser power and the number of total captured images (4 images/h maximum). Images can be acquired more frequently in case of shorter kinetics.
28. Image analysis can be performed on 2D acquisitions or on projections of 3D acquisitions.
29. The quantification procedure cannot be easily automated due to cell movement during long-term imaging.

Acknowledgments

We are grateful to Olivier Renaud (PICT-IBiSA, Curie Institute, Paris, France) for critical reading of the manuscript. Work in S.E.P. lab is supported by the European Research Council under Grant ERC-2013-StG-336427 "EpIn"; "Who am I?" laboratory of excellence under Grant ANR-11-LABX-0071. S.E.P. is a member of the EMBO YIP. J.D. is the recipient of PhD fellowships from Paris Diderot University and the Fondation ARC pour la recherche sur le cancer, and A.F. is funded by an international PhD fellowship from Sorbonne Paris Cité University.

References

1. Luger K, Mäder AW, Richmond RK et al (1997) Crystal structure of the nucleosome core particle at 2.8 A resolution. Nature 389:251–260
2. Buschbeck M, Hake SB (2017) Variants of core histones and their roles in cell fate decisions, development and cancer. Nat Rev Mol Cell Biol 18:299–314
3. Bannister AJ, Kouzarides T (2011) Regulation of chromatin by histone modifications. Cell Res 21:381–395

4. Bonev B, Cavalli G (2016) Organization and function of the 3D genome. Nat Rev Genet 17:661–678
5. Allis CD, Jenuwein T (2016) The molecular hallmarks of epigenetic control. Nat Rev Genet 17:487–500
6. Dabin J, Fortuny A, Polo SE (2016) Epigenome maintenance in response to DNA damage. Mol Cell 62:712–727
7. Smerdon MJ (1991) DNA repair and the role of chromatin structure. Curr Opin Cell Biol 3:422–428
8. Polo SE, Almouzni G (2015) Chromatin dynamics after DNA damage: the legacy of the access-repair-restore model. DNA Repair 36:114–121
9. Luijsterburg MS, Lindh M, Acs K et al (2012) DDB2 promotes chromatin decondensation at UV-induced DNA damage. J Cell Biol 197:267–281
10. Kruhlak MJ, Celeste A, Dellaire G et al (2006) Changes in chromatin structure and mobility in living cells at sites of DNA double-strand breaks. J Cell Biol 172:823–834
11. Smeenk G, Wiegant WW, Marteijn JA et al (2013) Poly(ADP-ribosyl)ation links the chromatin remodeler SMARCA5/SNF2H to RNF168-dependent DNA damage signaling. J Cell Sci 126:889–903
12. Strickfaden H, McDonald D, Kruhlak MJ et al (2015) Poly(ADP-ribosyl)ation-dependent transient chromatin decondensation and histone displacement following laser microirradiation. J Biol Chem 291(4):1789–1802
13. Polo SE, Roche D, Almouzni G (2006) New histone incorporation marks sites of UV repair in human cells. Cell 127:481–493
14. Dinant C, Ampatziadis-Michailidis G, Lans H et al (2013) Enhanced chromatin dynamics by FACT promotes transcriptional restart after UV-induced DNA damage. Mol Cell 51:469–479
15. Adam S, Polo SE, Almouzni G (2013) Transcription recovery after DNA damage requires chromatin priming by the H3.3 histone chaperone HIRA. Cell 155:94–106
16. Luijsterburg MS, de Krijger I, Wiegant WW et al (2016) PARP1 links CHD2-mediated chromatin expansion and H3.3 deposition to DNA repair by non-homologous end-joining. Mol Cell 61:547–562
17. Adam S, Dabin J, Bai S-K et al (2015) Imaging local deposition of newly synthesized histones in UVC-damaged chromatin. Methods Mol Biol 1288:337–347
18. Loyola A, Bonaldi T, Roche D et al (2006) PTMs on H3 variants before chromatin assembly potentiate their final epigenetic state. Mol Cell 24:309–316
19. Adam S, Dabin J, Chevallier O et al (2016) Real-time tracking of parental histones reveals their contribution to chromatin integrity following DNA damage. Mol Cell 64:65–78
20. Bodor DL, Rodríguez MG, Moreno N et al (2012) Analysis of protein turnover by quantitative SNAP-based pulse-chase imaging. Curr Protoc Cell Biol Chapter 8, Unit 8.8:1934–2616
21. Patterson GH, Lippincott-Schwartz J (2002) A photoactivatable GFP for selective photolabeling of proteins and cells. Science 297:1873–1877
22. Patterson GH, Lippincott-Schwartz J (2004) Selective photolabeling of proteins using photoactivatable GFP. Methods 32:445–450
23. Dinant C, de Jager M, Essers J et al (2007) Activation of multiple DNA repair pathways by sub-nuclear damage induction methods. J Cell Sci 120:2731–2740
24. Moné MJ, Bernas T, Dinant C et al (2004) In vivo dynamics of chromatin-associated complex formation in mammalian nucleotide excision repair. Proc Natl Acad Sci U S A 101:15933–15937

Chapter 14

CRISPR/Cas9 Gene Editing of Human Histone H2A Variant H2AX and MacroH2A

Justin W. C. Leung, Lara E. Emery, and Kyle M. Miller

Abstract

Histone H2A variants play important roles in maintaining the integrity of the genome. For example, the histone variant H2AX is phosphorylated on Ser139 (called γH2AX) at DNA double-strand breaks (DSB) and serves as a signal for the initiation of downstream DNA damage response (DDR) factor recruitment and DNA repair activities within damaged chromatin. For decades, genetic studies in human cells involving DNA damage signaling and repair factors have relied mostly on either knockdown by RNA interference (i.e., shRNA and siRNA) or the use of mouse embryonic fibroblasts derived from knockout (KO) mice. Recent advances in gene editing using ZNF nuclease, TALEN, and CRISPR/Cas9 have allowed the generation of human KO cell lines, allowing genetic models for studying the DDR, including histone H2A variants in human cells. Here, we describe a detailed protocol for generating and verifying KO of H2AX and macroH2A histone H2A variants using CRISPR/Cas9 gene editing in human cancer cell lines. This protocol allows the use and development of genetic systems in human cells to study histone variants and their functions, including within the DDR.

Key words Histone variant, H2A, H2AX, macroH2A, Chromatin, CRISPR/Cas9, Gene editing, DNA damage, DNA repair, Genome stability

1 Introduction

Histones are small, basic proteins that bind and organize nuclear DNA into chromatin in eukaryotes. Nucleosomes represent the fundamental structural unit of chromatin and are composed of ~146 bp of DNA wrapped around a histone octamer containing two copies each of the four core histones H2A, H2B, H3, and H4. In addition, variants of these core histones exist that are encoded by unique genes but are highly related to the core histone. For example, H2A is the largest and most diverse family of histones. H2A and related variants are encoded by twenty-six genes in the human genome [1]. Canonical histone H2A alone is represented by seventeen individual genes, while an additional nine genes encode H2A variants including H2AX, H2AZ, and macroH2A. These variants

Guillermo A. Orsi and Geneviève Almouzni (eds.), *Histone Variants: Methods and Protocols*, Methods in Molecular Biology, vol. 1832, https://doi.org/10.1007/978-1-4939-8663-7_14, © Springer Science+Business Media, LLC, part of Springer Nature 2018

participate in many crucial biological activities, including DNA repair and transcriptional regulation [2].

Histones regulate chromatin structure, function, and dynamics through several mechanisms, including those that involve post-translational modifications (PTMs) and histone variant deposition [2, 3]. Histones are highly modified by various PTMs including phosphorylation, methylation, acetylation, and ubiquitylation, which can affect nucleosome stability, structure, and chromatin binding of effector proteins to unique PTMs and/or histone variants [4, 5]. The presence of histone variants within chromatin also serves as a mechanism to alter the composition of chromatin, allowing variant-specific activities to occur within these specialized histone variant-containing chromatin environments. Collectively, these two mechanisms represent important regulatory pathways involved in orchestrating chromatin function across the genome.

H2A variant H2AX represents the premier example for the involvement of histone variants in regulating DNA damage response (DDR) pathways [6–8]. For instance, H2AX is phosphorylated at serine 139 (γH2AX) at DNA double-strand breaks (DSBs) and provides a binding interface for mediator of DNA damage checkpoint protein 1 (MDC1) to trigger the DDR [9–11]. Recent studies have further cemented the vital contribution of H2AX and other H2A variants in the DDR [12–16]. In addition to their involvement in basic mechanisms of DNA damage signaling and repair that maintain the integrity of the genome, aberrant H2A variant expression is observed in multiple types of human diseases including cancer [17–19]. Whether or not H2A variant dysfunction in cancer primarily alters genome stability and/or transcriptional regulation is unclear. Alterations in histone variants can also impact cancer therapies [18–20]. Therefore, new approaches to study human histone H2A variants, including genetically modified human cancer cell lines, would provide valuable experimental systems for basic and translational research involving this important class of histone variants.

Genetic studies of histone variants in human cells have been challenging due to the difficulties of obtaining histone variant knockouts (KOs) in human cells. Recent advances in gene editing techniques including zinc finger (ZNF) nucleases, transcription activator-like effect nuclease (TALEN), and clustered regulatory interspaced short palindromic repeats (CRISPR)/Cas9 have enabled scientists to generate genetic models in mammalian cells [21–24]. These techniques all rely on DNA sequence-directed nucleases for targeting a specific region of the genome for editing. Gene editing occurs through the generation of a DNA DSB by one of these nucleases at the targeted loci. These techniques exploit the engagement of the nonhomologous end-joining (NHEJ) DSB repair pathway, which is an error-prone DNA repair pathway that can create frame-shift mutations by introducing insertion/deletions (indel) at the break site. These mutations result in the formation of premature stop-codons and/or unstable transcripts that result in the loss of a functional gene product. ZNF nucleases and CRISPR/Cas9 have been successfully used to generate human cell lines lacking H2AX [25–28]. Rapid and efficient deletion of histone H2A variants in an isogenic background is essential for obtaining comparable and reproducible data involving H2A variants. The use of histone variant KO cell lines will not only provide an advantage over RNA interference approaches to create clean genetic KO cells, but will also provide the use of gene complementation studies to control for off-target effects and to also systematically and definitively decipher the function of these proteins, including their modifications, using structure/function analyses of wild type (WT) and mutant genes that are reintroduced into complemented KO cell lines.

In this chapter, we describe a method for generating H2AX and macroH2A KO human cells by CRISPR/Cas9 gene editing techniques. We also provide an optimized screening and verification platform for histone H2A variant KO cells using western blotting, immunofluorescence, and genomic PCR-coupled sequencing. The successful implementation of this method provides important experimental cell-based systems that can be utilized to genetically decipher the role of histone H2A variants in chromatin biology, epigenetics, transcription, and DNA repair.

2 Materials

2.1 Cloning

1. pSpCas9(BB)-2A-Puro (pX459, Addgene plasmid ID: 48139) (*see* **Notes 1** and **2**).
2. Two complementary oligonucleotides with overhangs (complimentary bases are in bold) are used to generate guide RNA sequences that can be cloned into pSpCas9(BB)-2A-Puro and used for targeting genes of interest using CRISPR/Cas9

Oligos	Sequence (5′-3′)
H2AX sgRNA 1	CACC**GGGTGGCCTTCTTGCCGCCCG** (top)/ AAAC**CGGGCGGCAAGAAGGCCACCC** (bottom)
H2AX sgRNA 2	CACC**GCGCCAACGCGCTCGGCGTAG** (top)/ AAAC**CTACGCCGAGCGCGTTGGCGC** (bottom)
macroH2A sgRNA 1	CACC**GCCACTCCAATCCTGTACTTG** (top)/ AAAC**CAAGTACAGGATTGGAGTGGC** (bottom)
macroH2A sgRNA 2	CACC**GTTTGGCAGACCTGGACGTCT** (top)/ AAAC**AGACGTCCAGGTCTGCCAAAC** (bottom)

3. T4 PNK buffer, 10×.
4. T4 Polynucleotide kinase.
5. Tango Buffer, 10×.
6. Dithiothreitol (DTT, 10 mM).
7. ATP (10 mM).
8. Restriction enzyme, BbsI.
9. T7 ligase.
10. PlasmidSafe buffer (10×) and exonucleases.
11. Stbl3 chemically competent cells.
12. Super optimal broth (S.O.C.) medium.
13. LB agar plates with ampicillin (100 μg/ml).
14. 1× LB medium with ampicillin (100 μg/ml).
15. Mini prep kit (*see* **Note 3**).
16. Thermocycler PCR machine.
17. A 37 °C incubator and shaker for bacteria growth.
18. U6 sequencing forward primer—5′ GAGGGCCTATTTCC CATG 3′.

2.2 Cell Culture

1. Dulbecco's Modified Eagle's Medium (DMEM) supplemented with 10% fetal bovine serum.
2. Penicillin–Streptomycin-Glutamine (used as 1×).
3. Opti-MEM.
4. U2OS osteosarcoma cells or other adherent cell line of interest (*see* **Note 4**).
5. 0.25% Trypsin.
6. Phosphate-buffered saline.
7. Fugene HD transfection reagent.
8. HA-Flag-H2AX and HA-Flag-macroH2A mammalian expression vectors.
9. A 37 °C incubator with 5% CO_2 for growing human cells.
10. Puromycin.
11. 50 ml reagent reservoir (*see* **Note 5**).

2.3 Antibodies

1. Anti-H2AX (Millipore, 07-627).
2. Anti-MacroH2A (Abcam, ab376240).
3. Anti-Flag (Sigma, 1804).
4. Anti-Beta tubulin (Abcam, ab6046).
5. Anti-rabbit HRP-linked secondary antibody.
6. Anti-mouse HRP-linked secondary antibody.

7. Alexa Fluor 488 goat anti-rabbit IgG.
8. Alexa Fluor 594 goat anti-rabbit IgG.
9. Alexa Fluor 594 goat anti-mouse IgG.

2.4 Western Blotting

1. 1× SDS-PAGE sample buffer: 50 mM Tris–HCl (pH 6.8), 100 mM Dithiothreitol (DTT), 2% SDS, 0.1% Bromophenol blue, 10% Glycerol.
2. Running buffer: 25 mM Tris–HCl, 200 mM Glycine, 0.1% SDS.
3. Transfer buffer: 25 mM Tris–HCl, 200 mM Glycine, 20% Methanol (v/v).
4. TBS-T: 150 mM NaCl, 50 mM Tris base, 0.1% Tween 20.
5. Blocking buffer: 1× TBS-T + 3% Bovine Serum Albumin (BSA, w/v).
6. Nitrocellulose Blotting membrane.
7. Chromatography paper (2.5 in × 4 in, thickness—0.34 mm).
8. Primary antibody solution: 1:1000 (Subheading 2.3, **antibodies 1** or **2** in 5% bovine serum albumin with 0.1% sodium azide, reusable).
9. Secondary antibody solution: 1:10,000 (Subheading 2.3, **antibodies 4** or **5** in TBS-T solution, non-reusable).
10. Enhanced chemiluminescence (ECL) western blotting detection reagent (Solution A- luminol; Solution B—hydrogen peroxide, prepare fresh by mixing solution A and solution B in a 1:1 ratio) (*see* **Note 6**).

2.5 Immunofluorescence

1. Microscope coverslips (12 mm Round, thickness 0.02 mm), Bellco Glass, Inc.).
2. Microscope slides.
3. Phosphate-buffered saline (PBS).
4. Fixative: 5% formalin solution.
5. Permeabilization buffer: 0.5% (v/v) Triton in PBS.
6. Blocking solution: PBS + 3% BSA.
7. Anti-fade mounting medium—Vectashield (Vector Laboratories, Inc.).
8. 4′,6-Diamidino-2-phenylindole (DAPI).
9. Primary antibody solution: Subheading 2.3, **antibodies 1** and **2**, or **3** (1:500 in 3% BSA), non-reusable.
10. Secondary antibody solution: Subheading 2.3, **antibodies 7** and **8**, or **9** (1:2000 in 3% BSA with 10 μg/ml DAPI), non-reusable.

2.6 Genotype Verification

1. DNeasy Blood & Tissue Kit.
2. KOD Hot start DNA polymerase.
3. PCR and sequencing primers.

Primers	Sequence
H2AX PCR-forward/ sequencing primer	ATGTCGGGCCGCGGCAAGAC
H2AX PCR-reverse	TTAGTACTCCTGGGAGGCCT
macroH2A PCR-forward/ sequencing primer	TCATGTGGCATAATGGCAAG
macroH2A PCR-reverse	TGCATCTTAAGCCAGTGGTG

3 Methods

3.1 Cloning H2AX/ macroH2A KO sgRNA Targeting Constructs

3.1.1 Annealing the sgRNA oligos

1. Dissolve the sgRNA oligos (100 μM) and prepare the reaction as following

Components	Volume (μl)
T4 PNK buffer, 10×	1
T4 PNK	1
sgRNA top (100 μM)	1
sgRNA bottom (100 μM)	1
ddH_2O	6
Total	**10**

2. Complementary oligos are annealed using the following thermocycler setting: 37 °C for 30 min; 95 °C for 5 min; ramp down to 25 °C at 5 °C/min.
3. Dilute the annealed oligos 1:200 by adding 1 μl of oligo to 199 μl of DNase/RNase-free water.

3.1.2 Inserting sgRNA oligos into the pSpCas9 (BB) Vector

1. Prepare the ligation reaction as follows:

Components	Volume (μl)
Tango buffer, 10×	2
Diluted annealed oligos from Subheading 3.1.1	2
pSpCas9(BB)-2A-Puro, 100 ng/μl	1
DTT, 10 mM	1
ATP, 10 mM	1
FastDigest BbsI	1

Components	Volume (μl)
T7 ligase	0.5
ddH_2O	11.5
Total	**20**

2. Incubate in the thermocycler: 37 °C for 5 min and 21 °C for 5 min (6 cycles).

3.1.3 Removing linearized DNA

1. Incubate the Following Reaction at 37 °C for 30 min Followed by 70 °C for 30 min.

Components	Volume (μl)
PlasmidSafe buffer, 10×	1.5
Ligation product from Subheading 3.1.2	11
ATP, 10 mM	1.5
PlasmidSafe exonuclease	1
Total	**15**

3.1.4 Plasmid transformation.

1. Add 2 μl of the final product from Subheading 3.1.3 into a tube containing 100 μl Stbl3 (or other ex.DH5α) chemically competent cells. Transform DNA into bacteria by incubating on ice for 20 min, followed by heat shock at 42 °C for 40 s and recovery on ice for 5 min.
2. Add 400 μl S.O.C. medium and shake at 37 °C for 20 min. Plate entire volume of medium and bacteria onto LB plate containing 100 μg/ml ampicillin.
3. Incubate the plate overnight at 37 °C.
4. Pick individual colonies and inoculate each colony in 5 ml LB medium containing 100 μg/ml ampicillin and shake overnight at 37 °C.
5. Isolate plasmid DNA using QIAprep spin miniprep kit following the manufacturer's protocol.
6. Verify the clone by Sanger DNA sequencing using the U6 sequencing forward primer.

3.2 CRISPR/sgRNA Transfection

Day 0: Seeding target cells

1. Seed target cells in a 6-well plate with 50% confluency.
   Day 1: Transfecting the targeting construct

2. Remove the medium from target cells.
3. Add 1.5 ml regular medium without Penicillin/Streptomycin.
4. Prepare two separate tubes: one containing 2 μg of targeting construct and 100 μl of Opti-MEM, the other containing 6 μl of Fugene transfection reagent and 100 μl of Opti-MEM. Incubate both tubes for 5 min at room temperature.
5. Combine and mix both tubes and incubate for 15 min at room temperature.
6. Add the transfection mixture dropwise onto the target cells, mix by gently swirling the place, and put the cells back into the incubator.
7. Remove medium containing transfection mixture after 6 h.
8. Add 2 ml of regular DMEM medium with 10% FBS and Penicillin–Streptomycin–Glutamine and place cells back into the incubator.

Day 2: Selecting cells with target construct

9. Replace with regular medium containing 2 μg/ml puromycin.
10. Incubate at 37 °C for 2 days.

Day 4: Seeding single cells into 96-well plates

11. Remove the medium and wash the cells with 100 μl 1× PBS.
12. Add trypsin to the cells and allow the cells to incubate at 37 °C until detached from plate. Resuspend detached cells with 2 ml of fresh, warmed medium, and spin down cells at 400 × *g*, 21, 130 × *g* for 3 min. Remove the supernatant by aspiration and resuspend pelleted cells in 2 ml of fresh medium. Count cells using a cell counter.
13. Add 500 cells into 40 ml of warmed, regular medium in a 50 ml reagent reservoir. Mix thoroughly and then add 200 μl of this mixture to each well of the first plate using a 100–1200 μl multi-channel pipette. Dilute and mix the remaining cell suspension thoroughly with 20 ml of additional regular medium (~40 ml cell suspension). Add 200 μl of this mixture to each well of the second and third 96-well plates to obtain single cell suspensions in each well (*see* **Note 7**).
14. Place plates into the incubator and allow the cells to grow for 10–14 days, or until single colonies are formed (*see* **Note 8**).

3.3 Single Clone Isolation and Screening

1. Once a colony has formed (at approximately day 10), mark the wells containing a single colony (*see* **Note 8**).
2. Remove the medium from wells that contain a single colony and wash with 100 μl 1× PBS.
3. Add 50 μl trypsin for 5 min or until the cells are detached at 37 °C.

4. Resuspend the cells with 1 ml of normal medium by pipetting up and down gently several times to ensure the cells are resuspended thoroughly. Transfer cells from each single colony to an individual well in a 24-well plate.
5. Allow cells to grow until 80% confluency is reached.
6. Screen each clone for successful KO of the histone variant by western blotting [29] and immunofluorescence [30], as shown in Fig. 1, using a specific antibody that recognizes unmodified H2AX (1:1000 for western blotting and 1:500 for immunofluorescence) or macroH2A (1:1000 and 1:500 for immunofluorescence). All positive clones detected by these methods should also be verified by PCR-coupled sequencing using primers described in genotype verification (*see* Subheading 2.6) to validate and determine the mutation that has occurred within the targeted gene.

3.4 Western Blotting

1. Harvest ~3 × 10^6 cells grown from each clone by adding 100 μl trypsin to each individual well of a 24-well plate. Put cells back in the incubator and allow cells to detach (~5 min). Next, mix cells and transfer 80 μl into a 1.5 ml eppendorf containing 80 μl of normal medium. Centrifuge 15,000 rpm for 30 s and remove the medium. Directly resuspend the cell pellet in 50 μl 1× SDS-PAGE sample buffer and boil for 10 min. Add 1 ml of normal medium to the well containing the remaining cells and allow to grow and expand for future use.
2. Load 5 μl of sample (or about 10 μg of cell lysate) into a well of a 15% polyacrylamide gel for each colony to be tested.
3. Run the gel using 1× running buffer (*see* **Note 9**).
4. Transfer of the gel onto a nitrocellulose membrane using 1× transfer buffer by a wet transfer system (ex. BioRad) (*see* **Note 9**).

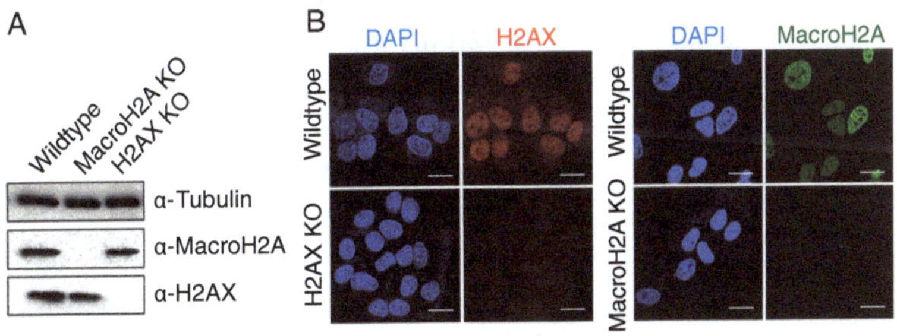

Fig. 1 Validation of histone H2A variant KO in human U2OS cells. (**a**) Western blot analysis of U2OS WT, H2AX KO, and macroH2A KO cells using an antibody against H2AX and macroH2A. Tubulin was used as a loading control to ensure equal loading of the samples. (**b**) Immunofluorescence analysis of H2AX and macroH2A in WT and KO cell lines. Lack of signal indicates the cells that are KO for the specified gene. Scale bar 20 μm

5. Thoroughly rinse the membrane in 1× TBS.
6. Incubate the membrane with 3% BSA in TBS-T for 1 h at room temperature.
7. Remove blocking buffer and incubate the membrane in primary antibody solution at 1 h room temperature or 4 °C overnight (*see* **Note 10**).
8. Wash the membrane with TBS-T for 5 min three times.
9. Incubate the membrane in TBS-T containing secondary antibody at room temperature for 1 h.
10. Wash the membrane with TBS-T for 5 min three times.
11. Incubate the membrane with 1 ml of ECL solution for 1 min at room temperature.
12. Remove excess ECL solution and analyze the membrane by chemiluminescence using a ChemiDoc XRS+ system (ex. BioRad) or film.

3.5 Immunofluorescence

1. Seed cells in regular medium onto a UV-sterilized glass coverslip that has been placed in a 24-well plate for ~24 h (*see* **Note 11**).
2. Once the cells have reached 50% confluency, remove the medium and wash the cells with 1 ml of 1× PBS briefly.
3. Fix the cells by incubating the coverslip in Fixative for 10 min at room temperature followed by 3 washes in 1 ml 1× PBS for 30 s each.
4. Add 1 ml of Permeabilization buffer and incubate for 5 min at room temperature.
5. Wash with 1× PBS three times for 30 s each.
6. Incubate the coverslip in Blocking solution for 30 min.
7. Remove the coverslip from the plate using a needle to pry the coverslip from the bottom of the dishes if needed and incubate the coverslip facedown on a small drop of primary antibody (~50 μl) for 1 h at room temperature or 4 °C overnight in a humidified chamber (*see* **Note 12**).
8. After incubation with the primary antibody, return the coverslip to a new 24-well plate and wash the coverslip with 1× PBS three times for 30 s.
9. Incubate the cells in secondary antibody solution for 1 h at room temperature followed by washing the cells with 1× PBS three times for 30 s.
10. Mount coverslips on a microscope slide by adding 2–3 μl of mounting medium on a slide, place the coverslip with the cell side against the mounting medium and slide. Gently press the coverslip until the mounting medium is dispersed evenly across

the bottom of the coverslip. Remove any excess mounting medium with a Kimwipe and use clear nail polish to seal the entire edge of the coverslip.

11. Once the nail polish has dried, the slides can be analyzed by fluorescence microscopy. For long-term storage, the slides should be stored in −20 °C and protected from light since DAPI and the fluorescent secondary antibodies are light sensitive.

3.6 Verification of KO Cells by Sequencing

1. Harvest ~1 × 10^6 cells (U2OS) from each well of a 6-well plate for each potential KO cell clone. Add 200 μl typsin and incubate in 37 °C for 5 min. Transfer the cells into 1.5 ml eppendorf containing 200 μl. Centrifuge 15,000 rpm for 30 s and resuspend in 200 μl PBS.
2. Extract genomic DNA using a DNeasy Blood & Tissue Kit according to the manufacturer's instructions (*see* **Note 13**). Perform PCR reaction with 100 ng genomic DNA and specific primers (*see* Subheading 2.6) using KOD Hot Start DNA polymerase, according to the manufacturer's instructions, to amplify the genomic region containing the sgRNA target site (*see* **Note 14**).
3. Sequence the amplicon by Sanger sequencing using gene-specific primers, which flank the sgRNA targeted region (*see* Subheading 2.6) that will amplify the sgRNA target site to verify the presence of a mutation.

3.7 Complementation of KO Cells Using WT and Mutant Histone Variant Derivatives

Due to the potential for off-target cleavage events by CRISPR/Cas9 that can result in mutations and cellular effects that are not related to loss of the targeted gene, complementation assays are required to rule out these potential off-target effects. Complementation assays are performed by re-introducing WT, as well as mutant genes that express the appropriate protein to ensure that the phenotype present in the histone variant KO cells is a result of the loss of the histone variant specifically. Complementation analyses are performed as follows:

1. Transfect 15 μg of a plasmid containing WT H2AX or macroH2A (ex. HA-Flag-H2AX or HA-Flag-macroH2A) using 10 ml of Opti-MEM without antibiotics and 40 μl Fugene HD transfection reagent into the appropriate KO cell line (density ~3 × 10^6) in a 10 cm dish and allow the cells to incubate with the transfection reagent for 6 h at 37 °C (as in Subheading 3.2, **step 1–7**) (*see* **Note 15**).
2. Replace the plate of cells with fresh regular medium to allow the cells to recover for 24 h.
3. Select stable cell lines that contain these plasmids using regular medium with 2 μg/ml puromycin.

4. Allow the cells to grow in a 10 cm dish for at least 5 days or until confluent, harvest and validate the reconstitution of the cells with the plasmid by analyzing the expression of the histone variant by western blotting (Fig. 2) and immunofluorescence using Flag antibody (for tagged proteins) (as in Subheading 3.5). H2AX and macroH2A specific antibodies can also be used (*see* **Note 16**).
5. Once cell lines are validated as complemented with a specific histone variant, WT, KO, and KO complemented cell lines can be used to confirm that the mutant phenotype is specific.

4 Notes

1. There are different types of CRISPR/Cas9 vectors available. We use pSpCas9 (BB)-2A-puro (pX459) because the puromycin resistance gene improves screening efficiency by eliminating non-transfected cells by selection in puromycin-containing medium.
2. This protocol is designed for U2OS cells. The transfection efficiency can be further optimized, for example in other cell lines, using fluorescence microscopy to identify plasmid-containing cells using pSpCas9 (BB)-2A-GFP (pX458, Addgene plasmid ID: 48138).
3. QIAprep Spin Miniprep kit is used in the current study. We have also used the Invitrogen PureLink Quick Plasmid Miniprep Kit, which is also acceptable for purifying plasmid DNA.
4. U2OS cells are adherent and this protocol is relevant to gene editing of histone variants in adherent cells only. If this protocol is to be used for cells in suspension, several modifications will need to be made to take into account this variation in cell growth conditions.
5. Corning™ Costar™ sterile disposable reagent reservoirs were used in this study. Other brands or even a 50 ml conical tube would also work.
6. We used GE Healthcare Amersham™ ECL™ Prime Western Blotting Detection Reagent in our current study, but different brands of enhanced chemiluminescence (ECL) will also work.
7. The diluted cells containing medium in the 50 ml reagent reservoir should be mixed thoroughly every time before dispensing into the 96-well plate by pipetting up and down to make sure the cells remain suspended evenly.

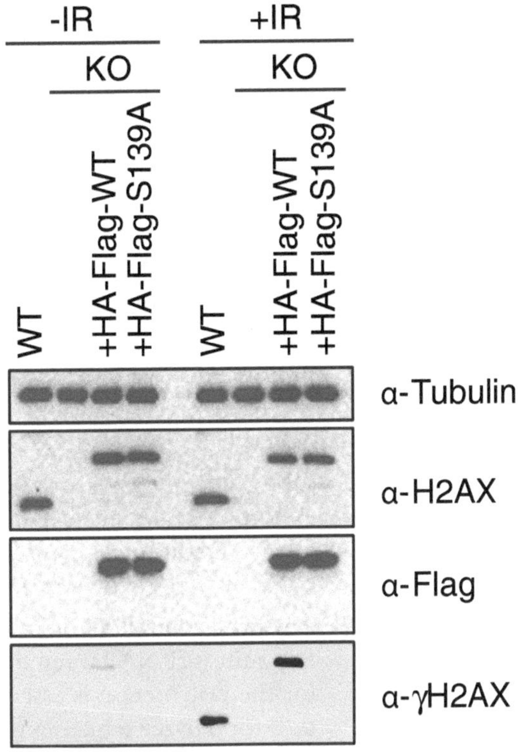

Fig. 2 Western blotting analysis of HA-Flag-WT and S139A mutant H2AX in H2AX KO reconstituted cells. Flag antibody is used to detect tagged H2AX WT and S139A mutant proteins. H2AX antibody is used to detect the endogenous and reconstituted H2AX WT and S139A mutant proteins. γH2AX antibody is used to detect the phosphorylated form of H2AX at S139 residue, which is induced by DNA damage from ionizing radiation (IR, 10 Gy). Tubulin is used as a loading control. Note the decreased mobility of tagged WT and mutant H2AX proteins compared to endogenous H2AX

8. Although several methods can be used to visualize single clones in 96-well plates, we recommend the use of a light microscope for this technique. Use a 4× or 10× objective of a light microscope to observe single colonies in each individual 96-well plate. The medium should be left in the well to ensure the healthy maintenance of the cells. This technique allows the visualization of cell colonies, which is important for determining if a well contains one single colony. This is critical since taking more than one colony from the same well can result in a mix population of cells. It is paramount that gene KOs are isogenic for genetic analysis and for drawing clear conclusions of phenotypes associated with the loss of a specific gene.

9. Western blots are performed using 15% SDS-PAGE gel under constant low voltage (~150 V) to achieve a better resolution for histone proteins. The gel was preferably transferred with wet transfer system in constant amperage (~300 A per blot).
10. Loading controls (ex. tubulin, core histones) should be included for Western blot screening to ensure equal loading between samples.
11. Glass coverslips were sterilized by exposing under the UV light in the tissue culture hood for >60 min.
12. Incubation of primary and secondary antibodies for immunofluorescence were performed in a humidified chamber, which can be created by adding a wetted paper towel with water in a closed box.
13. DNeasy Blood & Tissue Kit was preferably used in extracting genomic DNA to achieve consistent and highly purified DNA. Traditional phenol/chloroform method or other kits could also be used.
14. Any other high fidelity PCR polymerase can be used for amplifying the sgRNA targeted region. The proofreading property for the polymerase is essential to provide error-free amplification for Sanger sequencing results.
15. The amount of transfected cells can be scaled either down or up depending on experimental conditions (ex. transfection efficiency and cell death). The transfection efficiency can be verified by immunofluorescence staining using Flag antibody. *See* Subheading 3.3. If there is a lot of cell death after transfection, optimize the amount of transfection reagent and plasmids to reduce toxicity to the cells.
16. The suggested antibodies (*see* Subheading 2.3, **steps 1** and **2**) were verified using western blotting and immunofluorescence. Alternative antibodies for H2AX (Abcam, ab11175 and Cell Signaling, 2595); macroH2A (Novus Biologicals, BBP1-53018) can also be used.

Acknowledgments

We thank Fade Gong from the Miller lab for insightful comments on the manuscript. The K.M.M. laboratory is supported by the NIH National Cancer Institute (R01CA198279 and RO1CA201268) and the American Cancer Society (RSG-16-042-01-DMC).

References

1. Khare SP, Habib F, Sharma R et al (2012) HIstome—a relational knowledgebase of human histone proteins and histone modifying enzymes. Nucleic Acids Res 40:D337–D342
2. Maze I, Noh KM, Soshnev AA et al (2014) Every amino acid matters: essential contributions of histone variants to mammalian development and disease. Nat Rev Genet 15:259–271
3. Talbert PB, Henikoff S (2017) Histone variants on the move: substrates for chromatin dynamics. Nat Rev Mol Cell Biol 18:115–126
4. Kouzarides T (2007) Chromatin modifications and their function. Cell 128:693–705
5. Musselman CA, Lalonde ME, Cote J et al (2012) Perceiving the epigenetic landscape through histone readers. Nat Struct Mol Biol 19:1218–1227
6. Scully R, Xie A (2013) Double strand break repair functions of histone H2AX. Mutat Res 750:5–14
7. Bonner WM, Redon CE, Dickey JS et al (2008) GammaH2AX and cancer. Nat Rev Cancer 8:957–967
8. Celeste A, Petersen S, Romanienko PJ et al (2002) Genomic instability in mice lacking histone H2AX. Science 296:922–927
9. Stucki M, Clapperton JA, Mohammad D et al (2005) MDC1 directly binds phosphorylated histone H2AX to regulate cellular responses to DNA double-strand breaks. Cell 123:1213–1226
10. Yuan J, Adamski R, Chen J (2010) Focus on histone variant H2AX: to be or not to be. FEBS Lett 584:3717–3724
11. Rogakou EP, Pilch DR, Orr AH et al (1998) DNA double-stranded breaks induce histone H2AX phosphorylation on serine 139. J Biol Chem 273:5858–5868
12. Khurana S, Kruhlak MJ, Kim J et al (2014) A macrohistone variant links dynamic chromatin compaction to BRCA1-dependent genome maintenance. Cell Rep 8:1049–1062
13. Sansoni V, Casas-Delucchi CS, Rajan M et al (2014) The histone variant H2A.Bbd is enriched at sites of DNA synthesis. Nucleic Acids Res 42:6405–6420
14. Timinszky G, Till S, Hassa PO et al (2009) A macrodomain-containing histone rearranges chromatin upon sensing PARP1 activation. Nat Struct Mol Biol 16:923–929
15. Tolstorukov MY, Goldman JA, Gilbert C et al (2012) Histone variant H2A.Bbd is associated with active transcription and mRNA processing in human cells. Mol Cell 47:596–607
16. Xu Y, Ayrapetov MK, Xu C et al (2012) Histone H2A.Z controls a critical chromatin remodeling step required for DNA double-strand break repair. Mol Cell 48:723–733
17. Monteiro FL, Baptista T, Amado F et al (2014) Expression and functionality of histone H2A variants in cancer. Oncotarget 5:3428–3443
18. Vardabasso C, Hasson D, Ratnakumar K et al (2014) Histone variants: emerging players in cancer biology. Cell Mol Life Sci 71:379–404
19. Buschbeck M, Hake SB (2017) Variants of core histones and their roles in cell fate decisions, development and cancer. Nat Rev Mol Cell Biol 18:299–314
20. Gruosso T, Mieulet V, Cardon M et al (2016) Chronic oxidative stress promotes H2AX protein degradation and enhances chemosensitivity in breast cancer patients. EMBO Mol Med 8:527–549
21. Bibikova M, Beumer K, Trautman JK et al (2003) Enhancing gene targeting with designed zinc finger nucleases. Science 300:764
22. Christian M, Cermak T, Doyle EL et al (2010) Targeting DNA double-strand breaks with TAL effector nucleases. Genetics 186:757–761
23. Cong L, Ran FA, Cox D et al (2013) Multiplex genome engineering using CRISPR/Cas systems. Science 339:819–823
24. Mali P, Yang L, Esvelt KM et al (2013) RNA-guided human genome engineering via Cas9. Science 339:823–826
25. Chen WT, Alpert A, Leiter C et al (2013) Systematic identification of functional residues in mammalian histone H2AX. Mol Cell Biol 33:111–126
26. Weyemi U, Redon CE, Choudhuri R et al (2016) The histone variant H2A.X is a regulator of the epithelial-mesenchymal transition. Nat Commun 7:10711
27. Shimada M, Goshima T, Matsuo H et al (2016) Essential role of autoactivation circuitry on Aurora B-mediated H2AX-pS121 in mitosis. Nat Commun 7:12059
28. Leung JW, Makharashvili N, Agarwal P et al (2017) ZMYM3 regulates BRCA1 localization at damaged chromatin to promote DNA repair. Genes Dev 31:260–274
29. Kurien BT, Scofield RH (2006) Western blotting. Methods 38:283–293
30. Strickfaden H, Hendzel MJ (2017) Immunofluorescence of histone proteins. Methods Mol Biol 1528:165–171

Part IV

Evolution, Development, and Diseases

Chapter 15

Studying the Evolution of Histone Variants Using Phylogeny

Antoine Molaro and Ines A. Drinnenberg

Abstract

Histones wrap DNA to form nucleosomes that package eukaryotic genomes. Histone variants have evolved for diverse functions including gene expression, DNA repair, epigenetic silencing, and chromosome segregation. With the rapid increase of newly sequenced genomes the repertoire of histone variants expands, demonstrating a great diversification of these proteins across eukaryotes. In this chapter, we are providing guidelines for the computational characterization and annotation of histone variants. We describe methods to predict the characteristic histone fold domain and list features specific to known histone variants that can be used to categorize newly identified histone fold proteins. We continue describing procedures to retrieve additional related histone variants for comparative sequence analyses and phylogenetic reconstructions to refine the annotation and to determine the evolutionary trajectories of the variant in question.

Key words Histone variants, H2A, H2B, H3, Rapid evolution, Homology predictions, Multiple sequence alignments, Phylogenetic inference, Annotation

1 Introduction

The fundamental repeating unit of chromatin in eukaryotes is the nucleosome, in which DNA is wrapped around an octameric histone core complex [1]. The majority of nucleosomes in the cell are assembled by canonical histones and consist of 2 copies of each of the canonical histones H2A, H2B, H3, and H4. Histones all share a common fold, the histone fold domain (HFD), composed of three helices that mediate both protein dimerization and DNA binding in the context of the nucleosome [2]. Canonical histones provide the primary structural scaffold for DNA wrapping during whole genome replication. As such, canonical histone expression and synthesis peaks during S-phase and their deposition occurs genome-wide [3]. In addition to canonical histones, unique histone variants can be incorporated into nucleosomes to confer specialized functions in specific genomic regions. Such functions include processes as diverse as transcription, chromosome segregation, DNA repair

Guillermo A. Orsi and Geneviève Almouzni (eds.), *Histone Variants: Methods and Protocols*, Methods in Molecular Biology, vol. 1832, https://doi.org/10.1007/978-1-4939-8663-7_15,

and recombination, chromatin remodeling, germline-specific DNA packaging, and even extra-nuclear acrosome formation [4]. To accomplish their specific chromatin functions, variants differ from canonical histones in their primary amino acid sequence. In addition to variations in the HFD, most variants have evolved N- or C-terminal amino acid extensions with specific sequence motifs or acquired entirely new domains. These sequence features are key to categorize variants accomplishing similar functions between organisms.

The evolutionary trajectories of histone variants are also unique from their canonical counterparts. First, unlike most canonical histones, which are encoded by genomic clusters containing the four core histone genes, "stand-alone" genes encode variants [3]. Second, while most canonical histones have an old evolutionary history that can be traced back to the last common ancestor to all eukaryotes, variants can have diverse evolutionary origins [4]. Some variants are broadly distributed throughout the eukaryote phylogeny, and likely arose early in eukaryotic evolution, while others originated from recent lineage-specific events [5]. In fact, with the increasing number of sequenced genomes from diverse eukaryotes, the repertoire of new histone variants or diverse homologs of existing variants continues to expand [6]. As some of these newly sequenced organisms are not easily accessible to genetic manipulations enabling experimental investigation, histone variants characterization solely relies on in silico comparisons to known and well-annotated examples. In this context, the phylogenetic investigation of variants provides a unique insight into the origins and selective constraints driving their specific biological functions.

In this chapter, we will provide guidelines to enable the systematic characterization and phylogenetic study of histone variants with an emphasis on H2A and H3 variants that are most abundant in model organisms. First, we will describe approaches used to identify histone variant candidates from publically available curated or non-curated genomic sequences. We will then provide instructions for the application of phylogenetic analyses to help distinguish them from their canonical counterparts. Because the combination of distinct sequence and phylogenetic features are strong predictors of their structure and function, we also provide a list of features characteristic of known histone variants that can be used to survey candidates.

2 Materials

Sequence sources: curated histone sequences can be found in browsable and searchable public databases such as Histone DB 2.0 [7], NCBI [8], and UCSC [9]. User provided sequences can also be used.

Analysis tools: all of the analysis presented here can be performed on a desktop computer. Alignment and other sequence analysis software referenced here are open source software that can be freely downloaded from the web or run online.

3 Methods

In the following section, we provide general guidelines to perform sequence homology searches against publically available genome, transcriptome, and protein databases (*see* Subheading 3.1). Following up on those, we review characteristic features and evolutionary relationships of all known histone variants to help identify and classify HFD containing proteins (*see* Subheading 3.2). Once the histone variant has been identified with some confidence, we provide the general guidelines to extend searches to closely related and more distantly related species (*see* Subheading 3.3). Finally, we explain how to apply phylogenetic analyses to determine their evolutionary trajectories (*see* Subheading 3.4). The general workflow outline to study the evolution of histone variants is illustrated in Fig. 1.

3.1 Commonly Used Search Tools and Databases

With an unknown query sequence in hand, the first step is to establish its relationship, or its identity, to known histone fold containing proteins. Two strategies can be applied depending on the degree of divergence of the query sequence:

1. Search algorithms like BLAST or PSI-BLAST that test for direct sequence homology.
2. Hidden-Markov-Model-based algorithms capable of identifying remote homologies.

3.1.1 BLAST and PSI-BLAST

1. BLAST (basic local alignment search tool, *see* **Note 1**) [10] is a commonly used and efficient suite of tools to discover the evolutionary relationship of a query sequence with, searchable, publicly available sequences or between multiple user-provided input sequences. There are two types of queries that are supported by BLAST: nucleotide and amino acid sequences. When running BLAST on nucleotide sequences (BLASTn) it is advised to start with a spliced mRNA or coding sequence (CDS); as intron and regulatory regions might complicate searches since they are less evolutionarily constrained. A successful BLAST search should result in a list of sequence hits with additional attributes (i.e., genome of origin, chromosome location, protein ID, etc.). These hits can be downloaded and further processed outside the BLAST environment.

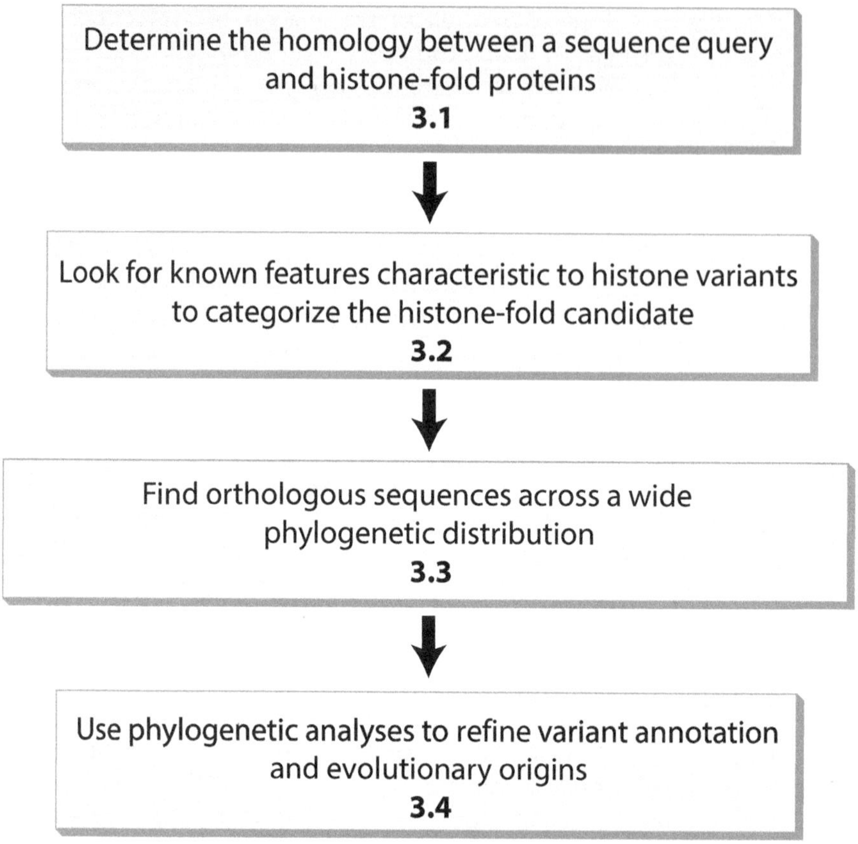

Fig. 1 Workflow to annotate and characterize new histone variants

2. Three parameters are used to describe a match: similarity, coverage, and expect value (*E* value). As general guidelines, a minimum of 30% similarity and at least 70% coverage domain between the query and the template are necessary for unambiguous alignment and potential structural models. The *E* value is a statistical estimate that describes the number of matches that can be expected by chance in a given database of a particular size. The lower the *E* value, the more "significant" the given match. Generally, *E* value less than 0.001 is considered to be significant to conclude homology.
3. When using protein queries, a more sensitive strategy than BLASTp (Protein BLAST) to detect more distant evolutionary relationships is position-specific iterated-BLAST (PSI-BLAST) [11]. Starting from a single sequence, a sequence profile or position-specific-scoring matrix (PSSM) of related proteins above a certain threshold (based on the *E* value) found using BLASTp is created. The PSSM is then used to search the protein database iteratively for several rounds. The matches are included to create another PSSM and the procedure is repeated

until no additional matches are found. By including all related proteins in the search, PSI-BLAST helps to uncover more distant homologs than ordinary BLASTp.

4. tBLASTn allows the user to search a nucleotide database (e.g., a whole genome) that has been in silico translated using a protein sequence query. This has the advantage of finding protein-coding nucleotide sequences that have no annotated protein, missed by BLASTp, or are highly diverged at the nucleotide level, missed by BLASTn. In addition, this allows the characterization of pseudogene sequences that have early stop codons or frame-shifts but still have significant homology to your query.
5. Finally, for highly diverged sequences, it might be better to run the search using the canonical histone sequence as a bait and focus on divergent matches (e.g., using H3 for cenH3 searches).

3.1.2 Hidden Markov Model-Based Predictions

While most histones can be detected using searches for direct sequence homology, some histone-fold proteins exhibit low levels of sequence conservation requiring alternative prediction tools to detect remote homologies [12]. In fact, it has been shown that two proteins can share a high degree of structural similarity despite the lack of detectable sequence similarity. Hidden Markov Model (HMM)-based techniques implemented in servers such as HHpred [13] are currently the most sensitive tools detecting remote protein homology.

1. In the first step, HHpred builds a multiple sequence alignment from the query sequence by multiple iterations of PSI-BLAST against the nr database from NCBI.
2. In the next step, a profile HMM is generated from this alignment that includes the information about predicted secondary structure.
3. This is then compared with a database of HMMs representing proteins with known structure (such as PDB [14]) and protein families (such as PFAM [15]). The result is a list of matches with pairwise alignments.
4. For the interpretation of the results, the most relevant statistical measure is the probability in percent that the database match is indeed homologous. This value takes the score of the secondary structure into account that helps to distinguish homologs from chance hits (*see* **Note 2**, for a discussion on false positives).

3.2 Characteristic Features Used to Distinguish Histone Variants and Their Evolutionary Trajectories

Histone variants are subject to strong selective pressures to perform their unique function. Consequently, many sequence features affecting both the HFD and additional, variant-specific, domains can be used to discriminate each variant from their canonical counterpart across broad phylogenetic distributions. It is important to keep in mind that canonical histones also display some degree of divergence across eukaryotes and, thus, only "conserved" substitutions should be used to specifically annotate variants. In Table 1 we provide an overview of the evolutionary origins and distinguishing features of histone variants.

3.3 Finding Orthologous Sequences

One key step in studying the evolution of histone variants, or any protein, is to compare it to homologous sequences across species to determine their evolutionary trajectories/origins.

To do so, a set of homologous sequences first needs to be collected keeping in mind that the phylogenetic distribution of the corresponding species used can greatly influence downstream analyses (*see* Subheading 3.4). In addition to surveying the group of organisms where orthologs of the corresponding histone variant are expected to be found, it is advised to include an homolog encoded by a more distantly related species in order to orient the comparison (i.e., root the phylogeny). To collect homologous sequences to the variant of interest, collection from public histone databases can be used as starting points. As an alternative and to further extend the set, BLAST searches can be performed to obtain homologs of additional, more closely related species (using the tools described in Subheading 3.1). We briefly describe both approaches.

3.3.1 HistoneDB 2.0

HistoneDB 2.0 [7] is a database classifying histone protein sequences by type and variant. It collects canonical histones and histone variants, provides protein sequence, structural and functional annotations to facilitate the performance of comparative analyses across organisms. In addition, one can input any histone protein to this database to find its most closely related histone annotation. HistoneDB 2.0 consists of two complementary parts.

1. First, each histone clade (H2A, H2B, H3, and H4) can be explored through a set of manually curated variants with annotations of their specific features and functions. This set is used to construct Hidden Markov Models in order to automatically extract collections of related proteins from the nonredundant database of protein sequence maintained by NCBI (nr).
2. Results for this search constitute the second part of this database. From these 2 sets of proteins HistoneDB also provides an estimated phylogenetic distribution for the variant.

Table 1
A summary table of histone variants and their characteristics. Note that frequently not all characteristics are common to all homologs of a particular variant and should therefore only be considered as general guidelines. Experimental validations are often needed to confirm the function of candidates. Furthermore, conservation across a specific group of organisms (e.g., Ophistokonts) does not imply that every eukaryote encodes this variant [4, 6, 31–55]

Variant	Conservation	Origin	Characteristics to canonical form[a]	References
H2A.Z	Eukaryotes	Monophyletic	60% protein sequence identity to H2A Under strong purifying selection Divergent C-terminal end—one a. a[b] insertion in loop L1 7 diverged residues within loop L1/α2 "DEELD" motif in the acidic patch—truncation of αC Duplication in vertebrates: H2A.Z.1 and 0.2–0.3 divergent a.a[b]: T/A15, S/T39 and V/A128	[4, 5, 34–36]
H2A.X	Eukaryotes	Polyphyletic	95% protein sequence identity to H2A along the HFD C-term SQ(E/D)-phi motif (phi: Hydrophobic residue—usually, Ile, Tyr or Phe)	[5, 7]
macroH2A	Metazoa	Monophyletic	~60% protein sequence identity to H2A along the HFD Extended C-term "macro domain" (Pfam:PF01661) Duplication in vertebrates: macroH2A.1 and macroH2A.2	[37–39]
Short H2As	Placental mammals	Monophyletic	Rapidly evolving, <50% identity with canonical H2A Diverged N-term residues and loss of acidic patch Truncated C-term tail with no Lys at position 119	[40, 53–56]
H2A.W	Seed bearing plants	Monophyletic	Paralogs have >80% protein sequence identity Diverged loop—"RY-S/A-K/Q" C-term motif "KSPK-K/S-A/K"	[6]
H2A.M	Moss, liverwort, lycophyte, angiosperms	Monophyletic	Similar to H2A.W Additional "KSPK" C-term motif Shorter N-term than H2A.W, at least 6 a.a[b]	[6]

(continued)

Table 1
(continued)

Variant	Conservation	Origin	Characteristics to canonical form[a]	References
H2A.J	Mammals, n.d.	Likely monophyletic	90% protein sequence identity to H2A Val at position 11 C-term "SQK" motif Variable position 4 a.a.[b] from the stop codon	[41]
H2A.1	Mammals, n.d.	Monophyletic	Ile at position 31 and 44, Ser at position 71 C-term "A/V/S/T-Q-S/A/T" motif Germline restricted	[4, 7, 43]
H2B.W	Mammals, n.d.	Likely monophyletic	<50% protein sequence identity to H2B 30 a.a. Extension of N-term Sperm restricted	[42, 44]
H2B.1	Mammals, n.d.	Monophyletic	Ile at position 43 many N-terminal substitutions Germline restricted	[4, 7, 55]
H2B.E	Mouse	n.d.	5 diverged residues Olfactory neuron restricted	[47]
H2B.Z	Apicomplexa	Likely monophyletic	>90% protein sequence identity to H2B Shorter N-term tail Variable residues within α2	[7, 45, 46]
H3.3	Eukaryotes	Polyphyletic	Under strong purifying selection Ser/Thr at position 31 Often additional differences at position 87, 89, and 90	[4, 48, 51]
cenH3	Eukaryotes	Possibly monophyletic (unclear)	50–60% protein sequence identity to H3 Rapidly evolving Extended N-terminal tail (20–200 aa) and loop1 region Lack of Gln at position 68; Trp at position 84 (instead of Phe); Ala, Cys or Ser at position 107 (instead of Thr)	[4, 5]
H3.5	Hominids	Monophyletic	Lack of Lys at position 37	[49, 52]
H3.T	Mammals	Monophyletic	Testis-restricted expression Val, met, Ser and Val at position 24, 72, 98, and 111	[31, 33]

(continued)

Table 1 (continued)

Variant	Conservation	Origin	Characteristics to canonical form[a]	References
H3.X	Primates	Monophyletic	Several mutations to canonical H3, almost identical to H3.Y	[50]
H3.Y	Primates	Monophyletic	Several mutations to canonical H3, almost identical to H3.X	[50]
CENP-T	Ophistokonts	Likely monophyletic	Rapidly evolving Large N-terminal domains with HFD at the C-terminus C-terminal 2-helical extension	[12, 32]
CENP-W	Ophistokonts	Likely monophyletic	Rapidly evolving	[12, 32]
CENP-S	Eukaryotes	Likely monophyletic	n.d.	[12, 32]
CENP-X	Eukaryotes	Likely monophyletic	n.d.	[12, 32]
subH2B	Mammals, n.d.	Likely monophyletic	Bipartite nuclear localization motif at N-term and C-term. Of the less than 50% identities with canonical H2B	[57]

[a]Positions are counted according to the canonical form
[b]Amino acid

3. This enables browsing through histone variant entries to analyze their characteristic features and predict their evolutionary trajectories (also *see* **Note 3**). HistoneDB 2.0 therefore represents a useful resource to obtain sets of histone variants for constructing multiple sequence alignments (MSA) and phylogenetic analyses as described in the following section.

3.3.2 BLAST Searches

The most thorough way of finding homologous sequences across species is to use BLAST searches using NCBI genomes and databases. One can also run BLAST on user complied datasets (genomes, transcriptomes, etc.). Through the GeneBank portal (https://www.ncbi.nlm.nih.gov/genbank/), one can explore the available datasets and their phylogenetic relationships.

1. We advise to begin by looking at the phylogeny (user determined or published) of species closely related to the query species—separated by a few My—and using this set of species to extend the search over distant clades (<100 My). With a few related sequences in hand, identifying distant orthologs becomes easier where conserved sequence features start to emerge.

2. As mentioned in Subheading 3.1 one should carefully decide between using amino acid or nucleotide sequences as input—generally, we advise to use tBLASTn and work with in silico translated protein sequences.
3. Finally, we advise to always retrieve the genetic context in which the homologs are found. Indeed, depending on the biology of the species as well as the evolutionary time span, certain histone variants are found within identical genomic context and this information can be crucial to help in finding the correct phylogenetic relationship between sequences (*see* Subheading 3.4). More generally if the gene/protein is annotated in any assembly, then BLAST searches can be combined with synteny searches using a comparative genomics online platform named Genomicus (http://www.genomicus.biologie.ens.fr/genomicus-88.01/).

3.4 Sequence Alignments and Phylogenetic Reconstructions

When studying the evolution of histone variants several questions can arise: when did this histone variant appear in genomes and how is it related to another variant? For example, while H2A.Z variants are expected to have deep roots in the most recent common ancestor of eukaryotes, the phylogeny of cenH3 variants is currently unresolved. It was proposed that cenH3 variants have evolved several times independently in multiple eukaryotic lineages from H3 [5], yet with limited statistical support for this hypothesis. Instead, the more parsimonious model of a single origin for cenH3 in an early eukaryotic ancestor appears more likely due to its conserved presence at the eukaryotic centromere. Are there new sequence features that arose in a group of species? For example, H2A.W acquired a unique sequence motif that is unique to plants, while the motif in H2A.X might have arisen several times during eukaryotic evolution. How diverged is this variant compared to other histone genes? While canonical histones as well as H2A.Z and H3.3 are deeply conserved, the protein sequences of other H2A and cenH3 variants are divergent among species.

Phylogenetic analyses use information derived from sequence alignments to infer the possible evolutionary path that led to extant gene/protein sequences. It allows the reconstruction of ancestral sequences and the assessment of orthologies between sequences. Finally, phylogenetic trees convey more information than sequences alignments alone. Following up on the retrieval of histone variants homologous sequences, we provide general guidelines to perform multiple sequence alignments (MSA) and to build meaningful phylogenies. However, since these are general tools that are applied outside the study of histone variants, we refer to individual chapters dedicated to both alignments and phylogenies [16, 17].

3.4.1 Methods for Multiple Sequence Alignments

Since all of the input sequences are histone-related genes/proteins one should attempt to anchor the alignment around the HFD, possibly using the appropriate canonical histone as a reference. Although current algorithms are very accurate, alignments should be inspected visually, especially when dealing with distantly related variants. At first glance the MSA can inform on highly constrained residues in the alignment or, to the contrary, reveal unexpected regions of variability. In addition, MSA will allow the user to trim potentially retained noncoding intronic sequences when using genomic sequences as input or to identify annotations errors such as mis-annotated start codons (*see* **Note 4**).

In the following, we review a few MSA tools:

1. *MUSCLE*: MUSCLE stands for stands for *MU*ltiple *S*equence *C*omparison by *Lo*g-*E*xpectation [18, 19]. This software that can be run both on an online server (http://www.ebi.ac.uk/Tools/msa/muscle/) and locally (the source code is freely available under http://www.drive5.com/muscle). One of the advantages of MUSCLE is that it can handle both DNA and protein sequences and is suitable for large datasets (>100 sequences).
2. *Clustal Omega*: Clustal Omega (http://www.clustal.org) is part of the widely used Clustal alignment suite [20]. Clustal Omega also achieves fast execution times for large datasets by implementing the co-called mBed algorithm [21] for the fast construction of guide trees to produce sequence alignments.
3. *DNA versus protein alignments*: Protein phylogenies are very well suited to inform on the patterns and rates of substitution that occurred between sequences, which will help in determining the origin and relatedness of the variant. On the other hand, nucleotide phylogenies can be more robust since they contain ~3 times as many positions to be considered in the MSA than proteins (3 nucleotides per amino acid). In addition, nucleotide phylogenies will better inform on the nature of the selective forces that have shaped the histone variant, since one can measure and compare the rates of synonymous (silent) and non-synonymous codon substitutions across the phylogeny [22]. In contrast to protein alignments that can be used when studying histone variants over deep evolutionary time scales (e.g., >100 My for mammals), alignments between nucleotide sequences are restricted to closely related species (e.g., <50 My for primates), but it largely depends on the mutation rate and generation time of the species studied since the total number of synonymous substitutions can quickly saturate and become uninformative.

3.4.2 Inferring the Phylogenies of Histone Variants

The final and last step to study histone variant evolution is the construction and visualization of a phylogeny. There are 3 major methods to infer phylogenies from sequence alignments:

1. *Maximum parsimony*: Phylogenies based on parsimony will treat each aligned position as a separate "character" and will attempt to build a relationship tree that minimizes the total number of steps required to explain the distribution of the "characters" in the MSA. A lot of platforms support parsimony tree building, one of the most commonly used is PAUP* (Phylogenetic Analysis Using Parsimony * and other methods) [23]. However, maximum parsimony methods are computationally heavy since they explore all possible trees before assessing the "most parsimonious" one, so they are best suited for a small number of sequences that are not too distantly related (e.g., all carnivores H2A.Zs).
2. *Neighbor joining*: This method is widely used since it is suitable for large datasets, and runs quickly locally or online. Neighbor joining phylogenies use identities in the MSA to assign distances between sequences. These distances become the branch length in the tree and the algorithm finds the "best phylogeny" by minimizing the total branch length of the tree [24]. This method can also be used with bootstrapping—which applies random sampling of branches to assign statistical significance to the tree topology. However, since this method assumes a constant rate of substitution across the MSA it will not resolve very deep nodes in the phylogeny, e.g., the base nodes of all land plants.
3. *Maximum Likelihood*: This is the most statistically robust method to build phylogenies. Computationally, building ML phylogenies is as intense as building maximum parsimony phylogenies, however it has the advantage of using a substitution model that evaluates the probability of mutations across the MSA. Since varying evolutionary rates are permitted along the tree, ML phylogenies are really efficient at dealing with deeply branching phylogenies and should be the preferred method when studying old histone variants. In addition, ML phylogenies should be preferred over other methods when attempting to measure selective pressure that acted upon a set of histone variants as it will provide a robust assessment of topology (substitution pattern) and branch length (distance between sequences). Substitution models vary between sequence types and overall divergence [25] and many online tools exist to evaluate the best model to use—such as ModelOMatic [26] or prottest [27]. To run ML phylogenies the reference software is phyML [28] which is an open access software and can be run online on a multitude of platforms.

Visualization of Phylogenetic Trees

The most commonly used output file formats for phylogenetic trees include nexus, phylip, and newick. These formats are suitable to the visualization of trees using a number of platforms as most contain information relative to the tree length, bootstrap support of topology, substitution rate (usually the branch scale), and additional information regarding the alignment.

1. When visualizing phylogenies, the user should decide if the tree should be rooted on a sequence, or a group of sequences, known to be the most "distantly related" histone in the MSA.
2. Rooted trees are preferred when attempting to resolve the evolutionary trajectories of the histone variant from its origins to its patterns of diversification. When unsure about which sequence to use, one should start with canonical histones from a similar set of species as well as from a true outgroup species for which there are no histone variant sequences being used. For example, when studying H2A.1 (known to be restricted to mammals) one can use mammalian canonical H2As and/or bird canonical H2As as outgroups. The assumption being that canonical H2A in these species arose before H2A.1.
3. Unrooted trees are most useful when studying deep phylogenies with many distantly related histone variants. These will allow one to infer the monophyletic (one origin or polyphyletic (multiple independent origins) nature of the studied variant (e.g., a phylogeny with representative H3s from most eukaryotes will group canonical sequences together and different variants as separated groups, see the next section).

3.4.3 Interpretation of the Obtained Phylogeny—The Evolution of Histone Variants

With these phylogenies in hand, it is now possible to trace back the evolutionary origin of the variant, assess its evolutionary rates within specific lineages, and even determine the selective forces that shaped its evolution. In this last section, we discuss specific histone variant examples to illustrate possible interpretations.

If the studied histone variant sequences all share a common ancestor they are expected to group together with a single node separating them from other, e.g., canonical, histone sequences. In most cases, histone variants are derived from their canonical counterpart via duplication or retroposition so these should constitute their closest relative in the tree. However, while most previously studied histone variants arose once and thus follow this topology, H3.3 and H2A.X variants have been proposed to have multiple evolutionary origins. These 2 variants diverge from their canonical counterparts several times in a phylogenetic tree consistent with either multiple independent origins of the variant or of canonical H2A or H3, or both [4, 5]. For H2A.X, considering its deep eukaryotic origin, it is still up for debate if H2A.X could have preceded the evolution of canonical H2As. Overall, these observations further highlight the necessity of phylogenetic approaches to raise important yet unresolved questions about histone ancestry.

Additional information that can be gained from the phylogeny is the rate at which the variant sequences accumulate mutations (evolutionary rate). This is an indicator of the selective forces shaping its evolution and can provide insight into functional constraints that structural analysis alone cannot uncover. The length of branches in a phylogenetic tree can convey such information. While some trees have branch lengths proportional to the total length of the tree (indicative of relative distances between sequences), others illustrate actual mutation rates in the form of substitution per site (either amino acids or nucleotides).

Across a wide phylogenetic distribution, variants such as macroH2As, H2A.Z, H2A.1, H3.3 as well as canonical histones display very short branch lengths indicative of the strong purifying selection acting on the sequences of these genes to maintain their function. However, branches connecting these histone clades can be long and are indicative of the major evolutionary transitions that have shaped the variant's sequence/function (*see* **Note 5** for caveats associated with long branches in phylogenies). In addition, these transitions can also be limited to a group of species within the variant phylogeny and indicate lineage-specific innovations that can be investigated further (e.g., H2A.M or H2A.J). On the other hand, histone variants such as cenH3 and short H2As have extremely long branches and their phylogeny does not necessarily match the expected species tree (*see* **Note 6**). In these cases, rapid sequence changes are hypothesised to be the result of diversifying/adaptive selection related to their special function.

Signatures of selection (purifying or adaptive) can be further determined when analyzing phylogenies based on nucleotide sequences. Nucleotide substitutions can be categorized into synonymous (S) or non-synonymous (N) codon substitutions. Assuming that an appropriate substitution model and maximum likelihood phylogeny was used, one can compare the rates of dS and dN along the tree. A small dN is usually the sign of purifying selection while dN > dS can indicate diversifying selection. Equal dN and dS rates indicate that the sequences are not subject to selection at all. There are excellent online tools to perform such analysis and provide some statistical assessment of the selective forces at play (e.g., the Datamonkey server http://www.datamonkey.org/).

4 Notes

1. *BLAST and related tools*: BLAST is a free software, if one wants to run jobs locally, or can be run remotely through its host at the NCBI (https://blast.ncbi.nlm.nih.gov/Blast.cgi). When

using BLAST one can specify the database that should be searched. The nonredundant protein/nucleotide sequence databases (nr database) are the most comprehensive and include both curated and non-curated sequences [29]. For comparative structural modeling of protein queries, the protein databank [14] that contains sequences associated with experimental structures is appropriate. BLASTp and BLASTn retrieve sequences based on similarity to the input sequence. This means that the alignment between the query and the hit must be sufficient to be reported, and works well with closely related sequences, but has clear limitations with increased divergence. Because protein databases used for BLASTp can be incomplete with annotations of certain variants missing and coding sequences can accumulate synonymous substitutions that limit BLASTn searches, these search algorithms might not return the true orthologs of the query. Instead, given that related histone genes are also often present in the genomes, BLASTp and BLASTn might return hits that only look closely related leading to wrong conclusions. We advise to systematically run translated nucleotide BLAST (tBLASTn) to take into account these limitations.

2. *False-positives in HMM-based predictions*: In contrast to the *E* value in the BLAST output, no general rule exists which probability value is considered to be sufficiently high to conclude true homology. Instead, the alignment needs to be manually inspected using criteria like similarity of the secondary structure, hydrophobicity profiles, and potential gaps.

   While these algorithms are currently the most sensitive approaches to reveal remote homologies, the false-positive rate among hits can be notably high. Therefore, additional criteria are often needed to evaluate putative histone-fold candidates. These can include short sequence motifs specific to certain variants and the overall protein domain architecture. Still, putative homologues should generally be experimentally verified.

3. *Histone DB*: As mentioned by the curators of this resource, not all variants might be represented even though new variants are recurrently added.

4. *Informative MSA*: As the parameters and algorithms differ, it is generally advised to use at least two different programs to construct the MSAs. Also, when possible, one should try to perform and compare both DNA and protein sequence alignment since they inform on different levels of selective pressure (codons vs. amino acids). In order to pick the "right" number and type of sequences to use, one should consider the overall amount of diversity present in the alignment. Having too many nearly identical sequences will not provide

enough information to discriminate between alternative evolutionary scenarios. The same holds true if too little identity exists in the alignment.

5. *Long-branch attractions*: When dealing with highly diverged sequences, as it is the case for cenH3 and short H2As or over large evolutionary span, one should pay special attention to the branching patterns of the longest branches. Indeed, a classical caveat to phylogenetic trees is that sequences that have accumulated many mutations share little homology to the rest of the alignment, and are thus artificially brought together during tree building. They may wrongfully appear as sharing a common ancestry. Using an appropriate substitution model can greatly improve this issue. One should always try to use alternative methods (synteny, species phylogeny, etc.) to infer the best relationship of the sequences subject to long branch attraction.
6. *Gene conversion*: When interpreting the topology of a phylogeny, concerted evolution (the fact that paralogs share more identity than their true orthologs) can create wrong assumptions. In such a case, duplicates might cluster together by species suggesting independent origins instead of forming separated clades suggestive of independent evolution from an ancestral duplication. Mammalian short H2As and, ironically, canonical histones display such complex phylogenies. While canonical histone genes seem to be subject to a "birth-and-death" mode of evolution [30], short H2As duplicates are undergoing gene conversion, through non-homologous recombination, which homogenizes their sequences within lineages [56]. When confronted to such topologies, synteny can greatly inform on the ancestry of the duplication and help resolve confusing phylogenies and uncover interesting modes of evolution like the ones mentioned above.

Acknowledgments

We would like to thank Paul Talbert and members of the Drinnenberg lab for comments on the manuscript. I.A.D. receives salary support from the CNRS. This work is supported by the Labex DEEP ANR-11-LABX-0044 part of the IDEX Idex PSL ANR-10-IDEX-0001-02 PSL, the Institut Curie and funds from the Atip Avenir 2015 program. A. M. was supported by the Damon Runyon Cancer Research Foundation (DRG:2192-14).

References

1. Kornberg RD (1974) Chromatin structure: a repeating unit of histones and DNA. Science 184(4139):868–871
2. Luger K, Mader AW, Richmond RK et al (1997) Crystal structure of the nucleosome core particle at 2.8 A resolution. Nature 389(6648):251–260. https://doi.org/10.1038/38444
3. Marzluff WF, Wagner EJ, Duronio RJ (2008) Metabolism and regulation of canonical histone mRNAs: life without a poly(A) tail. Nat Rev Genet 9(11):843–854. https://doi.org/10.1038/nrg2438
4. Talbert PB, Henikoff S (2010) Histone variants—ancient wrap artists of the epigenome. Nat Rev Mol Cell Biol 11(4):264–275. https://doi.org/10.1038/nrm2861
5. Malik HS, Henikoff S (2003) Phylogenomics of the nucleosome. Nat Struct Biol 10(11):882–891. https://doi.org/10.1038/nsb996
6. Kawashima T, Lorkovic ZJ, Nishihama R et al (2015) Diversification of histone H2A variants during plant evolution. Trends Plant Sci 20(7):419–425. https://doi.org/10.1016/j.tplants.2015.04.005
7. Draizen EJ, Shaytan AK, Marino-Ramirez L et al (2016) HistoneDB 2.0: a histone database with variants--an integrated resource to explore histones and their variants. Database. https://doi.org/10.1093/database/baw014
8. Coordinators NR (2016) Database resources of the National Center for biotechnology information. Nucleic Acids Res 44(D1):D7–D19. https://doi.org/10.1093/nar/gkv1290
9. Kent WJ, Sugnet CW, Furey TS et al (2002) The human genome browser at UCSC. Genome Res 12(6):996–1006. https://doi.org/10.1101/gr.229102 Article published online before print in May 2002
10. Altschul SF, Gish W, Miller W et al (1990) Basic local alignment search tool. J Mol Biol 215(3):403–410. https://doi.org/10.1016/S0022-2836(05)80360-2
11. Altschul SF, Madden TL, Schaffer AA et al (1997) Gapped BLAST and PSI-BLAST: a new generation of protein database search programs. Nucleic Acids Res 25(17):3389–3402
12. Schleiffer A, Maier M, Litos G et al (2012) CENP-T proteins are conserved centromere receptors of the Ndc80 complex. Nat Cell Biol 14(6):604–613. https://doi.org/10.1038/ncb2493
13. Soding J, Biegert A, Lupas AN (2005) The HHpred interactive server for protein homology detection and structure prediction. Nucleic Acids Res 33(Web Server):W244–W248. https://doi.org/10.1093/nar/gki408
14. Berman HM, Westbrook J, Feng Z et al (2000) The Protein Data Bank. Nucleic Acids Res 28(1):235–242
15. Finn RD, Mistry J, Tate J et al (2010) The Pfam protein families database. Nucleic Acids Res 38(Database issue):D211–D222. https://doi.org/10.1093/nar/gkp985
16. Whelan S, Morrison DA (2017) Inferring trees. Methods Mol Biol 1525:349–377. https://doi.org/10.1007/978-1-4939-6622-6_14
17. Bawono P, Dijkstra M, Pirovano W et al (2017) Multiple sequence alignment. Methods Mol Biol 1525:167–189. https://doi.org/10.1007/978-1-4939-6622-6_8
18. Edgar RC (2004) MUSCLE: a multiple sequence alignment method with reduced time and space complexity. BMC bioinformatics 5:113. https://doi.org/10.1186/1471-2105-5-113
19. Edgar RC (2004) MUSCLE: multiple sequence alignment with high accuracy and high throughput. Nucleic Acids Res 32(5):1792–1797. https://doi.org/10.1093/nar/gkh340
20. Sievers F, Wilm A, Dineen D et al (2011) Fast, scalable generation of high-quality protein multiple sequence alignments using Clustal omega. Mol Syst Biol 7:539. https://doi.org/10.1038/msb.2011.75
21. Blackshields G, Sievers F, Shi W et al (2010) Sequence embedding for fast construction of guide trees for multiple sequence alignment. Algorithms Mol Biol 5:21. https://doi.org/10.1186/1748-7188-5-21
22. Yang Z (2005) The power of phylogenetic comparison in revealing protein function. Proc Natl Acad Sci U S A 102(9):3179–3180. https://doi.org/10.1073/pnas.0500371102
23. Wilgenbusch JC, Swofford D (2003) Inferring evolutionary trees with PAUP*. Curr Protoc Bioinformatics Chapter 6:Unit 6 4. https://doi.org/10.1002/0471250953.bi0604s00
24. Saitou N, Nei M (1987) The neighbor-joining method: a new method for reconstructing phylogenetic trees. Mol Biol Evol 4(4):406–425
25. Shapiro B, Rambaut A, Drummond AJ (2006) Choosing appropriate substitution models for the phylogenetic analysis of protein-coding sequences. Mol Biol Evol 23(1):7–9. https://doi.org/10.1093/molbev/msj021
26. Whelan S, Allen JE, Blackburne BP, Talavera D (2015) ModelOMatic: fast and automated

model selection between RY, nucleotide, amino acid, and codon substitution models. Syst Biol 64(1):42–55. https://doi.org/10.1093/sysbio/syu062

27. Darriba D, Taboada GL, Doallo R, Posada D (2011) ProtTest 3: fast selection of best-fit models of protein evolution. Bioinformatics 27(8):1164–1165. https://doi.org/10.1093/bioinformatics/btr088
28. Guindon S, Dufayard JF, Lefort V et al (2010) New algorithms and methods to estimate maximum-likelihood phylogenies: assessing the performance of PhyML 3.0. Syst Biol 59(3):307–321. https://doi.org/10.1093/sysbio/syq010
29. Coordinators NR (2017) Database resources of the National Center for biotechnology information. Nucleic Acids Res 45(D1):D12–D17. https://doi.org/10.1093/nar/gkw1071
30. Rooney AP, Piontkivska H, Nei M (2002) Molecular evolution of the nontandemly repeated genes of the histone 3 multigene family. Mol Biol Evol 19(1):68–75
31. Witt O, Albig W, Doenecke D (1996) Testis-specific expression of a novel human H3 histone gene. Exp Cell Res 229(2):301–306. https://doi.org/10.1006/excr.1996.0375
32. Nishino T, Takeuchi K, Gascoigne KE et al (2012) CENP-T-W-S-X forms a unique centromeric chromatin structure with a histone-like fold. Cell 148(3):487–501. https://doi.org/10.1016/j.cell.2011.11.061
33. Albig W, Ebentheuer J, Klobeck G et al (1996) A solitary human H3 histone gene on chromosome 1. Hum Genet 97(4):486–491
34. Chakravarthy S, Bao Y, Roberts VA et al (2004) Structural characterization of histone H2A variants. Cold Spring Harb Symp Quant Biol 69:227–234. https://doi.org/10.1101/sqb.2004.69.227
35. Eirin-Lopez JM, Gonzalez-Romero R, Dryhurst D et al (2009) The evolutionary differentiation of two histone H2A.Z variants in chordates (H2A.Z-1 and H2A.Z-2) is mediated by a stepwise mutation process that affects three amino acid residues. BMC Evol Biol 9:31. https://doi.org/10.1186/1471-2148-9-31
36. Faast R, Thonglairoam V, Schulz TC et al (2001) Histone variant H2A.Z is required for early mammalian development. Curr Biol 11(15):1183–1187
37. Pehrson JR, Fried VA (1992) MacroH2A, a core histone containing a large nonhistone region. Science 257(5075):1398–1400
38. Pehrson JR, Fuji RN (1998) Evolutionary conservation of histone macroH2A subtypes and domains. Nucleic Acids Res 26(12):2837–2842
39. Rivera-Casas C, Gonzalez-Romero R, Cheema MS et al (2016) The characterization of macroH2A beyond vertebrates supports an ancestral origin and conserved role for histone variants in chromatin. Epigenetics 11(6):415–425. https://doi.org/10.1080/15592294.2016.1172161
40. Eirin-Lopez JM, Ishibashi T, Ausio J (2008) H2A.Bbd: a quickly evolving hypervariable mammalian histone that destabilizes nucleosomes in an acetylation-independent way. FASEB J 22(1):316–326. https://doi.org/10.1096/fj.07-9255com
41. Contrepois K, Coudereau C, Benayoun BA et al (2017) Histone variant H2A.J accumulates in senescent cells and promotes inflammatory gene expression. Nat Commun 8:14995. https://doi.org/10.1038/ncomms14995
42. Boulard M, Gautier T, Mbele GO et al (2006) The NH2 tail of the novel histone variant H2BFWT exhibits properties distinct from conventional H2B with respect to the assembly of mitotic chromosomes. Mol Cell Biol 26(4):1518–1526. https://doi.org/10.1128/MCB.26.4.1518-1526.2006
43. Marzluff WF, Gongidi P, Woods KR et al (2002) The human and mouse replication-dependent histone genes. Genomics 80(5):487–498
44. Churikov D, Siino J, Svetlova M et al (2004) Novel human testis-specific histone H2B encoded by the interrupted gene on the X chromosome. Genomics 84(4):745–756. https://doi.org/10.1016/j.ygeno.2004.06.001
45. Dalmasso MC, Echeverria PC, Zappia MP et al (2006) Toxoplasma gondii has two lineages of histones 2b (H2B) with different expression profiles. Mol Biochem Parasitol 148(1):103–107. https://doi.org/10.1016/j.molbiopara.2006.03.005
46. Dalmasso MC, Sullivan WJ Jr, Angel SO (2011) Canonical and variant histones of protozoan parasites. Front Biosci 16:2086–2105
47. Santoro SW, Dulac C (2012) The activity-dependent histone variant H2BE modulates the life span of olfactory neurons. eLife 1:e00070. https://doi.org/10.7554/eLife.00070
48. Waterborg JH (2012) Evolution of histone H3: emergence of variants and conservation of post-translational modification sites. Biochem Cell Biol 90(1):79–95. https://doi.org/10.1139/o11-036

49. Schenk R, Jenke A, Zilbauer M et al (2011) H3.5 is a novel hominid-specific histone H3 variant that is specifically expressed in the seminiferous tubules of human testes. Chromosoma 120(3):275–285. https://doi.org/10.1007/s00412-011-0310-4
50. Wiedemann SM, Mildner SN, Bonisch C et al (2010) Identification and characterization of two novel primate-specific histone H3 variants, H3.X and H3.Y. J Cell Biol 190(5):777–791. https://doi.org/10.1083/jcb.201002043
51. Postberg J, Forcob S, Chang WJ, Lipps HJ (2010) The evolutionary history of histone H3 suggests a deep eukaryotic root of chromatin modifying mechanisms. BMC Evol Biol 10:259. https://doi.org/10.1186/1471-2148-10-259
52. Urahama T, Harada A, Maehara K et al (2016) Histone H3.5 forms an unstable nucleosome and accumulates around transcription start sites in human testis. Epigenetics Chromatin 9:2. https://doi.org/10.1186/s13072-016-0051-y
53. Govin J, Escoffier E, Rousseaux S et al (2007) Pericentric heterochromatin reprogramming by new histone variants during mouse spermiogenesis. J Cell Biol 176(3):283–294. https://doi.org/10.1083/jcb.200604141
54. Ferguson L, Ellis PJ, Affara NA (2009) Two novel mouse genes mapped to chromosome Yp are expressed specifically in spermatids. Mamm Genome 20(4):193–206. https://doi.org/10.1007/s00335-009-9175-8
55. Shaytan AK, Landsman D, Panchenko AR (2015) Nucleosome adaptability conferred by sequence and structural variations in histone H2A-H2B dimers. Curr Opin Struct Biol 32:48–57. https://doi.org/10.1016/j.sbi.2015.02.004
56. Molaro A, Young JM, Malik HS (2018) Evolutionary origins and diversification of testis-specific short histone H2A variants in mammals. Genome Res 28(4):460–473.
57. Aul RB, Oko RJ (2001) The Major Subacrosomal Occupant of Bull Spermatozoa Is a Novel Histone H2B Variant Associated with the Forming Acrosome during Spermiogenesis. Dev. Biol 239(2):376–387

Chapter 16

Characterization of Post-Meiotic Male Germ Cell Genome Organizational States

Jérôme Govin, Sophie Barral, Yuichi Morozumi, Naghmeh Hoghoughi, Thierry Buchou, Sophie Rousseaux, and Saadi Khochbin

Abstract

Dramatic and unique genome reorganizations accompany the differentiation of haploid male germ cells, characterized by a gradual loss of the vast majority of histones leading to a final tight compaction of the genome by protamines. Despite being essential for procreation and the life cycle, the mechanisms driving the transformation of nucleosomes into nucleoprotamines remain poorly understood. To address this issue, our laboratory has developed a number of specific approaches, ranging from the purification of spermatogenic cells at specific stages, the analysis of chromatin transitional states, the functional characterization of histone variants, histone-replacing proteins and their chaperones. This chapter will detail all related relevant techniques with a particular emphasis on methods allowing the functional studies of histone variants and the genome organizational states associated with the studied histones in spermatogenic cells undergoing histone-to-protamine exchange.

Key words Histone, Histone variants, Transition proteins, Protamines, Chaperones, Chromatin, Nucleosome, Spermatogenesis, Germline, Sperm

1 Introduction

During mammalian spermatogenesis, haploid male germ cells are continuously produced following successive meiotic divisions (occurring in spermatocytes). The early post-meiotic cells known as round spermatids undergo radical nuclear shape changes while becoming elongating spermatids, which themselves evolve into condensing spermatids and undergo the final transformations to produce mature spermatozoa. Mature sperm cells are capable of leaving the male reproductive organ and will have to face a challenging environment as they will "swim" toward the oocyte. The preparation of the final genome compaction virtually starts when spermatocytes/spermatid chromatin incorporate specific histone variants that favor the ulterior genome-wide histone eviction [1, 2]. The commitment of progenitor cells into meiosis is associated with

Guillermo A. Orsi and Geneviève Almouzni (eds.), *Histone Variants: Methods and Protocols*, Methods in Molecular Biology, vol. 1832, https://doi.org/10.1007/978-1-4939-8663-7_16, © Springer Science+Business Media, LLC, part of Springer Nature 2018

the accumulation of testis-specific histones of the H2A, H2B, and H3 types with particular structural properties that survive meiosis and are inherited by the post-meiotic spermatids [3–7]. Additionally, haploid cells synthesize their own sets of histone H2A and H2B variants with specific roles in the control of testis-specific gene expression as well as in chromatin remodeling and reorganization and histone-to-protamine replacement [3, 7].

The gradual stage-specific incorporation of these histone variants along with the occurrence of a sharp and dramatic gain of histone posttranslational modifications (PTMs) in elongating spermatids [8–11], mediate the replacement of histones by basic small non-histone proteins known as transition proteins (TPs) and protamines (Prms).

Until recently, the prevailing view was that histones are first replaced by TPs, which are then themselves replaced by Prms [12, 13]. However, a recent functional investigation of a specific H2A histone variant, H2A.L.2, expressed at the time of histone eviction in elongating spermatids showed that an exchange of H2A-H2B histones is actually coupled and concomitant with the action of TPs and Prms, the generation of transitional states and DNA compaction [7]. This particular research on H2A.L.2, and our current understanding of the molecular basis of post-meiotic mouse male germ cell genome reorganization in our laboratory relies on years of pilot studies and development of specific approaches [14–16] that will be detailed here.

Our global approach to the issue of post-meiotic genome reorganization is built first on the design, adaptation, and improvement of techniques to isolate different types of post-meiotic cells [17], and second on the adaptation and development of methods to study different organizational states of chromatin, discover and identify histones, non-histone basic proteins and their chaperones, and generate relevant mouse models for functional molecular studies. Here below we successively describe the techniques for post-meiotic cell fractionation, the identification of basic and acidic proteins (respectively enriched in nucleic acid-binding proteins and chaperones, and in histone and non-histone basic proteins), the identification of histones [18], the study of chromatin organizational states, and the generation of mouse models.

2 Materials

All solutions are stored at room temperature (RT) unless otherwise specified.

2.1 Isolation of Condensing Spermatids

1. Homogenization Medium 1 (HM1): 0.32 M Sucrose, 1.5 mM $CaCl_2$, 10 mM Tris pH 8, 1 mM Dithiothreitol (DTT) and protease inhibitors cOmplete EDTA-free (Roche).

2. Homogenization Medium 2 (HM2): HM1 supplemented with 0.1% Triton.
3. Homogenization Medium 3 (HM3): HM2 supplemented with 0.88 M Sucrose (final concentration).
4. Filter with 100 μm wide pore.
5. Dissection material.
6. Motor-driven Potter-Elvehjem homogenizer.

2.2 Isolation of Basic and Acidic Proteins from Condensing Spermatids

2.2.1 Basic Proteins (Histone and Non-histones)

1. Sulfuric acid (H_2SO_4).
2. Trichloro-acetic acid (TCA).
3. Acetone.
4. Acetone supplemented with 0.05% HCl.

2.2.2 Acidic Proteins (TP Column)

1. TP1 cDNA cloned into pET28 plasmid for production as a recombinant protein (available upon request).
2. BL21 transformation-competent bacteria.
3. Isopropyl-beta-D-thiogalacto-pyranoside (IPTG).
4. Perchloric acid (PCA).
5. Trichloro-acetic acid (TCA).
6. Acetone.
7. Acetone supplemented with 0.05% HCl.
8. CNBr-activated sepharose 4b fast flow (GE Healthcare).
9. Coupling buffer: 0.5 M NaCl, 0.1 M $NaHCO_3$ pH 8.3.
10. Quenching buffer: 0.2 M Glycine pH 8.
11. Storage buffer: 0.1 M Tris pH 8, 0.5 M NaCl, 0.05% NaN_3.
12. Wash buffer 1: 0.1 M Na Acetate pH 4, 0.5 M NaCl.
13. Wash buffer 2: 0.1 M Tris pH 8, 0.5 M NaCl.
14. Extraction buffer 0 (EB0): 50 mM Tris pH 8, 1% NP-40, 0.5% Deoxycholic acid, 0.1% SDS, 1 mM DTT and protease inhibitors cOmplete EDTA-free (Roche).
15. Extraction buffer 1 (EB1): EB0 supplemented with 150 mM NaCl.
16. Extraction buffer 2 (EB2): EB0 supplemented with 1.2 M NaCl.

2.3 MNase Digestion of Condensing Spermatids Nuclei

1. Lysis buffer: 50 mM Tris pH 7.4, 300 mM NaCl, 0.1% NP-40, 0.1% Deoxycholic acid, 1 mM DTT and protease inhibitors cOmplete EDTA-free (Roche).
2. MNase digestion buffer: 10 mM Tris pH 7.5, 10 mM KCl, 1 mM $CaCl_2$.

3. Nuclease S7 Microccocale nuclease.
4. Chromatin Fiber Analysis Mix: 10 mM Tris pH 8, 1 mM EDTA, 0.1 mg/mL Proteinase K, 0.1% SDS.

2.4 Sucrose Gradient Fractionation of Nucleosomal and Sub-Nucleosomal Fragments

1. Gradient sucrose solution: 1 mM Phosphate Buffer pH 7.4 (prepared by mixing of 0.5 M NaH_2PO_4 pH 9 with 0.5 M Na_2HPO_4 pH 4.4 to reach pH 7.4), 80 mM NaCl, 0.2 mM EDTA, 20% Sucrose w/v, cOmplete EDTA-free (Roche), HDAC inhibitors (300 nM Tricostatin A).

2.5 TAP-Tag Histone Purification

1. Total germ cells or purified post-meiotic cells: Prepare as described in Subheading 3.1. They can be flash frozen in liquid nitrogen and stored at −80 °C.
2. Immunoprecipitation (IP) buffer: 20 mM HEPES pH 7.9, 300 mM NaCl, 10% Glycerol, 0.2 mM EDTA, 0.1% NP-40, 1 mM DTT and protease inhibitors cOmplete EDTA-free (Roche). Prepare freshly and cool on ice.
3. 0.45 μm wide pore membrane filters.
4. Anti-FLAG M2 Affinity Agarose Gel (Sigma-Aldrich).
5. Tris-buffered saline (TBS): 20 mM Tris pH 7.4, 150 mM NaCl.
6. 3× FLAG peptide (Sigma-Aldrich): Dissolve 3× FLAG Peptide in TBS at a concentration of 5 mg/mL as a stock solution. Since repeated freeze/thaw cycle is not recommended, make small aliquots of the stock solution and store at −20 °C.
7. Anti-HA antibody.
8. Dynabeads Protein G (Thermo Fisher Scientific).
9. Magnetic separation stand.
10. Conjugation buffer: 20 mM Phosphate buffer pH 7.4 prepared as described above (section 2.4), 150 mM NaCl.
11. Bis(sulfosuccinimidy)suberate (BS3) linker (Sigma-Aldrich, catalogue No. S5799): Dissolve BS3 in conjugation buffer at a concentration of 100 mM just prior to use.

2.6 Proteomic Analysis of Histone Variants

1. MS_HistoneDB files: download from reference 18 (Supp. Files 1 and 2).

3 Methods

A detailed protocol for the fractionation of spermatogenic cells by sedimentation has been published recently [17] (Fig. 1).

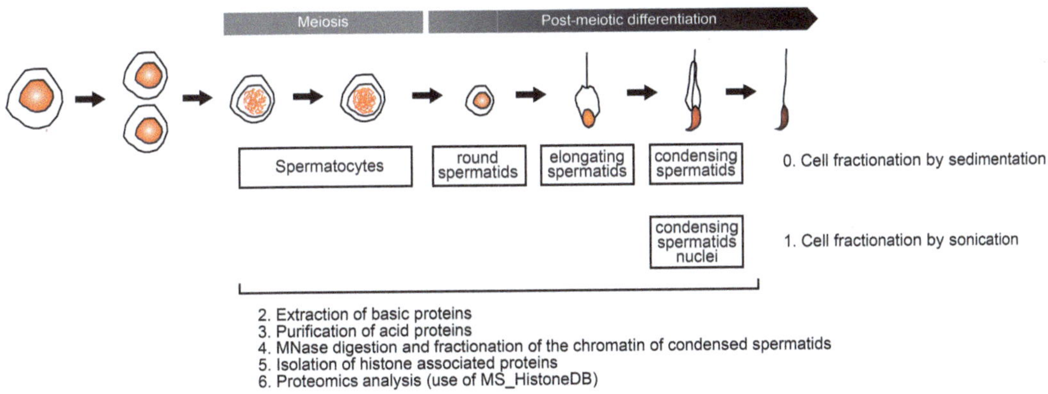

Fig. 1 Summary of the methods used to characterize the different chromatin organizational states during mouse spermatogenesis

3.1 Isolation of Nuclei from Condensing Spermatids

1. Start with 2 testes in cold 1× PBS (*see* **Note 1**).
2. Remove their tunica albuginea using fine tip tweezers.
3. Transfer all testicular material in a potter with 5 mL of Homogenizing Media 1 (HM1).
4. Disrupt mechanically using a motor-driven potter-homogenizer. Perform 10 strokes at 3000 rpm.
5. Complete the volume to 10 mL with Homegenizing Media 2 (HM2).
6. Centrifuge 10 min at 500 × *g* at 4 °C.
7. Resuspend the pellet in 3 mL of cold HM2.
8. Sonicate the tube on ice at 250 J.
9. Centrifuge for 10 min at 500 × *g* at 4 °C.
10. Resuspend the pellet in 3 mL of cold HM2.
11. Sonicate the tube on ice at 250 J.
12. Centrifuge for 10 min at 500 × *g* at 4 °C.
13. Filter the suspension with a 100 μm pore size filter.
14. Add HM2 to obtain a final volume of 10 mL.
15. Centrifuge for 10 min at 500 × *g* at 4 °C and resuspend the pellet in 10 mL of HM2.
16. Centrifuge for 10 min at 500 × *g* at 4 °C and resuspend the pellet in 10 mL of HM3.
17. Centrifuge for 10 min at 500 × *g* at 4 °C and resuspend the pellet in 1 mL of HM1.
18. Transfer in microfuge tubes 1.5 mL.
19. Centrifuge and flash freeze the dry pellets. They can be stored at −80 °C (*see* **Note 2**).

3.2 Isolation of Basic Proteins from Condensing Spermatids Nuclei

1. Resuspend a pellet of condensed spermatids in 500 μL of 0.2 M H_2SO_4. Basic proteins will be solubilized while other proteins will precipitate.
2. Sonicate briefly if necessary.
3. Incubate for 30 min on ice.
4. Centrifuge for 30 min at 21,000 × *g* at 4 °C and transfer the supernatant to a new prechilled tube.
5. Precipitate histones by adding 110 μL of Trichloroacetic acid (TCA) (20% v/v final concentration, *see* **Note 3**).
6. Incubate for 30 min on ice (*see* **Note 4**).
7. Centrifuge for 30 min at 21,000 × *g* at 4 °C. Precipitated histones should be visible on the side of the tube.
8. Discard the supernatant and wash the pellet with 500 μL of Acetone 0.05% HCl.
9. Centrifuge for 10 min at 21,000 × *g* at 4 °C.
10. Wash again in pure Acetone.
11. Centrifuge for 10 min at 21,000 × *g* at 4 °C.
12. Carefully remove the maximum of acetone using P200 and P10 pipettes. Dry the pellet at RT for 10–20 min (*see* **Note 5**).
13. Resuspend the pellet in H_2O (*see* **Note 6**).
14. Centrifuge 10 min at 21,000 × *g* at 4 °C to remove non-soluble proteins. The supernatant contains acid-soluble proteins.

3.3 Isolation of Acidic Proteins from Condensing Spermatids Nuclei

This protocol has been used to identify proteins able to interact with the basic transition protein TP1 [14].

3.3.1 Production and Purification of the Transition Protein 1 (TP1)

1. Transform TP1 cDNA cloned in pET 28 vector in BL21 competent bacteria.
2. The production of TP1 is induced in a 100 mL log phase culture (OD_{600} 0.7) for 1 h at 37 °C with 2 mM IPTG.
3. Wash the pellet in PBS and resuspend in 5 mL of 5% perchloric acid (*see* **Note 7**).
4. Incubate for 30 min incubation on ice.
5. Centrifuge at 21,000 × *g* during 10 min at 4 °C.
6. Add 1.1 mL of trichloroacetic acid (TCA) to reach a final concentration of 20%.
7. Incubate on ice for 30 min.
8. Centrifuge at 21,000 × *g* during 10 min at 4 °C.
9. Discard the supernatant and wash the pellet with 500 μL of 0.05% Acetone HCl.

10. Centrifuge during 10 min at 21,000 × *g* at 4 °C.
11. Wash again in pure Acetone. Centrifuge 10 min at 21,000 × *g* at 4 °C.
12. Carefully remove the maximum of acetone using P200 and P10 pipettes. Dry the pellet at RT for 10–20 min (*see* **Note 5**).
13. Resuspend the pellet in H_2O (*see* **Note 6**).
14. Centrifuge for 10 min at 21,000 × *g* at 4 °C to remove unsoluble proteins. The supernatant contains purified TP1, which is aliquoted and stored at −80 °C. Pure TP1 recombinant protein will be observed on Coomassie-strained SDS-PAGE gel.

3.3.2 TP1 Crosslinking on an Agarose Resin

TP1 is crosslinked to CNBr-activated agarose beads (GE Healthcare) following the manufacturer's instructions.

1. Resuspend 0.6 g of dried resin in 300 mL of 1 mM HCl.
2. Wash with 15 mL of coupling buffer.
3. Resuspend in 6 mL of coupling buffer.
4. Incubate half of the resin for 2 h at RT with 270 μg of TP1 whereas the other half is incubated with a similar amount of BSA.
5. Remove the buffer by centrifugation or filtering and add 2 mL of quenching solution.
6. Incubate for 2 h at RT.
7. Extensively wash the resin sequentially with 2 mL of wash buffers 1 and 2. Store the column at 4 °C in storage buffer.

3.3.3 Purification of TP1 Interacting Proteins

1. Resuspend condensing spermatids in 100 μL of cold Extraction Buffer 1 (EB1).
2. Sonicate briefly to disrupt the cells.
3. Incubate for 15 min at 4 °C.
4. Centrifuge 10 min at 4 °C at 10,000 × *g*.
5. Save the supernatant and resuspend the pellet in 100 μL of cold EB2.
6. Incubate for 15 min at 4 °C.
7. Centrifuge 10 min at 4 °C at 10,000 × *g*.
8. Pool the extracts obtained in EB1 and EB2 and add 700 μL of EB0 to adjust the concentration of NaCl to 150 mM.
9. Wash twice 60 μL of the BSA or TP1 crosslinked resins in 1 mL of EB1.
10. Incubate 200 μL of the protein extract with 60 μL of the resin for 2 h at 4 °C under agitation.
11. Wash three times the resins in 1 mL of EB1.

12. Elute proteins in 15 μL of loading buffer and boil for 5 min.
13. Purified proteins can be analyzed by SDS-PAGE or using standard mass spectrometry procedures (Subheading 3.7).

3.4 MNase Digestion of the Chromatin of Condensing Spermatids

1. Resuspend condensing spermatids (obtained as described in Subheading 3.1, using six testes) in 112.5 μL of Lysis buffer.
2. Incubate for 15 min at 4 °C.
3. Centrifuge for 10 min at 4 °C at 20,000 × *g*.
4. Save the supernatant.
5. Resuspend the pellet in 112.5 μL of lysis buffer.
6. Sonicate the tube on ice at 80 J.
7. Centrifuge for 10 min at 4 °C at 20,000 × *g*.
8. Save the supernatant.
9. Pool both supernatants.
10. Add 112.5 μL of MNase buffer and 11.5 Units of micrococcal nuclease S7.
11. Incubate for 2 min at 37 °C (*see* **Note 8**).
12. Stop the digestion by adding 5 mM EDTA (final concentration) and place on ice.
13. Analyze the length of DNA fragments: digest 5 μL of chromatin fibers with 2.5 μL of Chromatin Fibers Analysis Mix for 15 min at 50 °C. Analyze directly on a 1.5% agarose gel.

3.5 Separation of Nucleosomal Chromatin Fragments

1. Pour the gradient sucrose solution in the tubes and freeze at −20 °C (*see* **Note 9**).
2. The day before use, thaw the tubes slowly by placing them in at 4 °C. The gradient will be formed by differential freezing / thawing of the sucrose solution (*see* **Note 10**).
3. Load slowly the chromatin fibers obtained after **step 13** (Subheading 3.4) on the sucrose gradient. Equilibrate the tubes with EDTA 0.1 mM if necessary.
4. Centrifuge for 16 h at 4 °C in the SW28 rotor at 25,000 rpm (~82,000 × *g*, large tubes), or in the TLS-55 rotor at 32,000 rpm (~68,000 × *g*, small tubes).
5. Collect 500 or 100 μL fractions of the gradients for large or small tubes, respectively.
6. The length of DNA fragments from fractions can be analyzed by digesting 5-10 μL with 2.5 μL of Chromatin Fibers Analysis Mix for 15 min at 50 °C followed by migration on a 1.1% agarose gel.

3.6 Purification of Soluble Histone Associated Proteins in the Male Germline

Histone genes were tagged with a combination of three tags: His, Flag, and HA. The first affinity purification uses the Flag tag since the protein complexes can be recovered using a competitor Flag peptide under non-denaturing condition. The HA tag was used to perform a second affinity purification.

3.6.1 Anti-HA Antibody Cross-Linking to Dynabeads Protein G

Contamination of the purified complex by heavy and light chains of the antibody may interfere with subsequent analyses. For this reason, the antibody is covalently cross-linked to the beads. The following cross-linking protocol is suitable for 5 µg of HA antibody with 50 µL of Dynabeads Protein G (*see* **Note 11**). This procedure can be scaled up or down if necessary. Unless otherwise indicated, perform the whole procedure at RT.

1. Transfer 50 µL of Dynabeads Protein G in a microfuge tube. Place the tube on magnetic separation stand for 1 min and discard the supernatant.
2. Wash the beads with 200 µL of cold PBS.
3. Wash the beads twice with 200 µL of IP buffer.
4. Incubate the beads with 5 µg of the anti-HA antibody in 200 µL of IP buffer. Perform incubation under gentle rotation at 4 °C for 1 h.
5. Discard the supernatant and wash the beads twice with 200 µL of conjugation buffer. During the second wash, transfer the beads to a new microfuge tube.
6. Resuspend the beads in 250 µL of 5 mM BS3 cross-linker diluted in conjugation buffer and gently rotate for 30 min.
7. Add 12.5 µL 1 M Tris pH 7.5 to the beads and incubate for 15 min under rotation (*see* **Note 12**).
8. Wash the beads twice with 200 µL of IP buffer (*see* **Note 13**). During second wash, transfer the beads to a new tube.
9. Add 50 µL of IP buffer to the beads and store on ice until use.

3.6.2 Purification of Soluble Histones Using TAP-Tag

The entire protocol is performed at 4 °C unless otherwise specified.

1. Resuspend the required amount of frozen male germ cell or purified post-meiotic cell pellets prepared using histone TAP-tagged mice (*see* Subheading 3.1) in the appropriate volume of IP buffer (*see* **Notes 14** and **15**).
2. Extract soluble proteins by gently rotating the cell suspension at 4 °C for 30–60 min.
3. During **step 2**, transfer the required amount of Anti-FLAG M2 affinity beads in a new microfuge tube (*see* **Note 16**). Wash the beads once with 1 mL of 1× PBS, and twice with 1 mL of IP buffer. Add an equal volume of IP buffer to the beads to make 50% beads slurry.
4. Recover the clarified lysate by centrifugation at 21,000 × *g* for 10 min, and then filtrate the supernatant through a 0.45 µm wide pore membrane filter (*see* **Note 17**). Transfer 10–20 µL of the supernatant to a new microfuge tube (input fraction) and store at −20 °C.

5. Mix the lysate with the equilibrated Anti-FLAG M2 affinity beads and incubate on a rotator at 4 °C for 4 h (*see* **Note 18**).
6. During **step 5**, prepare the anti-HA antibody cross-linked Dynabeads Protein G (*see* Subheadig 3.5, **step 1** and **Note 11**).
7. Separate the lysate and the beads by centrifugation at 2000 × *g* for 2 min. Remove the supernatant and transfer 10–20 μL of the supernatant to a new microfuge tube (first unbound fraction) and store at −20 °C.
8. Wash the beads in 1 mL of IP buffer. Centrifuge at 2000 × *g* for 2 min and discard the supernatant. Repeat this wash step five times. During the fifth wash, transfer the beads to a new microfuge tube (*see* **Note 19**).
9. Resuspend the beads in an equal volume of 0.3 mg/mL 3× FLAG peptide diluted in IP buffer (*see* **Note 20**). Gently rotate on a rotator at 4 °C for 30 min.
10. Centrifuge the resin at 2000 × *g* for 2 min. Carefully transfer the supernatant to a new microfuge tube.
11. Repeat **steps 9** and **10** again, and combine the elution fractions.
12. To remove the possibly remaining small amount of beads in the combined eluate, spin the sample at 2000 × *g* for 2 min, and carefully recover the supernatant into a new microfuge tube. Transfer 10–20 μL of the supernatant to another microfuge tube (first elution fraction) and store at −20 °C.
13. Mix the clarified elution fraction with the appropriate amount of anti-HA antibody cross-linked Dynabeads Protein G, and incubate with gentle rotation at 4 °C overnight (*see* **Note 21**).
14. Place the tube on a magnetic stand for 1 min. Remove the supernatant and transfer 10–20 μL of the supernatant to a new microfuge tube (second unbound fraction) and store at −20 °C.
15. Add 1 mL of IP buffer to the beads and resuspend gently. After the beads settle down by placing the tube on the stand for 1 min, discard the buffer. Repeat this wash step three times. During the third wash, transfer the beads to a new microfuge tube.
16. Add 10–30 μL of 1× Laemmli sample buffer without reducing agent. Heat the beads at 95 °C for 5 min (*see* **Note 22**).
17. Place the tube on a magnetic stand and transfer the supernatant to a new microfuge tube. Store the sample at −20 °C for further experiments.

18. Analyze the collected samples at each purification step (**steps 4**, **7**, **12**, **14** and **16**) by western blot to estimate the efficiency of TAP-tag purification.

3.7 Proteomic Analysis of Histone Variants

Mass spectrometry is a powerful technique to analyze the composition of chromatin, the diversity of histones present in a specific chromatin fraction and their posttranslational modifications. Several published reviews and protocols already document this type of analysis [19–24]. In brief, histones are extracted and purified from the initial material. They are usually proteolyzed into smaller peptides using a specific enzyme and these peptides are subsequently analyzed by mass spectrometry. Raw data are mainly made of fragmentation spectra acquired on peptides from which it is possible to identify the amino acid sequences using search softwares, such as the widely used Mascot software (Matrix Science). Additional post-analysis treatments enable identification and quantification of the proteins and potentially of their posttranslational modifications present in the initial protein sample [25].

The interpretation of the raw spectra relies on public protein sequence databases, such as SwissProt or UniProtKD. Unfortunately, they are either incomplete or overly redundant with partial annotation, complicating the analysis of histone proteins and their variants. We recently developed a new resource, named MS_HistoneDB, which can be directly used into standard proteomic pipelines and greatly facilitate proteomic approaches dedicated to the analysis of histones [18]. >1000 entries have been collected and manually filtered to create a final list of ~80 unique histone entries, in human and mouse species.

4 Notes

1. Isolation of condensed spermatids is routinely performed on 2 mouse testes but it is possible to modulate the number of testes and adjust the volume of buffers proportionally. The testes can be collected from male mice upon availability in animal facility, briefly washed in cold PBS 1× and stored in microfuge tubes at −80 °C. Male mice have to be 8–9 weeks old in order to produce mature sperm: they reach puberty ~34–38 days after birth and another 28 days are required for the differentiation of sperm.
2. The quality of the preparation can be controlled observing an aliquot of each step with a microscope.
3. Alternatively, acid soluble extract can be buffered by adding 110 μL of Tris 2 M pH 8. The pH of the extract can be verified by pipetting 1 μL on a pH paper. Add more Tris solution if necessary. We recommend to incubate this solution for 30 min

on ice and centrifuge 30 min at 21,000 × *g* at 4 °C to remove any potential precipitate. This option is useful when high concentration of histones is not required.

4. TCA-precipitated histones can be stored overnight at −20 °C. However, a prolonged state in this strong acid could impact posttranslational modifications of histones. Long-term storage should be done on acetone washed pellets.
5. This step should remove the excess of acetone. Extensive drying will result in difficulties to resuspend the pellet in water.
6. Resuspension in water can be difficult. The pellet can be briefly sonicated to facilitate the resuspension of histones. Alternatively, the pellet can be resuspended in SDS-PAGE loading buffer.
7. The purification of TP1 takes advantage of the extreme basic property of this protein, which is soluble at very low pH. Acid soluble proteins are extracted directly from the cell pellet by solubilization in a strong acid (perchloric acid).
8. Conditions of MNase digestion can be optimized by running a time course experiment to obtain nucleosomal fibers of the desired length, evaluated on an agarose as described in Subheading 3.4, **step 13**.
9. Two volumes can be used: large gradients, formed with 16 mL of Sucrose Gradient solution in 16 × 102 mm tubes for SW28 Beckman-Coulter rotor (ref 337986), or small gradients, using 2.3 mL of the same solution in 11 × 34 mm tubes (ref 347357) to be used on a TLS-55 rotor.
10. Ideally in a cold room without any vibration (lab fridges are frequently opened and such vibrations will perturb the quality of the gradient).
11. The affinity of protein G for antibodies varies depending on the host species and antibody subclasses. Dynabeads Protein A is also available if the affinity of the antibody to protein G is not optimal.
12. BS3 crosslinker reacts on primary and secondary amines. Adding 1 M Tris pH 7.5 will provide them in excess and quench the cross-linking reaction.
13. Uncross-linked antibody may remain on the beads. An additional wash with 200 μL of Glycine 100 mM pH 3.0 will remove the uncross-linked antibody. This wash has to be brief to preserve the functionality of the antibody. Then wash the beads twice with 200 μL of IP buffer to remove traces of Glycine acidic solution.
14. The salt concentration of the IP buffer can be adapted depending on the purpose of the experiment. Weakly bound proteins will be lost when using higher salt concentrations.

15. The required amount of cell pellet is dependent on the expression level of the protein of interest and the scale of the experiment. Total male germ cells obtained from two or three mice would be applied to large-scale purification and the analysis of histone-associated proteins by mass spectrometry. In the case of purified post-meiotic cells, larger numbers of mice (more than ten) would be required. The volume of the IP buffer should be 10–20 times the volume of the cell pellet.
16. The required amount of beads varies depending on the scale of the experiment. Generally, 20 μL and 100–200 μL of beads are optimal for small- and large-scale purifications, respectively.
17. Particles (e.g., lipid) present in the cell lysate, which may nonspecifically bind to the beads and cause an increased background, will be removed at that step.
18. Longer time incubation may increase nonspecific binding of background proteins and disrupt weak protein interactions.
19. This step is important to eliminate the contamination of the components that have been absorbed on the surface of the tube wall during the incubation.
20. In the case of less than 50 μL of the beads, it is recommended to use 50 μL of the peptide solution to efficiently recover the eluate.
21. The necessary amount of antibody cross-linked beads depends on the affinity of the antibody toward the protein of interest. In our hands, 10–20 μL of the beads corresponding to 1–2 μg cross-linked antibody is sufficient for large-scale purification if a high-affinity antibody is used.
22. It is not recommended to use 1× Laemmli sample buffer containing reducing agents (DTT or β-mercaptoethanol). It may increase a contamination of the eluate by the light chains of the antibody, released by reducing agents.

Acknowledgments

SK laboratory is supported by a grant from "Foundation pour la Recherche Medicale (FRM)" "analyse bio-informatique pour la recherche en biologie" program as well as by ANR Episperm3 program and by INCa libre program (RPT13001CCA)). Additional supports were from Fondation ARC "Canc'air" project (RAC16042CLA), Plan Cancer (CH7-INS15B66) and Plan Cancer (ASC16012CSA), as well as the "Université Grenoble Alpes" ANR-15-IDEX-02 (LIFE and SYMER).

The group of J. Govin is supported by grants from the Agence Nationale de Recherche [ANR-14-CE16-0027-01 (FungiBET);

ANR-11-PDOC-011-01 (EpiGam); ANR-10-INBS-08 (ProFI); ANR-10-LABX-49-01 (GRAL)] and the EU FP7 Marie Curie Action (Career Integration Grant 304003). Additional supports were from the Univ. of Grenoble Alpes (Agir) and FINOVI from the Région Rhône Alpes (France).

References

1. Govin J, Caron C, Lestrat C et al (2004) The role of histones in chromatin remodelling during mammalian spermiogenesis. Eur J Biochem 271:3459–3469
2. Gaucher J, Reynoird N, Montellier E et al (2010) From meiosis to postmeiotic events: the secrets of histone disappearance. FEBS J 277:599–604
3. Govin J, Escoffier E, Rousseaux S et al (2007) Pericentric heterochromatin reprogramming by new histone variants during mouse spermiogenesis. J Cell Biol 176:283–294
4. Montellier E, Boussouar F, Rousseaux S et al (2013) Chromatin-to-nucleoprotamine transition is controlled by the histone H2B variant TH2B. Genes Dev 27:1680–1692
5. Shinagawa T, Huynh LM, Takagi T et al (2015) Disruption of Th2a and Th2b genes causes defects in spermatogenesis. Development 142(7):1287–1292
6. Ueda J, Harada A, Urahama T et al (2017) Testis-specific histone variant H3t gene is essential for entry into spermatogenesis. Cell Rep 18:593–600
7. Barral S, Morozumi Y, Tanaka H et al (2017) Histone variant H2A.L.2 guides transition protein-dependent protamine assembly in male germ cells. Mol Cell 66:89–101 e8
8. Goudarzi A, Shiota H, Rousseaux S, Khochbin S (2014) Genome-scale acetylation-dependent histone eviction during spermatogenesis. J Mol Biol 426:3342–3349
9. Tan M, Luo H, Lee S et al (2011) Identification of 67 histone marks and histone lysine crotonylation as a new type of histone modification. Cell 146:1016–1028
10. Dai L, Peng C, Montellier E et al (2014) Lysine 2-hydroxyisobutyrylation is a widely distributed active histone mark. Nat Chem Biol 10:365–370
11. Goudarzi A, Zhang D, Huang H et al (2016) Dynamic competing histone H4 K5K8 acetylation and Butyrylation are hallmarks of highly active gene promoters. Mol Cell 62:169–180
12. Lewis JD, Abbott DW, Ausió J (2003) A haploid affair: core histone transitions during spermatogenesis. Biochem Cell Biol 81:131–140
13. Meistrich ML, Mohapatra B, Shirley CR, Zhao M (2003) Roles of transition nuclear proteins in spermiogenesis. Chromosoma 111:483–488
14. Govin J, Gaucher J, Ferro M et al (2012) Proteomic strategy for the identification of critical actors in reorganization of the post-meiotic male genome. Mol Hum Reprod 18:1–13
15. Rousseaux S, Khochbin S (2012) Combined proteomic and in silico approaches to decipher post-meiotic male genome reprogramming in mice. Syst Biol Reprod Med 58:191–196
16. Rousseaux S, Boussouar F, Gaucher J et al (2011) Molecular models for post-meiotic male genome reprogramming. Syst Biol Reprod Med 57:50–53
17. Buchou T, Tan M, Barral S et al (2017) Purification and analysis of male germ cells from adult mouse testis. Methods Mol Biol 1510:159–168
18. El Kennani S, Adrait A, Shaytan AK et al (2017) MS_HistoneDB, a manually curated resource for proteomic analysis of human and mouse histones. Epigenetics Chromatin 10:2
19. Villar-Garea A, Israel L, Imhof A (2008) Analysis of histone modifications by mass spectrometry. Curr Protoc Protein Sci Chapter 14, Unit 14.10
20. Yuan Z-F, Arnaudo AM, Garcia BA (2014) Mass spectrometric analysis of histone proteoforms. Annu Rev Anal Chem 7:113–128
21. Sidoli S, Bhanu NV, Karch KR et al (2016) Complete workflow for analysis of histone post-translational modifications using bottom-up mass spectrometry: from histone extraction to data analysis. J Vis Exp 17(111):e54112
22. Sabari BR, Zhang D, Allis CD, Zhao Y (2017) Metabolic regulation of gene expression through histone acylations. Nat Rev Mol Cell Biol 18:90–101

23. Zhao Y, Garcia BA (2015) Comprehensive catalog of currently documented histone modifications. Cold Spring Harb Perspect Biol 7:a025064
24. Huang H, Lin S, Garcia BA, Zhao Y (2015) Quantitative proteomic analysis of histone modifications. Chem Rev 115(6):2376–2418 150217145638004
25. Ramus C, Hovasse A, Marcellin M et al (2016) Benchmarking quantitative label-free LC-MS data processing workflows using a complex spiked proteomic standard dataset. J Proteome 132:51–62

Chapter 17

An Animal Model for Genetic Analysis of Multi-Gene Families: Cloning and Transgenesis of Large Tandemly Repeated Histone Gene Clusters

Michael P. Meers, Mary Leatham-Jensen, Taylor J. R. Penke, Daniel J. McKay, Robert J. Duronio, and A. Gregory Matera

Abstract

Histone post-translational modifications (PTMs) are thought to participate in a range of essential molecular and cellular processes, including gene expression, replication, and nuclear organization. Importantly, histone PTMs are also thought to be prime candidates for carriers of epigenetic information across cell cycles and generations. However, directly testing the necessity of histone PTMs themselves in these processes by mutagenesis has been extremely difficult to carry out because of the highly repetitive nature of histone genes in animal genomes. We developed a transgenic system to generate *Drosophila melanogaster* genotypes in which the entire complement of replication-dependent histone genes is mutant at a residue of interest. We built a BAC vector containing a visible marker for lineage tracking along with the capacity to clone large (60–100 kb) inserts that subsequently can be site-specifically integrated into the *D. melanogaster* genome. We demonstrate that artificial tandem arrays of the core 5 kb replication-dependent histone repeat can be generated with relative ease. This genetic platform represents the first histone replacement system to leverage a single tandem transgenic insertion for facile genetics and analysis of molecular and cellular phenotypes. We demonstrate the utility of our system for directly preventing histone residues from being modified, and studying the consequent phenotypes. This system can be generalized to the cloning and transgenic insertion of any tandemly repeated sequence of biological interest.

Key words Epigenetics, Histones, Multi-gene families, Post-translational modifications, Cloning, Site-specific transformation, *Drosophila*

1 Introduction

Histone post-translational modifications (PTMs) are covalent additions of chemical groups to specific histone residues, which frequently reside in the unstructured N-terminal tails of the core histones. A remarkable diversity of histone PTMs has been uncovered over the years, the most prominent of which include lysine methylation and acetylation, arginine methylation, and serine and threonine phosphorylation [1]. The predominant approach for

Guillermo A. Orsi and Geneviève Almouzni (eds.), *Histone Variants: Methods and Protocols*, Methods in Molecular Biology, vol. 1832, https://doi.org/10.1007/978-1-4939-8663-7_17,

studying the functions of histone PTMs has been to ablate the function of the so-called "writer" enzymes that catalyze them or the "reader" complexes that bind them. However, it is possible that these factors play structural roles or have functions that are independent of histones. For example, there are several known non-histone substrates for histone PTM writers [2–6].

An attractive strategy for circumventing these difficulties is direct mutagenesis of histone genes, resulting in amino acid substitutions that render histone proteins refractory to particular writer enzymes. In the budding yeast *Saccharomyces cerevisiae*, there are two copies of each of the replication-dependent histone genes in the genome, meaning that direct histone mutagenesis can be employed with relative ease, *see* [7–9]. However, it is thought that histone PTMs have important roles as carriers of epigenetic information that can help implement the transcriptional programs that facilitate cell differentiation and development, among other processes that cannot be evaluated in a single-celled organism [10, 11]. To exhaustively analyze the importance of histone PTMs in development and differentiation, a metazoan model for histone mutagenesis is required. However, in metazoans, the highly repeated replication-dependent histone genes are typically localized at multiple clusters throughout the genome [12]. Moreover, these clustered histone genes are often interspersed with other genes, making a conventional mutagenesis strategy nearly impossible, even with CRISPR/Cas9 technology.

To overcome this problem, we and others have taken advantage of *Drosophila melanogaster*, in which all five of the replication-dependent histone genes (H1, H2A, H2B, H3, H4) reside at a single locus organized in repeating units without any other intervening genes [13]. This structure enables deletion of the entire histone locus and complementation by synthetic arrays of mutant histone genes contained in transgenes. This strategy has successfully enabled basic developmental, morphological, and phenotypic analyses of histone mutants [14–22].

Here, we detail a cloning, transgenesis, and validation strategy used to directly replace the endogenous replication-dependent histone genes in *Drosophila melanogaster* using a single transgene containing a tandem array of engineered histone genes. To accomplish this cloning procedure, we generated a vector, dubbed pMultiBAC, that facilitates cloning of a large array of histone gene repeats necessary to complement deletion of the endogenous histone genes [17]. As shown below, pMultiBAC has the capacity for efficient transgenesis using φC31-mediated recombination with an attP site in the *Drosophila* genome, for assembly of tandemly arrayed inserts via a simple restriction site-based cloning scheme, and for inducible expression of large-insert vectors at high copy number in *E. coli*.

We used pMultiBAC to assemble histone locus repeat arrays of varying sizes, and used the resulting constructs to evaluate the number of histone genes required for metazoan development and viability. As part of the initial characterization of the vector, we showed that 12 wild-type copies of transgenic histone genes rescued viability and fertility of *D. melanogaster* bearing a homozygous deletion of the endogenous histone gene cluster (which contains ~100 copies of each histone gene), and that histone and protein mRNA levels are comparable in both genotypes [17]. As a proof-of-principle, we generated *D. melanogaster* in which histone H3 lysine 36 (H3K36) is mutated to arginine (H3K36R) that cannot be modified by conventional H3K36 "writer" enzymes [17, 21]. Furthermore, we have used this system to successfully analyze the phenotypes of four additional histone genotypes: H3K27A, H3K27R, H3K9R, and H4K20A [17, 19]. Thus, we have confirmed the utility of our BAC-based system in analyzing histone mutant phenotypes in a multicellular organism.

The methods described herein are divided into three major parts. In Subheading 3.1, we describe the procedure for molecular cloning of tandemly repeated sequences using the pMultiBAC vector (Fig. 1), and the validation of sequence duplication at each step. In Subheading 3.2, we describe the protocol for inserting the artificial tandem array directly into the *Drosophila melanogaster* genome at a specific *attP* "landing site." In Subheading 3.3 we detail the procedure for visual screening of in vivo integrated transformants, and for the verification of proper insertion at the landing site.

2 Materials

2.1 Generation of Artificial Tandem Arrays

1. NotI restriction enzyme (*see* **Note 2**).
2. 2 SalI restriction enzyme.
3. XhoI restriction enzyme.
4. 10× Restriction Buffer.
5. 10× Bovine Serum Albumin (BSA) solution.
6. Low melt agarose.
7. Pulsed field gel apparatus.
8. UV Transilluminator.
9. 0.5× TBE solution.
10. T4 DNA Ligase.
11. EPI-300 electrocompetent cells (Lucigen).
12. CopyControl Autoinduction Solution (Lucigen).
13. 0.1 mm Electroporation Cuvettes.

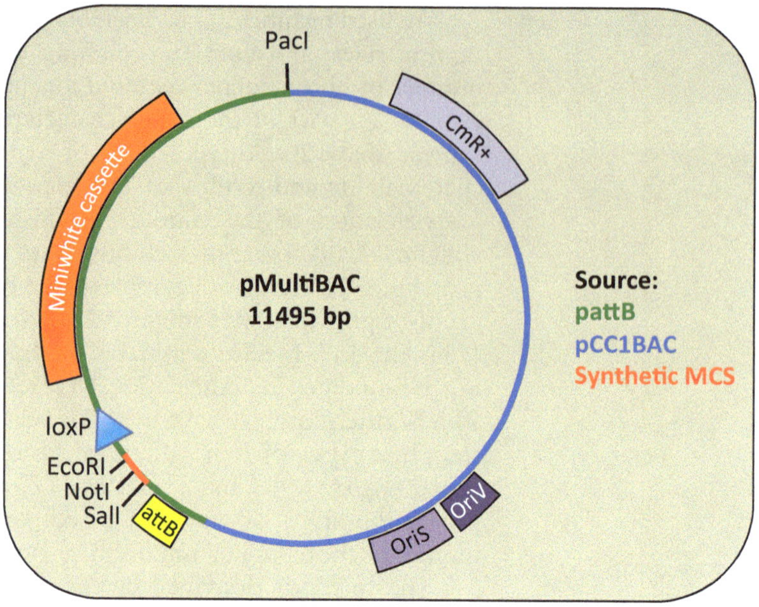

Fig. 1 pMultiBAC plasmid map. The vector was assembled by SalI-ScaI ligation of custom versions of pUASTattB (green) and pCC1BAC (blue) vectors that had been subjected to in vitro mutagenesis to remove all XhoI sites, and all but one SalI site (*see* [17] for additional details). The modified pCC1BAC vector contributed canonical (OriS) and inducible (OriV) replication machinery and chlroramphenicol selectable marker (CmR+). The modifed pattB vector contributed miniwhite cassette for visible screening, and attB site for site-specific integration into the *D. melanogaster* genome. A multiple cloning site (MCS, red) was introduced by ligation of SalI ends into SalI and XhoI sites at an extant MCS of pattB, regenerating a single SalI site and introducing a NotI site

14. Micropulser Electroporation System.
15. SOC Medium: 0.5% yeast extract, 2% tryptone, 10 mM NaCl, 2.5 mM KCl, 10 mM $MgCl_2$, 10 mM $MgSO_4$, 20 mM sterile glucose.
16. Luria Bertani (LB) broth.
17. 1 mg/mL Chloramphenicol solution.
18. LB Agar plates supplemented with 12.5 μg/mL Chloramphenicol.
19. Plasmid Miniprep Kit (Qiagen) or equivalent.
20. Sterile razor blades.
21. 10 mg/mL ethidium bromide stock.
22. 50× TAE: 2 M Tris, 1 M acetate, 50 mM EDTA, pH 8.6 (Adjusted with HCl).
23. 10× Tris/Borate/EDTA (TBE) buffer dry powder mix.

2.2 Screening for Potential Transgenic Animals

1. Injected embryos (*see* **Note 1**).
2. Balancer stock yw;+/+; MKRS/TM6B.
3. Fly food.
4. Squish Buffer: 10 mM Tris–HCl pH 8.2, 1 mM EDTA, 25 mM NaCL, 200 μg/mL Proteinase K, (added fresh).
5. Primers:

   attP forward primer: Ccttcacgttttcccaggt

   attP reverse primer: cgactgacggtcgtaagcac

   attB reverse primer: AGTGTGTCGCTGTCGAGATG
6. Taq polymerase and buffer (NEB M0273) or equivalent.
7. dNTPs.
8. Gel electrophoresis equipment.

2.3 Validation of Transgenic Fly Stocks

1. 30 isogenic flies.
2. Buffer A: 100 mM Tris–HCl pH 7.5, 100 mM EDTA, 100 mM NaCl, 0.5% SDS.
3. 1.5 mL centrifuge tubes designed for pestle grinding (e.g., Denville C-2170).
4. Disposable tissue pestle.
5. LiCl/KOAc solution: 1 volume 5 M potassium acetate: 2.5 volumes 6 M lithium chloride.
6. RNase A (10 mg/mL).
7. Phenol/Chloroform/Isoamyl alcohol.
8. Chloroform.
9. 3 M sodium acetate, pH 7.0.
10. 100% and 70% ethanol.
11. TE buffer: 10 mM Tris–HCl pH 8.0, 1 mM EDTA.
12. NotI restriction enzyme.
13. SalI restriction enzyme.
14. XhoI restriction enzyme.
15. Pulse field gel electrophoresis system.
16. Agarose.
17. Ethidium bromide 10 mg/mL.
18. 5× TBE: 1.1 M Tris, 900 mM Borate, 25 mM EDTA, pH 8.3.
19. Orbital shaker.
20. Gel imaging system (for ethidium bromide agarose gel).
21. Hybond N+ nylon membrane.
22. Denaturation solution: 1.5 M NaCl, 0.5 M NaOH.

23. Neutralization solution: 1.5 M NaCl, 0.5 M Tris–Cl pH 7.0.
24. Whatman filter paper.
25. 20× SSC: 3 M NaCl, 300 mM trisodium citrate, pH 7.0.
26. Plastic food wrap.
27. Paper towels.
28. UV crosslinker.
29. PCR product containing probe sequence of interest.
30. Random primers.
31. Other dNTPs: 1:1:1 mixture of 10 mM dATP, dTTP, and dGTP solutions.
32. Klenow enzyme and 10× buffer.
33. 0.5 M EDTA pH 8.0.
34. G50 column.
35. Scintillation counter.
36. Denhardt's solution.
37. 100× Denhardt's solution.
38. 10% SDS.
39. Formamide.
40. Hybridization Buffer: 6× SSC, 5× Denhardt's, 0.5% SDS, 50% formamide.
41. Hybridization oven.
42. Heat block.

3 Methods

3.1 Cloning of Tandemly Duplicated Repeat Sequences

1. Prepare two restriction digestion reactions on ice as follows (*see* **Notes 1** and **2**):
   BAC vector digestion:

pMultiBAC plasmid	500 ng
10× NEB buffer 3	2 μL
10× NEB BSA	2 μL
NEB NotI	0.5 μL
NEB SalI	0.5 μL
ddH_2O	q.s. To 20 μL

Insert digestion:

Insert DNA	500 ng
10× NEB buffer 3	2 μL
10× NEB BSA	2 μL
NEB NotI	0.5 μL
NEB SalI	0.5 μL
ddH_2O	q.s. To 20 μL

Insert DNA used can vary based on need (*see* **Note 3**).

2. Incubate reactions for 1 h in a 37 °C water bath.
3. Prepare a 1% low-melt agarose gel in 1× TAE buffer. Run for 1 h at 100 V (can be shorter or longer depending on size of insert) and place at 4 °C to harden gel before excising bands. Place excised agarose slice in 1.5 mL Eppendorf tube and incubate at 65 °C for 5 min to melt gel (*see* **Note 4**).
4. Add digested insert to digested BAC vector in roughly a 3:1 molar ratio with 10× ligation buffer and T4 DNA ligase in a 10 μL reaction (*see* **Note 5**).
5. Incubate at 16 °C overnight, add 1ul of the ligation reaction to 50 μL of Epi300 cells, and electroporate at 1.5 kV (*see* **Note 6**). Recover in 500 μL of Luria Broth (LB) for 1 h and plate on LB/agar plate supplemented with 12.5 μg/mL chloramphenicol overnight.
6. Pick colonies and incubate overnight with shaking at 37 °C in 1.5 mL of LB + 12.5 μg/mL chloramphenicol. Isolate DNA using a miniprep kit, and digest with SalI/NotI as in **step 1** (*see* **Note** 7).
7. For each overnight culture, prepare 4 mL of fresh LB supplemented with 12.5 μg/mL chloramphenicol and 1.2× CopyControl EPI300 induction solution (Epicentre). Add 1 mL of overnight culture to 4 mL fresh supplemented media and incubate with shaking at 37 °C for 5 h. Isolate DNA using a miniprep kit, and digest with SalI/NotI as in **step 1**.
8. Load digested samples in a 0.5× TBE, 1% agarose gel and run on a pulsed-field gel electrophoresis apparatus. We use a FIGE-mapper (BioRad) (*see* **Note 8**).
9. Stain the gel with 150 mL of water containing 7.5 μL of 10 mg/mL ethidium bromide for 15 min with gentle agitation on an orbital shaker. Destain for 15 min in water. Visualize gel on a UV trans-illuminator as shown in Fig. 2b.

Repeat **steps 1–9**, as desired, in order to generate larger insert clones (*see* Fig. 2a and **Note 9**).

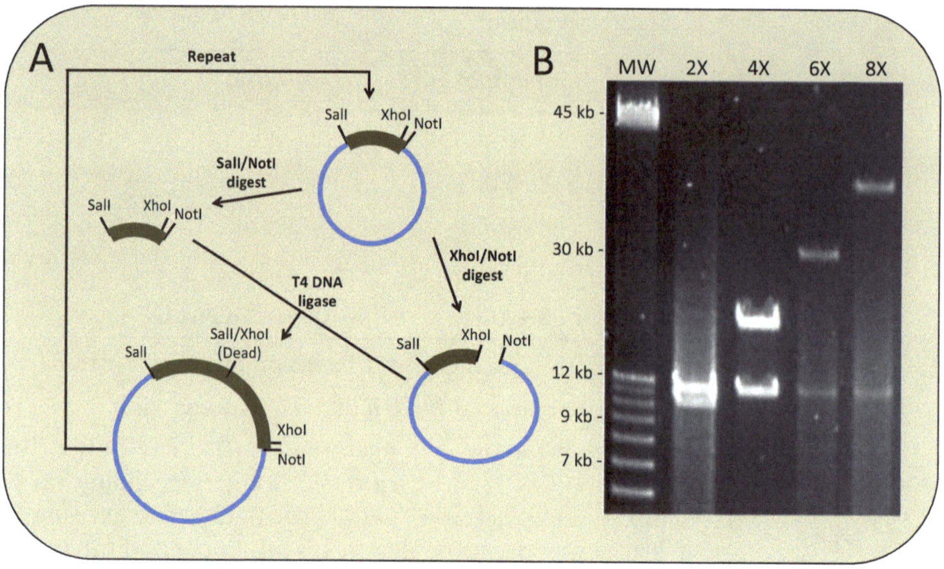

Fig. 2 Multimerization strategy for building tandem histone arrays. (**a**) Schematic of the multimerization strategy. The first step (not shown) is to insert a single repeat unit as single SalI/NotI fragment into the vector. The repeat unit should contain an XhoI site proximal to the NotI site as shown. To double the insert size, the insert is regenerated by SalI/NotI digestion (yellow), and the backbone is generated by XhoI/NotI digestion of a different aliquot of the BAC, which retains one copy of the insert (orange). T4 DNA ligase regenerates the NotI site and destroys the SalI-XhoI ligation junction, resulting in a directed tandem duplication of the insert. This process also regenerates the original pattern of restriction sites for cloning the next round. (**b**) Pulsed field gel of clones containing different sized tandem arrays of histone repeats, digested with SalI and NotI restriction enzymes. The pMultiBAC vector backbone runs at ~11.5 kb, whereas the inserts run at multiples of the 5 kb size of the histone repeat unit. Each lane contains a single BAC clone containing an insert of 2×, 4×, 6×, or 8× repeat units, as labeled

3.2 Screening for Potential Transgenic Animals by Eye Color and PCR

As outlined in Fig. 3a, after the animals have been screened for eye color, perform a PCR reaction to screen for animals with φC31 genomic landing sites (attP) that have properly recombined with the attB site on the BAC vector.

1. Inject at least 600 embryos with cloned BAC DNA (*see* **Notes 10** and **11**).
2. Cross survivors of injection to balancer stock (*see* **Note 12**). Single male survivors should be crossed with 3–5 virgin yw females; 1 or 2 females survivors should be crossed with 2–3 male balancer flies (*see* **Note 13**).
3. Allow cross to lay eggs for 3–5 days and then flip adults into a new vial for 2–4 days. Flip adults again into new vials for 2–4 days. Discard parents (*see* **Note 14**).
4. As flies eclose, sort through offspring to identify the transgenic (w+) flies by eye color (*see* **Note 10**). Anywhere from 0 to 5 vials out of 100 fertile crosses will produce w+ offspring (*see* **Note 15**).

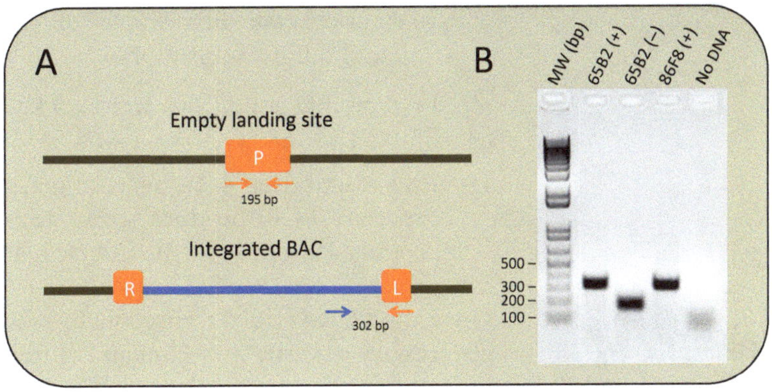

Fig. 3 PCR strategy for analyzing insertion of BAC transgenes into *Drosophila* genomic DNA. (**a**) Schematic of genome surrounding an attP (P) landing site used for φC31-mediated integration of transgenes. We have used two different chromosome 3 landing sites, one located at polytene band position 65B2 (called VK33) and another one located at 86F8 (called 86Fb). Primers flanking the attP site are used to interrogate the empty landing site, generating a 195 bp PCR product. Upon proper integration of the BAC transgene, the attP landing site is replaced by attR and attL sites flanking the transgene. Using a 5′ primer targeting the BAC insert along with the same 3′ primer used above, PCR amplification detects in a 302 bp fragment. (**b**) Agarose gel showing PCR products of positive (+) and negative (–) integrants

5. Cross individual w+ transgenic flies to a balancer stock (*see* **Note 16**). Make a balanced stock from the offspring; once a balanced stock has been established, freeze one w+ offspring for crude genomic DNA preparation and PCR verification.
6. Smash one frozen fly with P200 pipette tip containing 50 μL of Squish Buffer (SB), without expelling any liquid. Then expel the remaining SB into the tube. Breaking the fly into 2–4 pieces is sufficient to obtain enough DNA for PCR (*see* **Note 17**).
7. Incubate at 37 °C for 20 min. Heat inactivate Proteinase K at 85 °C for 10 min.
8. Use 0.5 μL of single fly squish genomic DNA in the following 25 μL PCR reaction:

10× NEB std. Taq buffer	2.5 μL
dNTPs	0.5
attP For	0.5
attP Rev	0.5
attB Rev	0.5
NEB taq	0.3
Water	*19.7*
	24.5

Make a cocktail sufficient for all reactions, aliquot, and then add 0.5 μL squish DNA.

9. PCR cycling conditions are: 1× 94 °C 5 min, 35× (94 °C 15 s, 55 °C 15 s, 68 °C 20 s), 1× 68 °C 5 min.
10. Run reaction on 1.2% agarose gel, as shown in Fig. 3b. The expected PCR product sizes are:(-) No integration, PCR product 195bp; (+) BAC inserted, PCR product 302 bp.

3.3 Validation of Transgenic Fly Stocks by Genomic Southern Analysis

Once the PCR-verified transgenic fly stocks have been obtained, it is necessary to verify they contain the full complement of tandemly repeated gene arrays by XhoI/SalI (or XhoI only) restriction digestion of genomic DNA followed by Southern blotting, *see* Fig. 4.

1. Collect 30 anesthetized flies in a 1.5 mL centrifuge tube and freeze at −80 °C.
2. Grind flies in 200 μL Buffer A with a disposable tissue pestle.
3. Add an additional 200 μL Buffer A and continue grinding until only cuticles remain. Incubate at 65 °C for 30 min.
4. Add 800 μL LiCl/KOAc Solution and incubate on ice for at least 10 min.
5. Spin for 15 min in a microcentrifuge at Vmax at 4 °C.
6. Transfer 1 mL of supernatant to a new 1.5 mL tube, avoiding floating debris. If debris transfers, repeat the spin and transfer to a clean tube.
7. Add 600 μL isopropanol, mix by inversion and spin 15 min at room temperature.
8. Pour off the supernatant. Carefully add 500 μL 70% ethanol. Pour off the supernatant. Spin 5 s and then pipette off the remaining ethanol.
9. Resuspend the pellet in 150 μL TE buffer.
10. Perform an RNaseA digestion by adding 1 μL RNAseA and incubating for 30 min at 37 °C.
11. Add 150 μL of Phenol/Chloroform/Isoamyl alcohol solution. Mix by flicking tube.
12. Spin 3 min at Vmax at room temperature in microcentrifuge.
13. Transfer upper (aqueous) phase to fresh 1.5 mL tube.
14. Add an equal volume of chloroform. Mix. Spin for 3 min.
15. Transfer upper (aqueous) phase to fresh tube.
16. Add 0.1 volume of 3 M sodium acetate. Mix. Add 2-volumes 100% ethanol. Mix. Incubate for 30 min at −80 °C. Thaw. Mix. Spin 15 min at 4 °C.

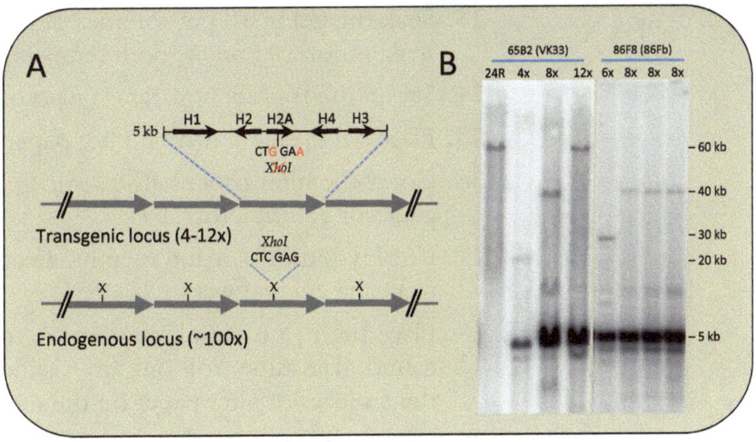

Fig. 4 Genomic Southern analysis of integrated BAC clones. (**a**) Restriction map of genomic DNA at the locus on chromosome 2 containing the tandemly arrayed Histone gene complex (endogenous locus), along with the map of the artificial tandem array of histone genes (transgenic locus). The transgenic arrays contain two translationally silent point mutations that abrogate an XhoI site present within each of the histone repeats at the endogenous locus. (**b**) Southern blots of genomic DNA from transgenic flies digested with XhoI and probed with a radio-labeled fragment as described in the Methods. The autoradiogram shows that the DNA from the endogenous histone locus, which contains roughly 100 copies of the 5 kb repeat unit, is cut by XhoI, whereas the transgenic histone arrays are uncut, and migrate at the appropriate sizes. BAC clones containing 4×, 6×, 8× or 12× histone repeats integrated at either 65B2 or 86F8 are analyzed, along with a 24× "rescue line" (24R) that contains no endogenous histone repeats (*see* ref. 17 for details)

17. Pour off the supernatant. Wash with 500 μL 70% ethanol. Pour off the supernatant. Spin 5 s and pipette off remaining ethanol. Air dry for 5 min.
18. Resuspend in 23 μL TE.
19. Restriction digest genomic DNA in 30 μL total volume. Use 3 μL 10× buffer, 3 μL 10× BSA, 0.5 μL each of SalI and NotI and as much DNA as possible, and at least 1 μg. Digest for 3 h at 37 °C. Digest 1 ng of the injected pMultiBac histone array containing vector as a positive control for size for the Southern.
20. Run the digest on a 0.5× TBE, 1% agarose gel on a pulsed-field gel electrophoresis unit. We use a FIGE-mapper (BioRad).
21. Stain the gel with 150 mL of water containing 7.5 μL of 10 mg/mL ethidium bromide for 15 min with gentle agitation on an orbital shaker. Destain the gel for 15 min in water and then image the gel. Note the exposure used.
22. Soak the gel in 10 gel volumes of denaturation solution with shaking for 20 min at room temperature. Repeat.

23. Soak the gel in 10 gel volumes of neutralization solution with shaking for 20 min at room temperature. Repeat.
24. Set up the transfer apparatus (*see* **Note 18**).
25. Place gel upside down on filter paper/transfer apparatus.
26. Cover the filter paper with plastic food wrap right up to the 4 edges of gel (*see* **Note 19**).
27. Cut Hybond N+ nylon membrane and 5 pieces of filter paper to fit the gel and wet in 20× SSC.
28. Place the nylon membrane on the top of the gel and using a round glass tube, roll out any visible air bubbles. Then place the 5 pieces of filter paper on the top of the membrane.
29. Cut paper towels to the size of the membrane and place on the top of the filter paper. Use 4 cm of paper towels.
30. Place a glass plate on the top of the paper towels and then a 400 g weight on the top of the glass plate. We use a large plastic bottle filled to 400 g with water. Allow DNA to transfer overnight.
31. Restain gel as described in **step 21** (see above) and take picture with same exposure to monitor transfer efficiency.
32. Submerge nylon membrane in distilled water for 10 min with shaking. Let the membrane air dry for 30 min.
33. Wrap the membrane in plastic wrap and UV crosslink DNA to nylon membrane. We use a Stratalinker 2400 with the autocrosslink setting and 60,000 μJ.
34. Generate a radioactively labeled probe by combining 25 ng of purified PCR product from the following reaction. Cycling conditions are 1× 94 °C 5 min, 35× (94 °C 15 s, 55 °C 30 s, 68 °C 1 min), 1× 68 °C 5 min. PCR mix is as follows:

10× NEB std Taq buffer	2.5 μL
dNTPs	0.5
For probe primer	0.5
Rev probe primer	0.5
NEB taq	0.3
Water	20.5
pMultiBac-histone plasmid	*0.2*

35. Mix 25 ng of PCR product with 10 μL random primers in a sterile 1.5 mL tube and bring to 34 μL with water.
36. Denature the DNA by heating the mixture in boiling water for 5 min. Centrifuge briefly.

37. Add the following components to the tube: 5 μL of other dNTPs, 5 μL of 10× Klenow buffer, 5 μL of labeled dCTP (50 uCi total), 1 μL Klenow exo$^-$.
38. Mix reaction thoroughly with pipette and incubate at 37 °C for 10 min.
39. Centrifuge briefly and then add 2 μL 0.5 M EDTA, pH 8.0, to stop the reaction.
40. Purify the reaction using a G50 column. Save 1 μL probe for pre-purification specific activity determination.
41. Follow the manufacturer's recommendations for the G50 column. Spin column once at 3000 rpm (~800 × *g*) for 2 min. Put column in a clean tube, add probe and spin again at 3000 rpm for 2 min.
42. Calculate the specific activity of the probe. Using a scintillation counter, measure the before and after column purification counts per million (CPM) of the reaction.
43. Equation and calculation using pre-G50 CPM of 1,896,098 and post-G50 CPM of 736,215.

$$SA = \frac{\left[50\,\text{uCi}\left(2.2\times10^{9}\right)\left(\frac{736,215}{1,896,098}\right)\right]}{\left[25+\left[1.3\times10^{3}\left(\frac{736,215}{1,896,098}\right)\left(\frac{50}{3000}\right)\right]\right]}$$

$$SA \sim 1.3\times10^{9}\ \text{dpm}/\mu g$$

44. Place membrane in a hybridization tube and incubate with 10 mL hybridization buffer for 2 h at 42 °C in hybridization oven.
45. Denatured labeled probe by a 5 min incubation in 100 °C heat block. Chill immediately on ice.
46. In a sufficient volume of hybridization buffer to cover the membrane (*see* **Note 20**) add radioactive probe and incubate overnight at 42 °C in a hybridization oven. For SA of 10^8 use probe at 10 ng/mL. For SA of 10^9 use probe at 2 ng/mL.
47. Wash the membrane twice with 50 mL of 2× SSC, 0.1% SDS for 10 min at room temperature in a hybridization oven.
48. Wash the membrane twice with 50 mL of 0.1× SSC, 0.1% SDS for 20 min at 50 °C in a hybridization oven.
49. Wrap membrane wrapped in plastic wrap and image.

4 Notes

Notes from cloning of tandemly duplicated repeat sequences

1. To generate an artificial tandem array, the sequence of interest (repeat unit) must be flanked on the 5′ end by SalI, and on the 3′ end by proximal XhoI-NotI restriction enzyme sites, as schematized in Fig. 2a. Note that the repeat unit must not contain any other SalI, XhoI, or NotI sites. The appropriate sequence can be generated by PCR amplification using primers with the following structure, where "X" represents the 5′ terminal sequence for the sequence of interest, and "Y" represents the reverse-complemented 3′ terminal sequence:

   Forward primer: NNNGTCGACXXXXXXXXXXXXXXX

   Reverse primer: NNNGCGGCCGCNNNNNNCTCGAGYYYYYYYYYYYYYY

2. Our protocol uses NEB cloning reagents, but any vendor that produces high quality restriction enzymes, Taq polymerase, and associated reagents can be used.
3. Although the amount of DNA used in the vector digestion should remain the same, the amount of insert DNA used should change based on maintaining a 3:1 ratio of insert-to-vector (*see* **step 4**). If a tandem array is being constructed via successive rounds of duplicative cloning, each subsequent round requires a reevaluation of the amount of insert DNA included in the ligation.
4. Low melt agarose gels are extremely fragile and liable to tear if perturbed, so handle with care. Ideally, gels should not be removed from their casting chambers at any point during the procedure. We run the low melt agarose gel in the same plastic cartridge in which it was cast, and then transfer that cartridge directly to a UV transilluminator after gel running to excise the gel slice.
5. Use the following equation to calculate the fold molar excess of insert to vector. Insert DNA mass (ng) = (fold-excess desired) × (insert DNA length/vector DNA length) × vector DNA mass (ng). Provided the ratio of insert-to-vector DNA is properly calibrated, the gel slices added from each reaction should be of roughly equal volume. It is also possible to optimize the ratio of insert-to-vector by setting up multiple ligation reactions from the same gel slices, in which each reaction contains the same volume of vector and a different volume of insert gel slice.
6. Epi300 cells can be diluted 1:4 with ice-cold water with little loss in efficacy. As constructs increase in size, the voltage can be

reduced, e.g., use 1.8 kV for a 10 kb insert, 1.5 kV for a 20 kb insert, and 1.3 kV for a 60 kb insert.

7. Once the ligated product exceeds 20 kb, it is recommended that the overnight cultures be used to seed a subsequent culture that contains EPI300 induction solution in order to induce pMultiBAC production to high copy number. The procedure for this step can be found in **sec 3.1**, **step 7**. At less than 20 kb, ligations can usually be transformed into standard bacterial cloning strains such as GC10 without induction.
8. It is highly recommended that the size of ligated products be verified after each round of cloning by SalI/NotI digestion and FIGE-mapper electrophoresis (or standard agarose gel electrophoresis in the case of products that are expected to be 10 kb or smaller). During cloning of tandemly repeated DNA, abnormalities have been known to arise even in DNA cloned from recombination-deficient bacterial strains. Such errors can be minimized by avoiding overgrowth of bacterial cultures beyond log-phase, especially during the CopyControl induction phase (**step 7**).
9. It is not always necessary to increase the size of the insert by 2^n repeat units. For example, to obtain a 12× insert add a 4× insert to a BAC vector containing an 8× repeat.

Notes from screening for potential transgenic animals by eye color and PCR

10. For injection of pMultiBAC clones, we often use landing site fly strains 86Fb or VK33, located on the third chromosome at band positions 86F8 and 65B2, respectively. Allow injected embryos to develop into adults. When integrated at VK33, the eyes are usually a pale yellow/orange in color.
11. We typically inject BAC DNA at a concentration 100 ng/μL and observe a 50–60% survival rate of injected embryos. We have found that changes in the concentration of injected DNA can improve both embryo survival and transgenesis rates. Thus this step can be carefully optimized for each individual clone. We use the Duke Model Systems Injection facility (modelsysteminjections@nc.rr.com, Durham, NC 27703, USA) for our transgenics, although other service providers can be used.
12. A balancer stock should be rigorously tested to ensure that it truly balances the landing site you are using. For instance, for our purposes, TM6B was the only available Chr3 balancer for which we observed 0% reversion and loss of the w^+ phenotype, indicative of positive VK33 transgenesis, in successive generations.
13. Because the health of small *Drosophila* cultures is partially dependent upon the number of females present, it is recom-

mended that either multiple transgenic females be pooled in a single culture, or else single females should be incubated on very hospitable, molasses-based food, and monitored carefully.

14. To stringently prevent cross-contamination of generations while also allowing enough time for healthy cultures to develop, vials should be flipped on the schedule recommended in **step 3**.
15. The percentage of positive transformants observed is dependent upon transgenesis efficiency, which itself is dependent upon the size of the clone, the concentration of DNA injected, and the PhiC31 "landing site" used. For our purposes, a ~71.5 kb BAC injected at 100 ng/μL into the VK33 landing site yielded roughly ~1–2% transgenesis from surviving injected lines. Smaller BAC clones are expected to yield higher transgenesis rates, and different landing sites may prove more efficient for different clones.
16. The cross can be schematized as follows:yw; +/+; VK33{transgene}/TM6Bxyw; +/+; MKRS/TM6B → yw; +/+; VK33{transgene}/TM6B
17. The same end can be achieved by grinding the fly in 50 μL of buffer with a plastic pestle that fits snugly into the bottom of an Eppendorf tube.

Notes from validation of transgenic fly stocks by genomic southern analysis

18. Inside a large Pyrex pan, place a pipet tip box lid and top with a glass plate large enough to support the gel. Cut 3 identical pieces of filter paper, wide enough to cover the glass plate and long enough to hang down on both sides and touch the bottom of the pan. Fill the pan with 20× SSC transfer buffer, making sure the edges of the filter paper are submerged.
19. Covering the filter paper insures the transfer solution moves through gel and not around it.
20. For an approximately 12 × 15 cm blot we use 10–15 mL of hybridation buffer in a rotating hybridization tube.

References

1. Rothbart SB, Strahl BD (2014) Interpreting the language of histone and DNA modifications. Biochim Biophys Acta 1839(8):627–643
2. Huang J, Berger SL (2008) The emerging field of dynamic lysine methylation of non-histone proteins. Curr Opin Genet Dev 18(2):152–158
3. Sims RJ 3rd, Reinberg D (2008) Is there a code embedded in proteins that is based on post-translational modifications? Nat Rev Mol Cell Biol 9(10):815–820
4. Biggar KK, Li SS (2015) Non-histone protein methylation as a regulator of cellular signalling and function. Nat Rev Mol Cell Biol 16(1):5–17
5. Carlson SM, Gozani O (2016) Nonhistone lysine methylation in the regulation of cancer pathways. Cold Spring Harb Perspect Med 6(11):a026435
6. Chen K, Liu J, Liu S et al (2017) Methyltransferase SETD2-mediated methyla-

tion of STAT1 is critical for interferon antiviral activity. Cell 170(3):492–506 e14

7. Park JH, Cosgrove MS, Youngman E et al (2002) A core nucleosome surface crucial for transcriptional silencing. Nat Genet 32(2): 273–279
8. Hyland EM, Cosgrove MS, Molina H et al (2005) Insights into the role of histone H3 and histone H4 core modifiable residues in Saccharomyces cerevisiae. Mol Cell Biol 25(22): 10060–10070
9. Tompa R, Madhani HD (2007) Histone H3 lysine 36 methylation antagonizes silencing in Saccharomyces cerevisiae independently of the Rpd3S histone deacetylase complex. Genetics 175(2):585–593
10. Strahl BD, Allis CD (2000) The language of covalent histone modifications. Nature 403(6765):41–45
11. Margueron R, Reinberg D (2010) Chromatin structure and the inheritance of epigenetic information. Nat Rev Genet 11(4):285–296
12. Marzluff WF, Wagner EJ, Duronio RJ (2008) Metabolism and regulation of canonical histone mRNAs: life without a poly(a) tail. Nat Rev Genet 9(11):843–854
13. Lifton RP, Goldberg ML, Karp RW et al (1978) The organization of the histone genes in Drosophila melanogaster: functional and evolutionary implications. Cold Spring Harb Symp Quant Biol 42(Pt 2):1047–1051
14. Günesdogan U, Jackle H, Herzig A (2010) A genetic system to assess in vivo the functions of histones and histone modifications in higher eukaryotes. EMBO Rep 11(10):772–776
15. Hödl M, Basler K (2012) Transcription in the absence of histone H3.2 and H3K4 methylation. Curr Biol 22(23):2253–2257
16. Pengelly AR, Copur O, Jackle H et al (2013) A histone mutant reproduces the phenotype caused by loss of histone-modifying factor Polycomb. Science 339(6120):698–699
17. McKay DJ, Klusza S, Penke TJ et al (2015) Interrogating the function of metazoan histones using engineered gene clusters. Dev Cell 32(3):373–386
18. Graves HK, Wang P, Lagarde M et al (2016) Mutations that prevent or mimic persistent post-translational modifications of the histone H3 globular domain cause lethality and growth defects in drosophila. Epigenetics Chromatin 9:9
19. Penke TJ, McKay DJ, Strahl BD et al (2016) Direct interrogation of the role of H3K9 in metazoan heterochromatin function. Genes Dev 30(16):1866–1880
20. Meers MP, Adelman K, Duronio RJ, et al (2018) Transcription start site profiling uncovers divergent transcription and enhancer-associated RNAs in Drosophila. BMC Genomics 19(1):157
21. Meers MP, Henriques T, Lavender CA et al (2017) Histone gene replacement reveals a post-transcriptional role for H3K36 in maintaining metazoan transcriptome fidelity. eLife 6:e23249
22. Penke TJ, McKay DJ, Strahl BD, et al. (2018) Functional redundancy of variant and canonical histone H3 lysine 9 modification in Drosophila. Genetics 208(1):229–244.

Chapter 18

Imaging and Quantitation of Assembly Dynamics of the Centromeric Histone H3 Variant CENP-A in *Drosophila melanogaster* Spermatocytes by Immunofluorescence and Fluorescence In-Situ Hybridization (Immuno-FISH)

Caitríona M. Collins and Elaine M. Dunleavy

Abstract

This chapter describes a method used to assay the cell cycle dynamics of the centromeric histone H3 variant CENP-A in meiosis using Drosophila males as the experimental system. Specifically, we describe a method that combines Immunofluorescence (IF) and Fluorescence in-situ Hybridization (FISH) protocols, performed on fixed Drosophila testes. An advantage of this protocol is the ability to localize individual centromeres on the four Drosophila homologous chromosomes that form distinct nuclear territories in spermatocytes. We also describe a method to quantify CENP-A focal intensities using Image J software.

Key words *Drosophila melanogaster*, Centromere, Chromosome segregation, CENP-A, Meiosis, Prophase I, Spermatogenesis

1 Introduction

At centromeres, the canonical histone H3 is replaced by the centromeric histone H3 variant CENP-A [1]. Centromeres are not determined by DNA sequence, and instead it is CENP-A that epigenetically defines centromere location on chromosomes [2, 3]. Correct levels of CENP-A at the centromere are essential for chromosome segregation in both mitotic and meiotic cell division cycles. To ensure this function, CENP-A nucleosomes at centromeres must be replenished each cell cycle. Unlike canonical histone H3 that is assembled at DNA replication, the de novo assembly of CENP-A at centromeres is a DNA replication independent process occurring outside of S phase [4]. Strikingly, investigations so far indicate that the cell cycle timing of CENP-A assembly differs between mitosis and meiosis. In mitosis in most somatic cells, CENP-A is assembled at late telo-

Guillermo A. Orsi and Geneviève Almouzni (eds.), *Histone Variants: Methods and Protocols*, Methods in Molecular Biology, vol. 1832, https://doi.org/10.1007/978-1-4939-8663-7_18,

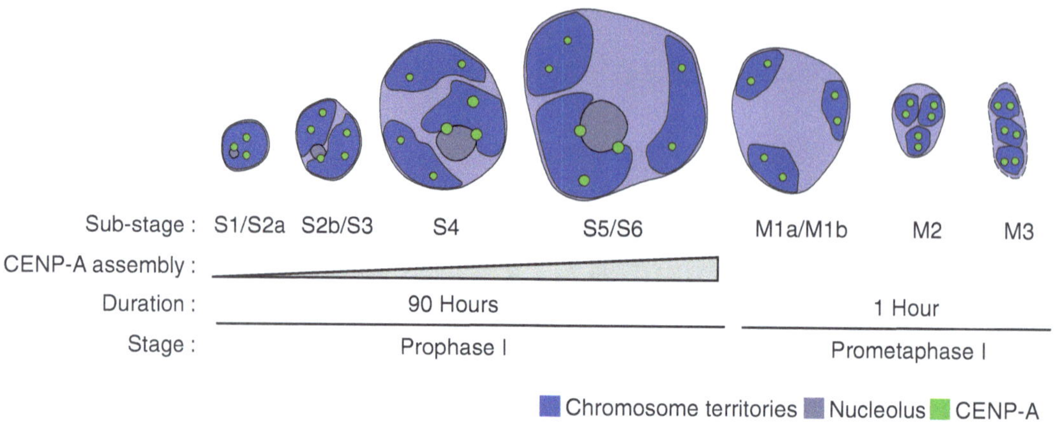

Fig. 1 Schematic illustrating nuclear morphology, centromere organization, and CENP-A dynamics during meiotic prophase and prometaphase I in Drosophila spermatocytes. At S1/S2a stages homologous and non-homologous centromeres cluster. Between S2b and S5/S6 stages homologous centromeres separate and distinct nuclear territories form. Widespread chromatin decondensation results in the formation of DAPI-dense and -light regions. At prometaphase I (stages M1 to M3) chromosome territories condense and centromeres align at the metaphase plate. Sister centromeres remain cohesed throughout

phase/early G1 phase [5]. Yet, in meiosis in germ cells, CENP-A is assembled at prophase I in the first meiotic cell division cycle [6, 7]. Here, we describe a method to assay CENP-A assembly and localization dynamics in Drosophila cell meiosis.

Drosophila melanogaster males are an excellent model system in which to study centromere (CENP-A) dynamics in meiosis and development. Advantages of this model system include the ease of genetic manipulation and the sequential progression of meiosis through the tubular shaped testes, which allows for easy identification of different cell stages. At the tip of Drosophila testes, germ-line stem cells divide asymmetrically giving rise to pre-meiotic primary spermatogonia. These cells are then amplified in number through a series of mitotic divisions producing a cyst of 16 cells, which synchronously undergo a single round of DNA replication prior to entering into meiosis I [8]. Drosophila males do not follow the standard meiotic prophase I script whereby homologous chromosomes pair, assemble a synaptonemal complex, recombine and are held together by chiasma until anaphase I [9]. Instead, male fruit flies opt for an alternative (and not very well understood) method of conjoining homologous chromosomes in distinct nuclear territories. Therefore, stages of meiotic prophase I have a unique classification (designated S1 to S6; Fig. 1) which is based on the formation of these chromosome territories [10]. At stage S6, the four Drosophila chromosomes are separated into three to four nuclear territories [9]. The second and third chromosomes each form a large territory; the X–Y chromosomes harboring the rDNA form a third territory adjacent to the nucleolus; the 4th chromosomes generally form a small territory close to the X–Y chromosomes. In this system, previous

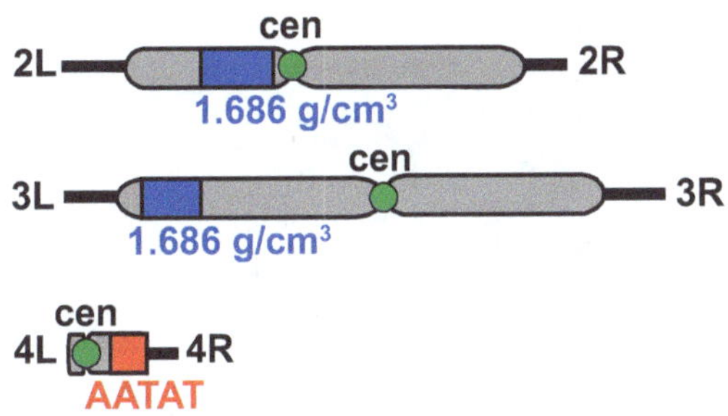

Fig. 2 Schematic illustrating the chromosomal locations of 1.686 g/cm^3 (blue) and AATAT (red) FISH probe targets

studies have shown that CENP-A is assembled at centromeres between S1 and S6 stages [6, 7, 11]. Moreover, different levels of CENP-A associate with different chromosomes at S6 [7, 11].

Here, we describe procedures for immuno-staining S1 and S6 stage spermatocytes for CENP-A, combined with chromosome specific FISH, to enable the tracking of centromere and CENP-A dynamics on each of the four Drosophila homologous chromosomes during meiosis. We use heterochromatic FISH probes recognizing distinct sites on the second and third autosomes (1.686 g/cm^3) and a homologue-pairing site on the fourth chromosome (AATAT) [12] (Fig. 2). Using Immuno-FISH it is possible to study CENP-A assembly dynamics at the centromeres of distinct chromosomes (Fig. 3). In addition, it is possible to use this method to study meiotic centromere and chromosome dynamics including (but not limited to); homologous centromere and chromosome pairing, centromere clustering, sister centromere and arm cohesion, chromosome segregation in meiotic divisions I or II and the correct formation of chromosome territories. This method could also be applied to study the dynamics of posttranslational modified versions of histones and histone variants that localize as discrete nuclear foci, e.g., gamma-H2AX.

2 Materials

2.1 Fly Culture

1. 25 mm polystyrene vials containing standard cornmeal medium preserved with 0.5% propionic acid and 0.1% Tegosept (methylparaben anti-fungal agent) (*see* **Note 1**).
2. Incubator (20 °C) under 12 h light-dark cycle.

2.2 Dissection

1. Stereomicroscope and light source.
2. One pair of ultra-fine dissecting forceps.
3. Translucent glass 1 ml dissecting dishes.
4. Tungsten needle (diameter 0.125 mm).
5. Dissection buffer: 1× PBS.

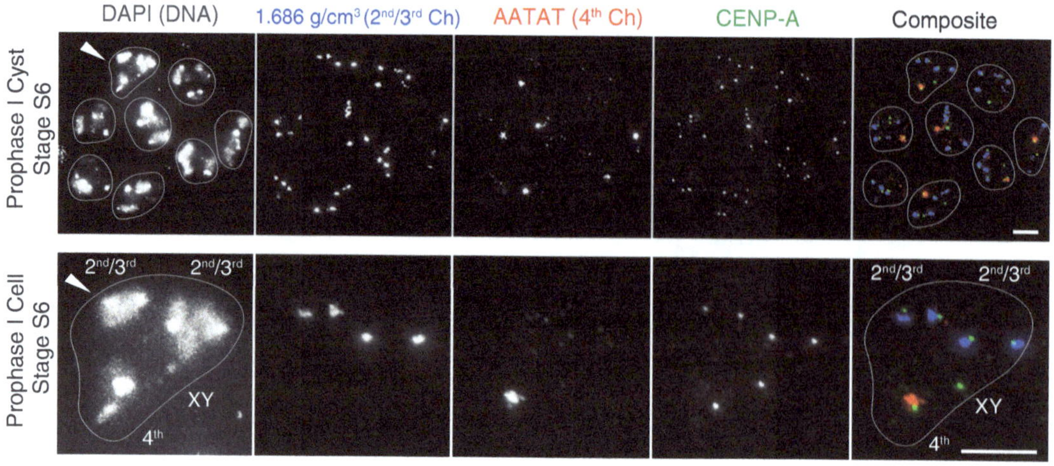

Fig. 3 Immuno-FISH on meiotic prophase I (S6 stage) spermatocytes. CENP-A (green) in combination with the 2nd/3rd chromosome probe (1.686 g/cm^3, blue) and the 4th chromosome probe (AATAT, red). DNA is stained with DAPI (gray). Nuclear membranes are outlined in white and chromosome-specific centromeres are illustrated. White arrowhead indicates one S6 stage cell from the prophase I cyst that is magnified in the bottom panel. Scale bar 5 μm

2.3 Fixation

1. Poly-L-lysine coated glass slides.
2. Liquid Nitrogen and 1 l Dewar container.
3. Cryogenic safety goggles and gloves.
4. Hydrophobic coverslips (*see* **Note 2**).
5. Sharp razor blade.
6. 4% paraformaldehyde.
7. Glass Coplin staining jars.
8. 70% ethanol (−20 °C).
9. 75%, 85%, and 95% ethanol at −20 °C.
10. Permeabilization buffer: PBS with 0.4% Triton-X (0.4% PBTX).

2.4 FISH

1. FISH wash buffer: 2× Saline-Sodium Citrate with 0.1% Triton X (0.1% SSC-TX).
2. Pre-hybridization buffer 1: 0.1% SCC-TX with 25% formamide (*see* **Note 3**).
3. Pre-hybridization buffer 2: 0.1% SCC-TX with 50% formamide, solutions at 23 °C and 37 °C (*see* **Note 4**).
4. Fluorescently labeled oligonucleotide probes. Oligonucleotides are synthesized and end labeled with Alexa fluorophores. Oligonucleotides recognizing the second and third chromosome 1.686 g/cm^3 site $(AATAACATAG)_3$ are labeled with Alexa-555, the fourth chromosome oligonucleotide $(AATAT)_6$

are labeled with Alexa-647 (*see* **Note 5**). Sequences are described in ref. [12].

5. Hybridization buffer: 3× SSC, 50% formamide and 10% dextran sulfate.
6. Rubber cement.
7. Hot plate at 95 °C.
8. Incubators at 37 °C and 25 °C.
9. Posthybridization wash buffer 1: 0.1% SCC-TX and 50% formamide (solutions at 23 °C and 37 °C).
10. Post-hybridization wash buffer 2: 0.1% SCC-TX and 25% formamide (solution at 23 °C).

2.5 Immunostaining

1. IF wash buffer: 0.4% PBTX.
2. Blocking buffer: 0.4% PBTX with 3% BSA.
3. Humid chamber.
4. Hydrophobic barrier (high-strength multipurpose adhesive or PAP pen).
5. Primary antibody: rabbit anti-*Dm*CENP-A, Active Motif, #39713.
6. Secondary antibody: Alexa-546 conjugated goat anti-rabbit.
7. DAPI: 1 μg/ml solution in 1× PBS.
8. 1× PBS.
9. Mounting medium: SlowFade® Gold antifade reagent (Life Technologies™) (*see* **Note 6**).
10. Coverslips (22 × 22 mm, 0.13–0.17 mm thick).
11. Sealing varnish.

2.6 Imaging, Image Quantification, and Processing

1. Fluorescent microscope.
2. Image J software (NIH).

3 Methods

3.1 Dissection

1. Sex larval or adult flies and transfer males to 1 ml of dissecting buffer in a dissecting dish. Dissect 12–16 testes per sample being careful to remove any excess fat and/or any associated tissues as well as testes associated seminal vesicles (*see* **Notes 7** and **8**).
2. Individually transfer testes using forceps (*see* **Note 9**) to a 10 μl drop of dissecting buffer on a poly-L-lysine coated slide.
3. Tear open testes using forceps and/or tungsten wire to release cell cysts from inside. Separate testes on the slide so that the tissues are not overlapping (*see* **Note 10**).

4. Place a hydrophobic coverslip (*see* **Note 2**) over the sample, being careful to avoid air bubbles. Squash the sample beneath the coverslip (*see* **Note 11**).

3.2 Fixation and Permeabilization

1. Immediately freeze the slides with squashed testes in liquid nitrogen using a large forceps.
2. Remove the frozen slide from the liquid Nitrogen, place on the bench on a piece of paper towel, and immediately remove the coverslip with a sharp blade (*see* **Note 12**).
3. While the slide is still frozen, pipette 500 μl of 4% paraformaldehyde over the samples and incubate at room temperature for 10 min.
4. Drain the paraformaldehyde from the slide into a waste container and immediately transfer the slide into 70% ethanol at −20 °C (*see* **Note 13**).
5. Pass slides through a series of ethanol concentrations at −20 °C in Coplin staining jars; 2 min in 75% ethanol, 2 min in 85% ethanol and 2 min in 95% ethanol.
6. Remove slides from ethanol and place on paper towel; tap excess ethanol from the slide and dry by evaporation for 2 min.
7. Incubate the sample at room temperature for 1 × 15 min in permeabilization buffer.

3.3 FISH

1. Transfer sample to FISH wash buffer and at room temperature carry out a 1× 10 min wash.
2. Incubate the sample at room temperature for 10 min in pre-hybridization buffer 1, followed by 10 min in pre-hybridization buffer 2, perform washes in the fume hood.
3. Incubate the sample at 37 °C for 2 h in pre-hybridization buffer 2.
4. Prepare 20 μl of hybridization buffer containing FISH probes per slide. Use 20 ng of DNA probe for the 1.686 g/cm^3 $(AATAACATAG)_3$ probe and 40 ng of DNA probe for 4th chromosome $(AATAT)_6$ per sample (Fig. 2).
5. Remove slides from pre-hybridization buffer (in the fume hood) and transfer to paper towel to dry slightly.
6. Pipette 20 μl of hybridization buffer onto a coverslip, invert the slide over the coverslip, and gently mount onto the slide being careful to avoid air bubbles.
7. Seal the edges of the coverslip with rubber cement and denature the slide on a hotplate at 95 °C for 4 min.
8. Incubate slides overnight at 20 °C in a humid chamber (*see* **Note 14**).
9. After probe hybridization, gently remove coverslip from the slide using a sharp blade. Wash the sample at 20 °C for 10 min in post-hybridization wash buffer 1 (*see* **Note 15**).

10. Incubate the sample for a further 2× 30 min at 20 °C in post-hybridization wash buffer 1.
11. Incubate the sample at room temperature for 10 min in post-hybridization wash buffer 2.
12. Finally, carry out 3× 10 min washes at room temperature in FISH wash buffer.

3.4 Immunostaining

1. After FISH protocol, transfer the sample to IF wash buffer and incubate at room temperature for 10 min.
2. Place a hydrophobic barrier (using high-strength multipurpose adhesive or a PAP pen) around the samples on the slide to create a well (*see* **Note 16**). Fill the well with 200 μl of blocking buffer and incubate in the dark at room temperature for 1 h in a humid chamber.
3. Drain blocking buffer from the slide and add 200 μl of anti-CENP-A primary antibody solution to the well created by the hydrophobic barrier. Incubate samples overnight at 4 °C in a humid chamber.
4. Following primary antibody incubation remove hydrophobic barrier from slides with a blade/sharp forceps and wash for 3× 15 min at room temperature in IF wash buffer.
5. Prior to secondary antibody staining, replace the hydrophobic ring around the samples and add 200 μl of the secondary antibody solution.
6. Incubate the samples in the dark for 1 h at room temperature in a humid chamber.
7. Remove hydrophobic ring and wash in the dark at room temperature for 3× 15 min in IF wash buffer.
8. Stain DNA with 1 μg/ml DAPI in the dark at room temperature for 10 min in a humid chamber (*see* **Note 17**).
9. Wash the samples in the dark for 10 min in 1× PBS.
10. To mount, allow the slides to dry slightly for 2–3 min and then place 20 μl of mounting medium on the coverslip and 20 μl on the sample and invert slide onto the coverslip being careful to avoid air bubbles (*see* **Note 18**).
11. Seal coverslips with varnish.

3.5 Imaging

Fluorescent imaging was carried out using a DeltaVision Elite wide-field microscope system (Applied Precision). The images were acquired with a 60× lens. As centromeres have a diameter of <5 μm images were acquired as z-stacks with a step size of 0.2 μm in order to ensure capture of all CENP-A foci. Raw data files were deconvolved using a conservative maximum intensity algorithm (10 cycles) and 3D z-stack images were represented in 2D by projection using SoftWorx (Applied Precision). RBG images were exported in TIFF format for further quantification analysis.

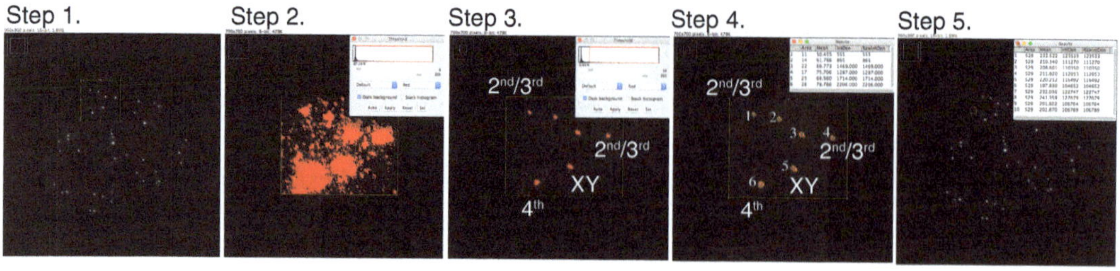

Fig. 4 Steps required to calculate the corrected total cellular fluorescence (CTCF) of centromeric foci using the image processing software Image J. Chromosome-specific centromeres are illustrated

3.6 Quantitation of Fluorescent Intensity

Focal fluorescent intensities were measured as corrected total cellular fluorescence (CTCF) using Image J software (NIH).

1. Export RGB image from microscope software in TIFF format.
2. Import TIFF image to Image J.
3. Split the channels of the composite RGB image. Image > color > split channels.
4. For quantification of CENP-A signal, select channel displaying CENP-A and select nucleus of interest (Fig. 4, **step 1**).
5. Copy and paste selected nucleus into a new 8-bit file.
6. Apply a threshold to the image (Fig. 4, **step 2**) and reduce the background reading to visualize the particles if interest (Fig. 4, **step 3**).
7. Set parameters to be analyzed. Analyze > set measurements > select area, integrated density and mean gray value > OK.
8. Analyze selected particles. Analyze > analyze particles > OK (Fig. 4, **step 4**).
9. Export results to a Microsoft Excel workbook.
10. To generate background readings return to original CENP-A image and select an area of fixed size. Take 10 background readings in the area surrounding your particles of interest. Analyze > measure (Fig. 4, **step 5**).
11. Export background readings to a Microsoft Excel workbook.
12. Calculate the corrected total cellular fluorescence (CTCF) using the following formula:

    $\text{CTCF} = \text{Integrated Density}_{(\text{CENP-A})} - (\text{Area}_{(\text{CENP-A})} \times \text{Average Mean}_{(\text{background})})$.
13. For the identification of CENP-A levels on a specific chromosome identify the chromosome from the composite FISH image and extract the CTCF pertaining to that centromere. To determine the average centromeric CENP-A sum the CTCF of each centromere per nucleus.

4 Notes

1. In humid climates/conditions the addition of low concentrations of methylparaben (0.1% Tegosept) helps to prevent growth of contaminating mounds in fly stocks. Tegosept is most commonly used in Drosophila food media.
2. To reduce sticking of the coverslip during squashing, coverslips (22 mm × 22 mm) are treated with a hydrophobic substance such as rain repellent. The coverslips are coated with thin layer of the rain repellent and allowed to dry at room temperature. Hydrophobic coverslips can be stored for later use.
3. SAFETY: formamide (CAS 75-12-7). Formamide is characterized as CMR (carcinogenic, mutagenic, and reprotoxin). To avoid exposure wear appropriate PPE and prepare all solutions and carry out all incubations in a fume hood. When removing Coplin staining jars from the fume hood seal lids with plastic paraffin film to avoid leakage/spillage.
4. Pre-hybridization buffers containing formamide can be collected after use and stored overnight at 4 °C for reuse during washing steps.
5. The $(AATAT)_6$ probe also recognizes sites on the X and Y chromosomes, however the most intense hybridization signal is visible at h61 on the 4th chromosome. h61 is a site of stable homologue pairing and is visible throughout meiotic prophase I as a single bright focus that colocalizes with the condensed DNA signal of the 4th chromosome. The 1.686 g/cm^3 probe recognizes non-centromeric heterochromatic regions h37 on chromosome 2L and h48 on chromosome 3L. Homologue pairing at these sites is not stable, however sister chromatid cohesion is and thus two h37 foci are observed per second chromosome territory and two h48 foci are observed per third chromosome territory. DNA probes to identify the X and Y chromosomes, as well as specific sites on the second and third chromosomes, are further described in ref. [12].
6. SlowFade® Gold antifade reagent preserves signal across the entire spectrum and causes little or no quenching of the initial fluorescent signal.
7. For the analysis of the early stages of spermatogenesis including early and late prophase I, dissect testes from third instar larval testes. To enrich for the later stages of meiosis including late prophase, meioses I and II and all stages of spermatid differentiation dissect testes from newly eclosed adult males (< 1 day old).
8. To cleanly remove larval testes; identify the testes, which are visible on the ventral surface toward the aboral end of the larvae.

Place one set of dissecting forceps on each side of the testes and gently tease apart the larvae. The released testes are surrounded by fat tissue; using the forceps remove as much of this as possible.

For the dissection of the whole adult testes; place anaesthetized adult males in 500 μl of dissecting buffer with the dorsal surface facing upward. Secure the fly by placing one forceps on the upper third of the abdomen; the testes are removed from the abdomen by gently removing the epandrium. Remove the attached accessory glands and seminal vesicles from the testes.

9. Transferring testes individually prevents loss of sample, as the testes are prone to sticking to the inside of the plastic pipette tip. For transfer by pipette, this can be minimized by coating the inside of the pipette tip with 1% BSA in 1× PBS.
10. During transfer of testes additional buffer can carry over to the slide, additional buffer should be removed with a pipette to prevent over spreading of the sample when squashed.
11. To squash, place the slide on the bench with a folded sheet of paper towel on top and give a relatively firmly squash. If the sample is not squashed enough, the cysts of cells remain inside the tissue and are difficult to visualize. Over-squashing leads to distorted nuclear morphology.
12. To avoid damaging/losing the sample the coverslip needs to be removed quickly, before the slide thaws, leaving the slide on paper towel insulates the slide and stops it from thawing immediately. To remove the coverslip, firmly hold the slide down (being careful to avoid cold burn) and quickly flick off the coverslip with a sharp blade.
13. At this point, slides can be stored in 70% ethanol at −20 °C until all samples are collected or until further processing.
14. Incubating slides over a period of 48 h (12 h for FISH plus 12 h for IF) in a humid chamber can lead to bacterial growth on the slide, to avoid this ensure to replace water, moist tissue and wash down the chamber surfaces with 70% ethanol prior to use.
15. Pre-warm buffers to 20 °C before washing steps. Washing at specific temperatures was found to be critical for probe retention on DNA.
16. Creating a well around the samples on the slide allows for full immersion of the sample in the primary antibody solution and leads to more uniform antibody staining.
17. Steps involved in the FISH protocol reduce the permeability of the sample to DAPI staining and this gives a blurry appearance when the cells are visualized by fluorescence microscopy. Steps using detergents that increase permeabilization, e.g., Sodium deoxycholate impairs DAPI staining of DNA in these cells.

18. Placing mounting medium on both surfaces (slide and coverslip) acts to prevent the formation of air bubbles.

Acknowledgments

This work is supported by a Science Foundation Ireland (SFI)-Health Research Board-Wellcome Trust Research Career Development Fellowship and SFI-President of Ireland Young Researcher Award (PIYRA) to EMD. CMC is funded by a National University of Ireland Galway, College of Science Scholarship and SFI-PIYRA.

References

1. Westhorpe FG, Straight AF (2015) The centromere: epigenetic control of chromosome segregation during mitosis. Cold Spring Harb Perspect Biol 7(1):a015818. https://doi.org/10.1101/cshperspect.a015818
2. Allshire RC, Karpen GH (2008) Epigenetic regulation of centromeric chromatin: old dogs, new tricks? Nat Rev Genet 9(12):923–937. https://doi.org/10.1038/nrg2466
3. Panchenko T, Black BE (2009) The epigenetic basis for centromere identity. Prog Mol Subcell Biol 48:1–32
4. Probst AV, Dunleavy E, Almouzni G (2009) Epigenetic inheritance during the cell cycle. Nat Rev Mol Cell Biol. https://doi.org/10.1038/nrm2640
5. Valente LP, Silva MCC, Jansen LET (2012) Temporal control of epigenetic centromere specification. Chromosom Res. https://doi.org/10.1007/s10577-012-9291-2
6. Dunleavy EM, Beier NL, Gorgescu W et al (2012) The cell cycle timing of Centromeric chromatin assembly in drosophila meiosis is distinct from mitosis yet requires CAL1 and CENP-C. PLoS Biol. https://doi.org/10.1371/journal.pbio.1001460
7. Raychaudhuri N, Dubruille R, Orsi GA et al (2012) Transgenerational propagation and quantitative maintenance of paternal centromeres depends on Cid/Cenp-a presence in drosophila sperm. PLoS Biol. https://doi.org/10.1371/journal.pbio.1001434
8. Fuller M (1993) The development of Drosophila Melanogaster, volume 1. Cold Spring Harbour Press, New York
9. McKee BD, Yan R, Tsai J-H (2012) Meiosis in male drosophila. Spermatogenesis. https://doi.org/10.4161/spmg.21800
10. Cenci G, Bonaccorsi S, Pisano C et al (1994) Chromatin and microtubule organization during premeiotic, meiotic and early postmeiotic stages of Drosophila melanogaster spermatogenesis. J Cell Sci 107:3521 LP–3534 LP
11. Kwenda L, Collins CM, Dattoli AA, Dunleavy EM (2016) Nucleolar activity and CENP-C regulate CENP-A and CAL1 availability for centromere assembly in meiosis. Development. https://doi.org/10.1242/dev.130625
12. Tsai J-H, Yan R, Mckee BD (2012) Homolog pairing and sister chromatid cohesion in heterochromatin in Drosophila male meiosis I Abbreviations CDI Centromere identifier FISH Fluorescence in situ hybridization MNM Modifier of Mdg4 in meiosis rDNA Ribosomal DNA Rsp Responder SC Synaptonemal

Chapter 19

Probing the Function of Oncohistones Using Mutant Transgenes and Knock-In Mutations

Dong Fang, Heping Wang, and Zhiguo Zhang

Abstract

Recently, frequent somatic mutations at histone genes have been detected in high grade pediatric brain tumor, chondroblastoma, and giant cell tumor of bone. These mutant histones are also termed oncohistones. Since oncohistone proteins co-exist with wild type histone proteins in cells, it is critically important to understand how they promote tumorigenesis. Here, we describe two methods to analyze the impact of these oncohistones on histone modification and epigenome, including the expression of oncohistone from a transgene and the utilization of CRISPR/Cas9 system to knock-in specific oncohistone mutations. The methods described are useful for the initial characterization of oncohistones. Other methods such as ChIP-seq and RNA-seq, which analyze the effect of oncohistone mutations genome wide, are not detailed in this protocol.

Key words Oncohistone, Histone marks, Acid extraction, Western Blot, Immunofluorescence, CRISPR/Cas9, ChIP-seq and RNA-seq

1 Introduction

In eukaryotic cells, DNA is assembled into chromatin. The basic repeat unit of chromatin is the nucleosome, which contains one histone H3-H4 tetramer and two H2A-H2B dimers wrapped around 147 bp of DNA [1, 2]. In human genome, there are 13 genes encoding canonical histone H3 (H3.1/H3.2), which are assembled into nucleosomes during the S phase of the cell cycle. In addition, two genes (*H3F3B* and *H3F3A*) encode histone H3 variant H3.3, which is assembled into nucleosomes in a replication-independent manner [3]. At the amino acid level, however, H3.1/H3.2 and H3.3 differ by only four or five amino acid residues. Therefore, it is interesting that these H3.1/H3.2 and H3.3 proteins are assembled into nucleosomes via distinct pathways.

Because histones are fundamentally important proteins for chromatin structure, it was surprising that histone H3 genes are mutated in various tumors. For instance, more than 75% of diffuse intrinsic pontine glioma (DIPG) cases contain heterozygous mutation at

Guillermo A. Orsi and Geneviève Almouzni (eds.), *Histone Variants: Methods and Protocols*, Methods in Molecular Biology, vol. 1832, https://doi.org/10.1007/978-1-4939-8663-7_19,

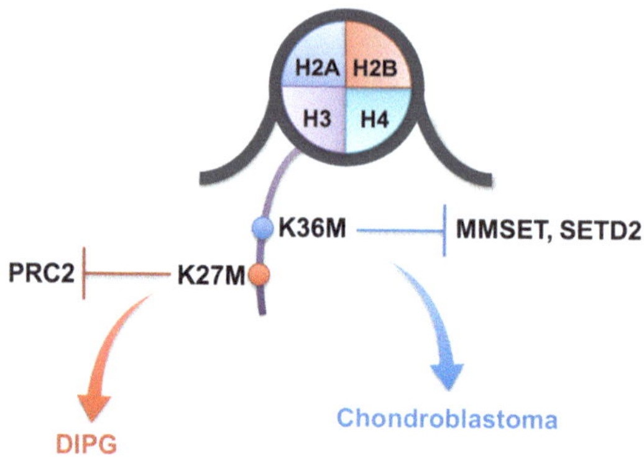

Fig. 1 Oncohistone H3.3K27M and H3.3K36M altered chromatin states in DIPG and chondroblastoma cases, respectively. H3.3K27M mutation proteins inbhibit the PRC2 in vitro. H3.3K36M mutation proteins inhibited MMSET and SETD2 in vitro. In cells, these mutant proteins trap the corresponding enzymes at genome loci

H3F3A, replacing lysine 27 to methionine in histone H3.3 (K27 M) [4–6]. In addition, *HIS1H3B* and *HIST1H3C*, which encoding H3.1, are also mutated in DIPG, but with a lower incidence [7–11]. In addition to the K27 to M mutation, the glycine 34 to arginine/valine (G34R/V) mutations occurring on *H3F3A* predominantly associate with pediatric glioblastoma multiform (GBM) [12–16]. Glycine 34 of H3.3 is also mutated to W/L in over 90% of giant cell tumors of bone, harboring heterozygous mutation at *H3F3A* [17–21]. Finally, about 95% of chondroblastomas contain heterozygous mutation at the *H3F3B* gene, replacing lysine 36 with methionine (K36 M) [8, 12, 21–23]. K36 M mutation at *H3F3B/HIST1H3C/HIST1H3E/HIST1H3G/HIST1H3I* was also detected in human papillomavirus (HPV)-negative head and neck squamous cell carcinomas (HNSCCs) [24]. In all these cases, only one allele of 15 histone H3 genes, including those encoding H3.1 and H3.3, is mutated in various tumors. Therefore, it is interesting and challenging to determine how mutant proteins promote tumorigenesis in cells with abundant wild type histone proteins.

We and others have previously studied the functions of oncohistones, H3.3K27M and H3.3K36M, in DIPG and chondroblastomas, respectively (Fig. 1). The H3.3K27M mutant protein dominantly drives global loss of H3K27me3 on wild type histones. This is most likely due to inhibition of PRC2, the H3K27 methyltransferase [25–27]. Similarly, we and others previously observed that H3K36me2/me3 are reduced globally in chondroblastomas and chondrocytes harboring H3.3K36M mutations, due to inhibition of at minimum two H3K36 methyltransferases, MMSET and SETD2 [28, 29]. Surprisingly, in addition to the global loss of

H3K27me3, we also observed a retension and/or gain of H3K27me3 at hundreds of gene promoters. Remarkably, compared to reference neural stem cells, genes with retension/gain of H3K27me3 are enriched in pathways associated with tumorigenesis [25]. More recently, it has been shown that Ezh2 is required for tumorigenesis of DIPG, supporting our original hypothesis that in addition to global loss of H3K27me3, locus-specific rentation/gain of H3K27me3 is also important for tumorigenesis [30, 31]. We also show that WT1 is silenced through H3K27me3 in DIPG cells, and this silencing is required for the proliferation of DPIG cells, providing an explanation for retention/gain of H3K27me3 in the environment of global reduction of H3K27me3.

To analyze the effect of K to M mutant transgenes on histone modifications, we and others usually exogenously express the mutant transgenes in cells. However, through analysis of the effect of H3.3K36M mutant proteins expressed from exogenous transgene and from endogenous *H3F3B* gene with heterozygous mutation [32], we realize that different expression levels of K to M mutant histone proteins, potentially other oncohistones, will have profoundly differential effect on cellular phenotypes. While we and other utilize K to M mutant transgene to study the functions of histone lysine methylation [25, 26, 29, 33–35], it would be prudent to use CRISPR/Cas9 approach to engineer the same mutation found in tumor samples in order to study cancer-associated phenotypes in the future.

Here, we outline two methods used in our laboratory to analyze these K to M oncohistones. For instance, to analyze the dynamic changes in histone modifications after introducing oncohistone mutant transgene, we used the immunofluorescence to analyze changes in individual cells after overexpression of oncohistones by plasmids or virus. Second, we describe knock-in specific mutation found in patient samples using CRISPR/Cas9, and we think that this method would be better to analyze the impact of the mutations on cellular phenotypes as well as epigenetic changes. In addition, we also utilize ChIP-seq, ChIP-seq with Spik-in yeast or *Drosophila* chromatin normalization as well as RNA-seq to analyze changes in histone modification and gene expression globally. These genome-wide methods are not described in detail in this protocol, but detailed procedures can be found [16, 27–29, 35]. Instead, we outline routine methods to analyze the impacts of oncohistones on epigenome initiatially using mutant transgenes as well as the CRISPR/Cas9 system to knock-in H3.3K36M mutation at *H3F3B* gene.

2 Materials

All reagents are prepared and stored at room temperature (unless indicated otherwise). Consult institution's Environmental Health and Safety Office and the Material Safety Data Sheets for proper handling of hazardous materials.

2.1 Probe the Function of Oncohistones Using Mutant Transgene

1. pOZ-FH-C-puro (Addgene, Cat. # 32516) with the inserted *H3F3B* or *HIST1H3C* gene, which is tagged at the C-terminal with tandem FLAG and HA tag, as the clone vector for oncohistone transgenes [25].
2. pQCXIP (Clontech, Cat. # 631516).
3. PfuTurbo DNA polymerase and 10× Cloned Pfu DNA polymerase reaction buffer (Agilent, Cat. # 600250) or equivalent DNA polymerase.
4. dNTP mixture 2.5 mM each.
5. Site-directed mutagenesis primers: *see* Table 1 (*see* **Note 1**).
6. DpnI, BamHI, XhoI.
7. Alkaline Phosphatase, Calf Intestinal (CIP).
8. Chemically-competent cells.
9. QIAprep Spin Miniprep Kit (Qiagen, Cat. # 27106).
10. LB agar plates with ampicillin: Dissolve 10 g tryptone, 5 g yeast extract, and 10 g NaCl, 15 g agar in 950 ml deionized water and bring volume up to 1 l with deionized water. Autoclave for 20 min at 15 psi. Cool to around 55 °C and add ampicillin to 50 mg/l. Mix well and pour into petridishes. Invert and store plates at 4 °C when the LB become harden.
11. QIAquick Gel Extraction Kit (Qiagen, Cat. # 28706).
12. Isopropanol.
13. Ethanol.
14. iMEF cells.
15. Lipofectamine™ 3000 Transfection Reagent (Invitrogen, Cat. # L3000001).
16. Opti-MEM™ I Reduced Serum Medium (Gibco, Cat. # 31985070).
17. PFA: 3% paraformaldehyde. Weight 1.5 g PFA and add it to 30 ml warm PBS, add 20 μl 10 M NaOH, warm up to 65 °C. Mix until complete dissolve. Add 1 g sucrose, shake until dissolve, adjust to 50 ml with PBS. Store at 4 °C in the dark (*see* **Note 2**).
18. 0.5% Triton X-100 solution: 1 ml 1 M Hepes buffer pH 7.4, 0.5 ml 5 M NaCl, 150 μl 1 M $MgCl_2$, 25.7 ml 20% sucrose, 250 μl Triton X-100, adjust to 50 ml with H_2O.

Table 1
Oligos used

Oligo name	Sequence, 5′ to 3′	Comments
H3.3K27M-H3F3B-F	CGAAAGCCGCCAGGATGAGCGCTCCCTCTACC	Site-directed mutagenesis primers for H3.3K27M mutation at human *H3F3B*
H3.3K27M-H3F3B-R	GGTAGAGGGAGCGCTCATCCTGGCGGCTTTCG	Site-directed mutagenesis primers for H3.3 K27 M mutation at human *H3F3B*
H3.3G34R-H3F3B-F	AGCGCTCCCTCTACCAGAGGGGTGAAGAAGCCT	Site-directed mutagenesis primers for H3.3 G34R mutation at human *H3F3B*
H3.3G34R-H3F3B-R	AGGCTTCTTCACCCCTCTGGTAGAGGGAGCGCT	Site-directed mutagenesis primers for H3.3 G34R mutation at human *H3F3B*
H3.3K9M-H3F3B-F	CAAGCAGACTGCTCGTATGTCCACCGGTGGGAAAG	Site-directed mutagenesis primers for H3.3 K9M mutation at human *H3F3B*
H3.3K9M-H3F3B-R:	CTTTCCCACCGGTGGACATACGAGCAGTCTGCTTG	Site-directed mutagenesis primers for H3.3 K9M mutation at human *H3F3B*
H3.3K36M-H3F3B-F	CTCTACCGGCGGGGTGATGAAGCCTCATCGCTAC	Site-directed mutagenesis primers for H3.3K36M mutation at human *H3F3B*
H3.3K36M-H3F3B-R	GTAGCGATGAGGCTTCATCACCCCGCCGGTAGAG	Site-directed mutagenesis primers for H3.3K36M mutation at human *H3F3B*

(continued)

Table 1 (continued)

Oligo name	Sequence, 5′ to 3′	Comments
H3.3K79M-H3F3B-F	GATCGCGCAGGATTTCATGACCGACCTGAGGTTTC	Site-directed mutagenesis primers for H3.3K79M mutation at human *H3F3B*
H3.3K79M-H3F3B-R	GAAACCTCAGGTCGGTCATGAAATCCTGCGCGATC	Site-directed mutagenesis primers for H3.3K79M mutation at human *H3F3B*
H3.1K27M-HIST1H3C-F	CTAAAGCAGCCCGTATGAGCGCTCCGGCCACC	Site-directed mutagenesis primers for H3.1K27M mutation at human *HIST1H3C*
H3.1K27M-HIST1H3C-R	GGTGGCCGGAGCGCTCATACGGGCTGCTTTAG	Site-directed mutagenesis primers for H3.1K27M mutation at human *HIST1H3C*
H3.1G34R-HIST1H3C-F	GCTCCGGCCACCGGTAGAGTGAAGAAACCTCAT	Site-directed mutagenesis primers for H3.1 G34R mutation at human *HIST1H3C*
H3.1G34R-HIST1H3C-R	ATGAGGTTTCTTCACTCTACCGGTGGCCGGAGC	Site-directed mutagenesis primers for H3.1 G34R mutation at human *HIST1H3C*
H3.1K9M-HIST1H3C-F	GAAGCAAACAGCTCGCATGTCTACCGGCGGCAAAG	Site-directed mutagenesis primers for H3.1 K9M mutation at human *HIST1H3C*
H3.1K9M-HIST1H3C-R	CTTTGCCGCCGGTAGACATGCGAGCTGTTTGCTTC	Site-directed mutagenesis primers for H3.1 K9M mutation at human *HIST1H3C*

H3.1K36M-HIST1H3C-F	CCACCGGTGGCGTGATGAAACCTCATCGCTAC	Site-directed mutagenesis primers for H3.1 K36 M mutation at human *HIST1H3C*
H3.1K36M-HIST1H3C-R	GTAGCGATGAGGTTTCATCACGCCACCGGTGG	Site-directed mutagenesis primers for H3.1 K36 M mutation at human *HIST1H3C*
H3.1K79M-HIST1H3C-F	GAAATCGCCCAGGACTTCATGACCGACCTGCGTTTCCAG	Site-directed mutagenesis primers for H3.1 K79 M mutation at human *HIST1H3C*
H3.1K79M-HIST1H3C-R	CTGGAAACGCAGGTCGGTCATGAAGTCCTGGGCGATTTC	Site-directed mutagenesis primers for H3.1 K79 M mutation at human *HIST1H3C*
sgRNA-F	CACCGAGGAAAAGCGCTCCCTCTAC	sgRNA for CRISPR/Cas9 cutting at *H3F3B* gene in mouse
sgRNA-R	AAACGTAGAGGGAGCGCTTTTCCTC	sgRNA for CRISPR/Cas9 cutting at *H3F3B* gene in mouse
Donor ssDNA	ACCGGTGGGAAAGCCCCCCGCAAACAGCTGGCCACGAAAGCCGCCAGGAAAAGCGCTCCC**AG**TACCGGCGGGGTG<u>ATG</u>AAGCCTCATCGCTACAGGTAGGTCGGGCGGGGGAACAATGGCCCGGCGGTGGCCGGCTTTGTGCGGCAGCGTCCG	Donor ssDNA to generate K36 M mutation at *H3F3B* in mouse The underline indicates the K36 M mutation and the bold nucleotide indicated the synonymous mutations to avoid cutting by CRISPR/Cas9
H3F3B-forward	TCGAGGAAGGGAAGTGACTCCT	To PCR the *H3F3B* gene in mouse
H3F3B-reverse	AGAGCCGCACTATTAATCCCA	To PCR the *H3F3B* gene in mouse

19. 5% (w/v) normal goat serum in PBS.
20. Antibodies: H3K36me2 (Cell Signaling Technology, Cat. # 2901), H3K36me3 (Active Motif, Cat. # 61101), H3K27me3 ((Cell Signaling Technology, Cat. #9733), H3K27 M (Millipore, Cat. # ABE419), H3K36 M (RevMAb Biosciences, Cat. # 31-1085-00), FLAG (Sigma, Cat. # F1804), Alexa fluor 594-conjugated anti-rabbit secondary antibodies, and Alexa fluor 488-conjugated anit-mouse secondary antibodies.
21. PBS-T: 1× PBS with 0.1% Tween 20.
22. DAPI staining solution: dilute DAPI in deionized water to make a 300 μM stock solution. 1:1000 dilute the stock solution in PBS to make a 300 nM DAPI staining solution.
23. ProLong™ Gold Antifade reagents (Invitrogen, Cat. # P36934).

2.2 Probe the Function of Onco-Histone Using Knock-In Mutation (see Note 3).

1. Plasmids: pSpCas9(BB)-2A-Puro V2.0 (Addgene, Cat. # 62988, PX459 V2.0).
2. BbSI-HF (NEB, Cat. # R3539S).
3. Cutsmart buffer (NEB, Cat. # B7204S).
4. DNA Ligation Kit Ver. 2.1 (TaKaRa, Cat. # 6022).
5. Chemical competent *E. coli* cells.
6. LB: Dissolve 10 g tryptone, 5 g yeast extract, and 10 g NaCl in 950 ml deionized water and bring volume up to 1 l with deionized water. Autoclave for 20 min at 15 psi. Store at room temperature.
7. LB agar plates with ampicillin: Dissolve 10 g tryptone, 5 g yeast extract, and 10 g NaCl, 15 g agar in 950 ml deionized water and bring volume up to 1 l with deionized water. Autoclave for 20 min at 15 psi. Cool to around 55 °C and add ampicillin to 50 mg/l. Mix well and pour into petridishes. Invert and store plates at 4 °C when the LB become harden.
8. Transfection reagent: use ideal transfection reagent for the cells. Nucleofector Kit V (Lonza, Basel, Switzerland) is used for T/C28a2 cells according to the manufacturer's instruction.
9. PCR reagent: Ex Taq, 10× Ex Taq reaction buffer, dNTP (10 mM), DMSO.
10. Oligos: *see* Table 1.
11. Antibody: H3K36M (RevMAb Biosciences, Cat. # 31-1085-00), Alexa fluor 488-conjugated anti-rabbit secondary antibodies (Jackson ImmunoResearch, Cat. # 11-545-003).
12. Gentra Puregene Cell Kit (Qiagen, Cat. # 158767).
13. Isopropanol.
14. 70% ethanol.
15. QIAquick PCR Purification Kit (Qiagen, Cat. # 28106).

3 Methods

Carry out all the steps at room temperature unless otherwise indicated.

3.1 Probe the Function of Oncohistones Using Mutant Transgene

1. Construct the expression vector of oncohistone mutant transgene (*see* **Note 4**).
2. Site-directed mutagenesis is performed with oncohistone mutation primers as described above.
3. The PCR reaction is set up as: 10 ng pOZ vector with inserted *H3F3A* or *HIST1H3C* as templet, 2.5 μl of 10 Cloned Pfu DNA polymerase reaction buffer, 2 μl of 10 mM dNTP, 0.5 μl of 10 mM primer onchohistone mutation forward primer, 0.5 μl of 10 mM primer onchohistone mutation Reverse primer, 0.5 μl of DMSO, 0.5 μl PfuTurbo DNA polymerase and add H_2O to 25 μl.
4. PCR steps: Denature at 95 °C 3 min, 18 cycles of 95 °C 30 s, 55 °C 30 s, 68 °C 18 s, extend the DNA at 72 °C for 10 min.
5. After the PCR reaction, directly add 1 μl DpnI to the samples, mix well.
6. Incubate at 37 °C for 4 h.
7. Add 10 μl of the sample to 200 μl of freshly thawed competent *E. coli* cells.
8. Incubated on ice for 30 min.
9. Heat shock at 42 °C for 90 s.
10. Incubate on ice for 2 min.
11. Add 800 μl LB and incubate at 37 °C for 1 h.
12. Spin at 1500 × g for 1 min and aspirate 800 μl supernatant.
13. Resuspend the cells and spread to LB plates with Ampicillin.
14. Incubate at 37 °C overnight.
15. Pick two clones and send for sanger sequencing to confirm the site-mutation of histones.
16. Sub-clone the oncohistone mutations to expression vector pQCXIP.
17. Digest 2 μg of pOZ-oncohistone plasmid with BamHI and XhoI for 2 h at 37 °C: 2 μg plasmid, 0.5 μl BamHI, 0.5 μl XhoI, 5 μl 10X buffer 3.1 (NEB), and add H_2O to 50 μl.
18. Run all the samples in 1.2% agarose and purify the lower histone gene band (around 500 bp) by QIAquick Gel Extraction Kit as follows.
19. Excise the DNA from the agarose gel and weigh the gel slice.
20. Add 3 volumes of Buffer QG from the QIAquick Gel Extraction Kit to 1 volume of gel (1 mg gel is around 100 μl volumn).

21. Incubate at 50 °C for 10 min with vortexing every 2 min until the gel has been completely dissolved.
22. Add 1 gel volume of isopropanol, mix by pipetting, and transfer the samples to QIAquick column (*see* **Note 5**).
23. Spin at 16,000 × *g* for 1 min and discard the flow-through.
24. Wash with 0.75 ml of Buffer PE in the the QIAquick Gel Extraction Kit (*see* **Note 6**), spin at 16,000 × *g* for 1 min, and discard the flow-through.
25. Spin for an additional 1 min at 16,000 × *g*.
26. Place the column in a new 1.5 ml microcentrifuge tube.
27. Add 30 μl H_2O to the center of column and incubate for 1 min.
28. Spin at 16,000 × *g*, 1 min to collect the digested insertion of oncohistones.
29. Digest 2 μg of pQCXIP plasmid with BamHI and XhoI for 2 h at 37 °C: 2 μg plasmid, 0.5 μl BamHI, 0.5 μl XhoI, 5 μl 10X buffer 3.1 (NEB), and add H_2O to 50 μl.
30. Add 1 μl CIP, 37 °C, 1 h.
31. Purify the digested plasmids by QIAquick PCR Purification Kit.
32. Add 500 μl of Buffer PB in the QIAquick PCR Purification Kit to the 20 μl of digestion reaction.
33. Load the samples to the QIAquick spin column and spin at 16,000 × *g*, 30 s.
34. Discard flow-through and add 0.75 ml Buffer PE in the QIAquick PCR Purification Kit to the column.
35. Spin at 16,000 × *g*, 30 s and discard flow-through.
36. Spin the column for additional 1 min at 16,000 × *g*.
37. Place the column in a new 1.5 ml microcentrifuge tube.
38. Add 30 μl H_2O to the center of column and incubate for 1 min.
39. Spin at 16,000 × *g*, 1 min to collect the digested plasmid.
40. Set up the ligation reaction: 50 ng digested pQCXIP plasmid, 100 ng digested insertion, 5 μl Solution I in DNA Ligation Kit Ver. 2.1 (TaKaRa), and add H_2O to 10 μl.
41. Incubate at 16 °C for 30 min. Transform to chemical competent *E. coli* cells as follows.
42. Add ligation reaction to 200 μl of freshly thawed competent *E. coli* cells.
43. Incubated on ice for 30 min.

44. Heat shock at 42 °C for 90 s.
45. Incubate on ice for 2 min.
46. Add 800 μl LB and incubate at 37 °C for 1 h.
47. Spin at 1500 × *g* 1 min and aspirate 800 μl supernatant.
48. Resuspend the cells and spread to LB plates with Ampicillin.
49. Incubate at 37 °C overnight.
50. Pick two clones and send for sanger sequencing to confirm the insertion of oncohistones.
51. Seed 2×10^5 iMEF cells to a 6-well plate 24 h before transfection (*see* **Note 7**).
52. Warm up Opti-MEM™ medium and thaw plasmids.
53. Make the DNA mix with 2 μg oncohistone expression plasmids, 4 μl P3000 reagent, and 125 μl Opti-MEM™ medium.
54. Make the lipofectamine mix with 5 μl lipofectamine 3000 reagent and 125 μl Opti-MEM™ medium.
55. Combine the DNA mix and lipofectamine mix, and incubate at room temperature for 10 min.
56. Add the mix to iMEF cells drop by drop and mix well.
57. After 6 h of transfection, seed iMEF cells to 24-well plates with cover slips and then perform Immunofluorescence staining every 24 h.
58. Discard the culture medium, wash the cells with PBS, twice.
59. Fix the cells in 200 μl of 3% PFA for 12 min, then wash with 500 μl of PBS twice (*see* **Note 8**).
60. Permeabilize cells in 200 μl of 0.5% Triton-X solution for 5 min, then wash with PBS twice.
61. Blocked with 200 μl of 5% (w/v) normal goat serum in PBS for 1 h.
62. Add 50 μl of primary antibodies with appropriate dilutions (H3K36me3 (1:1000), H3K36me2 (1:2000), H3K36M (1:500), H3K27M (1:500), H3K27me3 (1:500), FLAG (1:1000)) and incubate at 4 °C overnight.
63. Aspirate the primary antibody, wash with PBS-T for 3 times, 5 min each.
64. Probe with 1:1000 diluted Alexa fluor 594 anti-rabbit and 488-conjugated anti-mouse antibodies for 1 h in the dark (*see* **Note 9**).
65. Aspirate the secondary antibody, wash with PBS-T for 3 times in the dark, 5 min each.
66. Counterstain DNA with DAPI staining solution for 5 min in the dark. Wash with PBS 3 times in the dark, 5 min each.

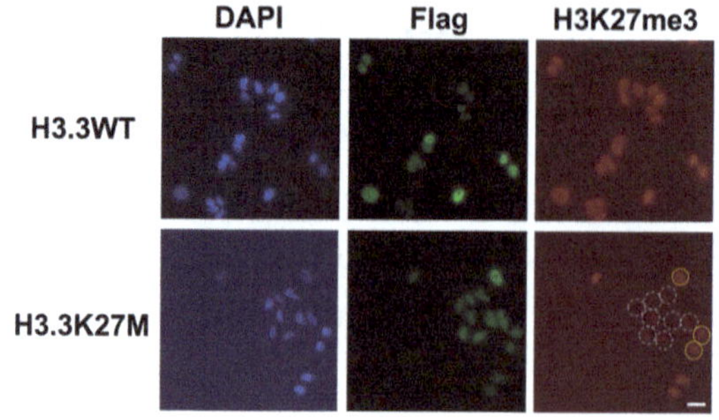

Fig. 2 Loss of H3K27me3 in astrocyte cells expressing H3.3K27M mutant proteins is dynamic among cells. Immunofluorescence analysis of H3K27me3 and Flag-H3.3 (H3.3WT) or Flag-H3.3K27M mutant proteins when plasmids were transfected into astrocyte cells for 2 days. Solid yellow circles indicate cells expressing H3.3K27M proteins but maintaining high levels of H3K27me3. Dash lined white circles identify cells with H3.3K27M mutant protiens and loss H3K27me3 after 2 days. Bar, 10 μm

67. Mount coverslip upside down onto slides with ProLong™ Gold Antifade reagents or equivalent antifade reagents. Seal the coverslip with invisible nail polish.
68. Examine the slides under fluorescence microscope. The histone marks will be stained with Alexa fluor 594 and the exogenous expressed oncohistone will be stained with Alexa fluor 488. The cells without the exogenous expressed oncohistone which are Alexa fluor 594 negative can be used as the negative control for the analysis (Fig. 2).

3.2 Probe the Function of Onco-Histone Using Knock-In Mutation

1. Insert the gRNA oligos into the pSpCas9(BB)-2A-Puro V2.0 plasmid according to the manufacturer's protocol.
2. Digest 1 μg of pSpCas9(BB)-2A-Puro V2.0 plasmid with BbsI for 30 min at 37 °C: 1 μg plasmid, 1 μl BbsI (NEB), 2 μl 10× CutSmart buffer (NEB), and add H_2O to 20 μl.
3. Purify by QIAquick PCR Purification Kit.
4. Add 100 μl of Buffer PB in the QIAquick PCR Purification Kit to the 20 μl of digestion reaction.
5. Load the samples to the QIAquick spin column and spin at 16,000 × *g*, 30 s.
6. Discard flow-through and add 0.75 ml Buffer PE in the QIAquick PCR Purification Kit to the column.
7. Spin at 16,000 × *g*, 30 s and discard flow-through.
8. Spin the column for additional 1 min at 16,000 × *g*.
9. Place the column in a new 1.5 ml microcentrifuge tube.
10. Add 30 μl H_2O to the center of column and incubate for 1 min.
11. Spin at 16,000 × *g*, 1 min to collect the digested plasmid.

12. Anneal the sgRNA oligos.
13. Mix the reagent: 1 μl sgRNA-F (100 mM), 1 μl sgRNA-R (100 mM), 1 μl 10× CutSmart buffer (NEB), 7 μl H_2O.
14. Anneal in a thermocycler: 95 °C 5 min and then ramp down to 25 °C at 5 °C/min.
15. Dilute the annealed oligos by 200 fold in H_2O.
16. Set up the ligation reaction: 50 ng digested pSpCas9(BB)-2A-Puro V2.0 plasmid, 1 μl diluted sgRNA oligo, 5 μl Solution I in DNA Ligation Kit Ver. 2.1 (TaKaRa), and add H_2O to 10 μl.
17. Incubate at 16 °C for 30 min.
18. Transform to chemical competent *E. coli* cells.
19. Add ligation reaction to 200 μl of freshly thawed competent *E. coli* cells.
20. Incubate on ice for 30 min.
21. Heat shock at 42 °C for 90 s in a heating plate.
22. Incubate on ice for 2 min.
23. Add 800 μl LB and incubate at 37 °C for 1 h.
24. Spin at 1500 × *g* for 1 min and aspirate 800 μl supernatant.
25. Resuspend the cells and spread to LB plates with Ampicillin.
26. Incubate at 37 °C overnight.
27. Pick one clone and send for sanger sequencing to confirm the insertion of sgRNA oligo.
28. Transfect the constructed pSpCas9(BB)-2A-Puro V2.0 vector expressing gRNA and Cas9 endonuclease along with donor ssDNA into targeting cells with appropriate reagent. Use 2 μg pSpCas9(BB)-2A-Puro V2.0 vector and 4 μl 10 mM donor ssDNA for 10^6 T/C28a2 cells.
29. After 24 h of transfection, select the cells by puromycin with appropriate concentration. 1 μg/ml puromycin is used for T/C28a2 cells.
30. 72 h after transfection, seed cells for single clone formation. Seed 2000 cells in 10 cm dish and do a serial 3-fold dilution for 4 plates.
31. When clones grow up, pick clones to 96-well plates from the well separated dishes under the microscope (*see* **Note 10**).
32. Expand the clones and collect cells for Western Blot to check the integration of H3.3K36M. Perform Western Blot or immunofluorescence using H3K36 M antibody to determine the knock-in of H3.3K36M.
33. Extract genomic DNA from positive clones using Gentra Puregene Cell Kit according to the manufacturer's instruction (*see* **Note 11**).
34. Trypsinize the cells and collect 10^6 cells in PBS.

35. Spin at 13,000 × *g* for 5 s to pellet the cells.
36. Carefully discard the supernatant and leave around 20 μl PBS to resuspend the cells.
37. Add 300 μl Cell Lysis Solution in the Gentra Puregene Cell Kit and vortex for 10 s.
38. Add 100 μl Protein Precipitation Solution in the Gentra Puregene Cell Kit and vortex for 20 s.
39. Spin at 16,000 × *g* for 1 min to precipitated the protein (*see* **Note 12**).
40. Transfer the supernatant to a new 1.5 ml microcentrifuge tube and add 300 μl isopropanol.
41. Precipitate the genomic DNA by gently inverting the tube 50 times.
42. Spin at 16,000 × *g* for 1 min to precipitate the genomic DNA.
43. Aspirate the supernatant.
44. Wash the DNA by adding 300 μl of 70% ethanol and invert the tube several times.
45. Spin at 16,000 × *g* for 1 min.
46. Aspirate the supernatant and invert the tube on absorbent paper for 5 s.
47. Air dry for 5 min.
48. Add 100 μl DNA Hydration Solution in the Gentra Puregene Cell Kit and pipet to mix.
49. Incubate at 65 °C for 1 h to dissolve the genomic DNA.
50. PCR amplify the surrounding sequence of targeting site. Set up the PCR reaction as follows: genomic DNA 500 ng, 2.5 μl of 10× Ex Taq reaction buffer, 0.5 μl of 10 mM dNTP, 0.5 μl of 10 mM primer H3F3B-Forward, 0.5 μl of 10 mM primer H3F3B-Reverse, 0.5 μl of DMSO, 0.2 μl Ex Taq and add H_2O to 25 μl.
51. PCR steps: Denature at 95 °C 3 min, 32 cycles of 95 °C 30 s, 57 °C 15 s, 72 °C 30 s, extend the DNA at 72 °C for 10 min.
52. Run 2 μl of the PCR reaction by 1.2% agarose gel to confirm the PCR product (690 bp).
53. Purify the rest 23 μl of the PCR product by QIAquick PCR Purification Kit.
54. Add 115 μl of Buffer PB in the QIAquick PCR Purification Kit to the rest 23 μl of the PCR product.
55. Load the samples to the QIAquick spin column and spin at 16,000 × *g*, 30 s.
56. Discard flow-through and add 0.75 ml Buffer PE in the QIAquick PCR Purification Kit to the column.

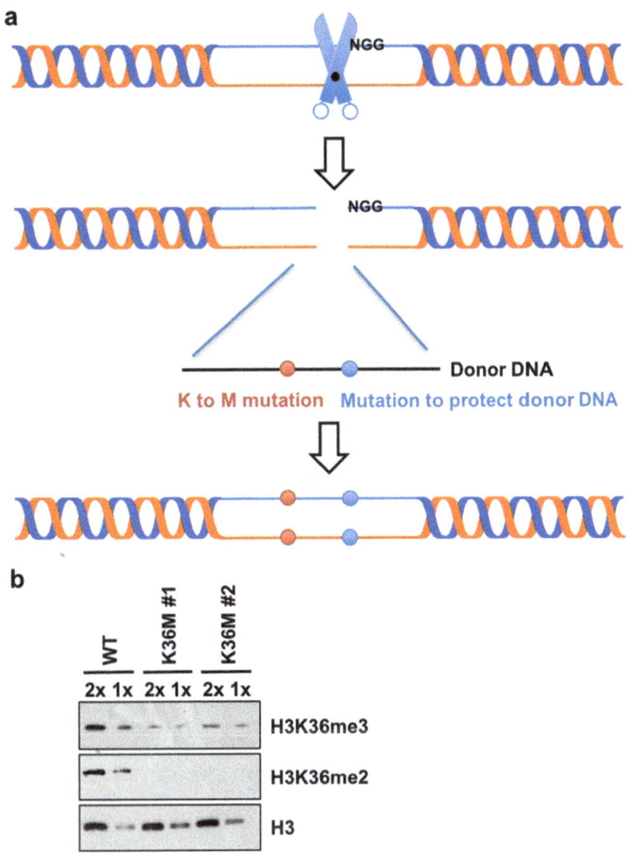

Fig. 3 Analysis of impact of H3.3K36M mutation on H3K36 methylation using knock-in oncohistone mutation by CRISPR/Cas9. (**a**) Schemata of CRISPR/Cas9 knock-in process. Donor DNA is designed to contain the onchohistone mutation as well as the synonymous mutations that prevent cutting by the CRISPR/Cas9. (**b**) Western blot analysis of H3K36me2 and H3K36me3 in two independent clones with H3.3K36M knock-in T/C28a2 cells

57. Spin at 16,000 × *g*, 30 s and discard flow-through.
58. Spin the column for additional 1 min at 16,000 × *g*.
59. Place the column in a new 1.5 ml microcentrifuge tube.
60. Add 50 μl H_2O to the center of column and incubate for 1 min.
61. Spin at 16,000 × *g*, 1 min to collect the purified PCR product.
62. Sequence the purified PCR product using PCR primer H3F3B-Forward by sanger sequencing to confirm the genotype at the *H3F3B* locus.
63. Save at least two independent clones each for heterozygous or homozygous mutations (Fig. 3) (*see* **Note 13**).

4 Notes

1. We use *H3F3B* and *HIST1H3C* as examples for H3.3 and H3.1. Site-directed mutagenesis can also be performed on other histone H3 genes. The design of site-directed mutagenesis primers is based on the requirement as: Set the mutation site in the middle of the forward primer, extend each side of the mutation for other 15 to 17 bps as the templete, and the reverse primer is the reverse-complement of forward primer. By using this strategy, we have made all K4M, K9M, K27M, K36M, and K79M mutant transgenes in both H3.1 and H3.3, respectively [33].
2. Use within 2 weeks.
3. We use H3.3K36M mutation at *H3F3B* as the example for the CRISPR knock-in. Other mutations at different genes can be done in a similar way.
4. We use pOZ-FH-C-puro (Addgene, Cat. # 32516) as the clone vector for oncohistone transgenes. The H3F3B or HIST1H3C gene was inserted into the expression site and tagged with a tandem FLAG and HA tag at the C terminal. Site-directed mutagenesis is performed using this vector and then sub-clone to pQCXIP (Clontech, Cat. # 631516) for better expression.
5. Do not spin the samples before load to the column.
6. Make sure ethanol is added to the Buffer PE before use.
7. We describe the usage of iMEF cells here to monitoring the dynamics of histone marks with the expression of oncohistones. Other cells can also be used to do the similar analysis with modified protocol of transfection.
8. Do not over-fix the cells which will destroy the epitope of antigen.
9. Keep samples in the dark, while and after probing with secondary antibody.
10. Usually 96 clones were randomly collected. Select the rounded or oval shapped ones with a bright edge and dark necrotic center. Avoid the flat ones.
11. Other methods can also be used. Alternatively, you can collect around 1000 cells in PCR tube and wash once with PBS. Then add 1 μl of 10X Extaq buffer (or other PCR buffer used in the following PCR step), 0.5 μl of 10 mg/ml Proteinase K, and 8.5 μl H_2O, incubate at 65 °C for 1 h to extract the genomic DNA and 95 °C for 15 min to inactive the Proteinas K. The DNA is ready for PCR. We usually use 2 μl of the product for the following PCR step.

12. If the protein pellet is not tight enough to take the supernatant, incubate samples on ice for 5 min and spin at 16,000 × *g* for 1 min.
13. The downstream analysis of histone marks could be done by acid extraction of histones and then Western blot or Mass spectrometry.

Reference

1. Bentley GA, Lewitbentley A, Finch JT et al (1984) Crystal-structure of the nucleosome Core particle at 16 a resolution. J Mol Biol 176:55–75
2. Luger K, Mader AW, Richmond RK, Sargent DF, Richmond TJ (1997) Crystal structure of the nucleosome core particle at 2.8 angstrom resolution. Nature 389:251–260
3. Tagami H, Ray-Gallet D, Almouzni G, Nakatani Y (2004) Histone H3.1 and H3.3 complexes mediate nucleosome assembly pathways dependent or independent of DNA synthesis. Cell 116:51–61
4. Schwartzentruber J, Korshunov A, Liu XY et al (2012) Driver mutations in histone H3.3 and chromatin remodelling genes in paediatric glioblastoma. Nature 482:226–231
5. Sturm D, Witt H, Hovestadt V et al (2012) Hotspot mutations in H3F3A and IDH1 define distinct epigenetic and biological subgroups of glioblastoma. Cancer Cell 22:425–437
6. Wu G, Broniscer A, McEachron TA et al (2012) Somatic histone H3 alterations in pediatric diffuse intrinsic pontine gliomas and non-brainstem glioblastomas. Nat Genet 44:251–253
7. Fontebasso AM, Papillon-Cavanagh S, Schwartzentruber J et al (2014) Recurrent somatic mutations in ACVR1 in pediatric midline high-grade astrocytoma. Nat Genet 46:462–466
8. Behjati S, Tarpey PS, Presneau N et al (2013) Distinct H3F3A and H3F3B driver mutations define chondroblastoma and giant cell tumor of bone. Nat Genet 45:1479–1482
9. Nikbakht H, Panditharatna E, Mikael LG et al (2016) Spatial and temporal homogeneity of driver mutations in diffuse intrinsic pontine glioma. Nat Commun 7:11185
10. Castel D, Philippe C, Calmon R et al (2015) Histone H3F3A and HIST1H3B K27M mutations define two subgroups of diffuse intrinsic pontine gliomas with different prognosis and phenotypes. Acta Neuropathol 130:815–827
11. Solomon DA, Wood MD, Tihan T et al (2016) Diffuse midline Gliomas with histone H3-K27M mutation: a series of 47 cases assessing the Spectrum of morphologic variation and associated genetic alterations. Brain Pathol 26:569–580
12. Lan F, Shi Y (2015) Histone H3.3 and cancer: a potential reader connection. Proc Natl Acad Sci U S A 112:6814–6819
13. Liu X, McEachron TA, Schwartzentruber J, Wu G (2014) Histone H3 mutations in pediatric brain tumors. Cold Spring Harb Perspect Biol 6:a018689
14. Gessi M, Gielen GH, Hammes J et al (2013) H3.3 G34R mutations in pediatric primitive neuroectodermal tumors of central nervous system (CNS-PNET) and pediatric glioblastomas: possible diagnostic and therapeutic implications? J Neuro-Oncol 112:67–72
15. Jones C, Baker SJ (2014) Unique genetic and epigenetic mechanisms driving paediatric diffuse high-grade glioma. Nat Rev Cancer 14:651–661
16. Bjerke L, Mackay A, Nandhabalan M et al (2013) Histone H3.3. Mutations drive pediatric glioblastoma through upregulation of MYCN. Cancer Discov 3:512–519
17. Presneau N, Baumhoer D, Behjati S et al (2015) Diagnostic value of H3F3A mutations in giant cell tumour of bone compared to osteoclast-rich mimics. J Pathol Clin Res 1:113–123
18. Amary F, Berisha F, Ye H et al (2017) H3F3A (histone 3.3) G34W immunohistochemistry: a reliable marker defining benign and malignant Giant cell tumor of bone. Am J Surg Pathol 41:1059–1068
19. Lee JC, Liang CW, Fletcher CD (2017) Giant cell tumor of soft tissue is genetically distinct from its bone counterpart. Mod Pathol 30:728–733
20. Koelsche C, Schrimpf D, Tharun L et al (2017) Histone 3.3 hotspot mutations in conventional osteosarcomas: a comprehensive clinical and molecular characterization of six H3F3A mutated cases. Clin Sarcoma Res 7:9

21. Kallappagoudar S, Yadav RK, Lowe BR et al (2015) Histone H3 mutations--a special role for H3.3 in tumorigenesis? Chromosoma 124:177–189
22. Amary MF, Berisha F, Mozela R et al (2016) The H3F3 K36M mutant antibody is a sensitive and specific marker for the diagnosis of chondroblastoma. Histopathology 69:121–127
23. Fontebasso AM, Gayden T, Nikbakht H et al (2014) Epigenetic dysregulation: a novel pathway of oncogenesis in pediatric brain tumors. Acta Neuropathol 128:615–627
24. Papillon-Cavanagh S, Lu C, Gayden T et al (2017) Impaired H3K36 methylation defines a subset of head and neck squamous cell carcinomas. Nat Genet 49:180–185
25. Chan KM, Fang D, Gan H et al (2013) The histone H3.3K27M mutation in pediatric glioma reprograms H3K27 methylation and gene expression. Genes Dev 27:985–990
26. Lewis PW, Muller MM, Koletsky MS et al (2013) Inhibition of PRC2 activity by a gain-of-function H3 mutation found in pediatric Glioblastoma. Science 340:857–861
27. Bender S, Tang Y, Lindroth AM et al (2013) Reduced H3K27me3 and DNA hypomethylation are major drivers of gene expression in K27M mutant pediatric high-grade gliomas. Cancer Cell 24:660–672
28. Fang D, Gan H, Lee JH et al (2016) The histone H3.3K36M mutation reprograms the epigenome of chondroblastomas. Science 352:1344–1348
29. Lu C, Jain SU, Hoelper D et al (2016) Histone H3K36 mutations promote sarcomagenesis through altered histone methylation landscape. Science 352:844–849
30. Mohammad F, Weissmann S, Leblanc B et al (2017) EZH2 is a potential therapeutic target for H3K27M-mutant pediatric gliomas. Nat Med 23:483–492
31. Piunti A, Hashizume R, Morgan MA et al (2017) Therapeutic targeting of polycomb and BET bromodomain proteins in diffuse intrinsic pontine gliomas. Nat Med 23:493–500
32. Fang D, Gan H, Wang H et al (2017) Probe the function of histone lysine 36 methylation using histone H3 lysine 36 to methionine mutant transgene in mammalian cells. Cell Cycle 16(19):1781–1789
33. Chan KM, Han J, Fang D et al (2013) A lesson learned from the H3.3K27M mutation found in pediatric glioma: a new approach to the study of the function of histone modifications in vivo? Cell Cycle 12:2546–2552
34. Herz HM, Morgan M, Gao X et al (2014) Histone H3 lysine-to-methionine mutants as a paradigm to study chromatin signaling. Science 345:1065–1070
35. Funato K, Major T, Lewis PW et al (2014) Use of human embryonic stem cells to model pediatric gliomas with H3.3K27M histone mutation. Science 346:1529–1533

Index

Guillermo A. Orsi and Geneviève Almouzni (eds.), *Histone Variants: Methods and Protocols*, Methods in Molecular Biology, vol. 1832, https://doi.org/10.1007/978-1-4939-8663-7, © Springer Science+Business Media, LLC, part of Springer Nature 2018

U

W

Y

FSC
www.fsc.org
MIX
Papier | Fördert gute Waldnutzung
FSC® C083411

Kolibri 360 GmbH
Ferdinand-Jühlke-Straße 7
99095 Erfurt, Deutschland
produktsicherheit@kolibri360.de